1983

ENERGY FOR A TECHNOLOGICAL SOCIETY

PRINCIPLES
PROBLEMS
ALTERNATIVES

SECOND
EDITION

JOSEPH PRIEST

Miami University

ENERGY FOR A TECHNOLOGICAL SOCIETY

PRINCIPLES
PROBLEMS
ALTERNATIVES

SECOND EDITION

ADDISON-WESLEY PUBLISHING COMPANY

Reading, Massachusetts • Menlo Park, California

London • Amsterdam • Don Mills, Ontario • Sydney

This book is in the
ADDISON-WESLEY SERIES IN PHYSICS

Cover photograph by
David Attie Stock Photographs

Library of Congress Cataloging in Publication Data

Priest, Joseph.
    Energy for a technological society.

    Includes bibliographies and index.
    1. Power resources.   2. Power (Mechanics)
I. Title.
TJ163.2.P74 1979   621   78-55829
ISBN 0-201-05936-3

ISBN 0-201-05936-3
ABCDEFGHIJK-MA-798

To Mary Jean, Mary Jo, John, Catherine, and Daniel
for their love and inspiration

Energy for a Technological Society is a second edition of a book published in 1975. By developing and applying the principles of physics, it treats those aspects of societal energy use in which a physical scientist has competence. These include

1. teaching physical principles as they relate to the societal uses of energy;
2. examining the consequences of working with finite energy resources;
3. examining the effects, extent, and control of by-products that are usually produced in energy conversion processes; and
4. evaluating the technology of energy converters and the trade-offs involved in their societal deployment.

Following this philosophy, the book is organized as follows:

Chapter 1 Energy and the Technological Society
Chapter 2 Fossil Fuels and Their Role in the Energy Picture
Chapter 3 Physical Basis for Energy
Chapter 4 Environmental Aspects of Burning Fossil Fuels
Chapter 5 Electric Energy and Power
Chapter 6 Thermodynamic Principles and the Steam Turbine
Chapter 7 Disposal of Rejected Heat from Steam Turbines
Chapter 8 Atmospheric Problems
Chapter 9 Nuclear Physics Principles
Chapter 10 Nuclear-fueled Electric Power Plants
Chapter 11 Solar Energy
Chapter 12 Other Energy Alternatives
Chapter 13 Nuclear Breeder and Nuclear Fusion Reactors
Chapter 14 The Potential and Importance of Energy Conservation

While this format is basically the same as in the first edition, a few organizational changes have been made. Recognizing the increasingly important role of solar energy, a chapter has been dedicated entirely to solar energy.

The nuclear breeder and nuclear fusion reactors are combined into a chapter following the solar energy presentation. This recognizes the importance of these nuclear technologies, but reflects some of the uncertainties that have emerged. The important concept of an energy efficiency based on the second law of thermodynamics is stressed throughout the revised edition.

Every effort has been made to incorporate all the features that are found in a good teaching book. Each chapter concludes with four categories of instructional aids designed to further student understanding. These are

1. *Topical Review.* This section asks questions that highlight the main topics and ideas in the chapter.

2. *References.* The references provide both an accurate source of information and a guide to the types of journals and books students should seek themselves.

3. *Suggestions for Further Understanding of Chapter* ____. Questions are posed that do not require a numerical answer.

4. *Numerical Problems.* Problems are posed that emphasize practical numerical calculations.

Experience using the first edition in a variety of schools has contributed to a clearer presentation of a number of topics. Overall, about 200 new entries have been included in the questions and problems section of this second edition. The glossary of important terms and ideas has been refined and expanded.

No previous exposure to physics is presumed in this text. Some familiarity with elementary algebra and numerical manipulation is assumed. Generally, the text is used in a course meeting three hours a week for 15 weeks. Course formats, sources of films and other teaching materials, and some suggested classroom demonstrations are presented in a teacher's guide that is available from the publisher. A student study guide is also available.

Oxford, Ohio
January 1979

J. P.

TO THE STUDENT

In late 1973, the Organization of Petroleum Exporting Countries (OPEC) imposed an oil embargo on the United States. The sudden loss of the oil imports forced our country into some effective energy conservation measures. However, these steps were gradually relaxed when the embargo was lifted. Gasoline shortages are but a memory. Accompanying the embargo was a significant increase in the price of crude oil. This aspect is not a memory and in all likelihood the prices will continue to increase. In 1977, $21.9 billion of a $26.6 billion trade deficit was due to oil imports. Since the oil embargo, the dependence on foreign oil imports has increased from 36% in 1973 to 42% in 1976. More than conservation will be needed if another embargo is imposed.

Pollution is a second major component of societal energy use. Following enactment of the 1970 Clean Air Act, some overall improvement has been made in overall air quality. That is not to say, however, that we have achieved a satisfactory condition in air quality. Much improvement is required.

Deciding on energy sources for our technological society is very difficult. We have an abundance of coal, but there are problems associated with mining it, transporting it, and burning it. Nuclear energy contributes significantly to the generation of electricity and many citizens would like to see its role amplified. But there are concerns with nuclear energy and it is difficult to decide the proper course. While solar energy will undoubtedly be a major factor in supplying our future energy needs, its technology is in its infancy and whether we can risk a total commitment to solar energy at this time is not at all clear. Although hopes are high for nuclear fusion energy, its ultimate development to commercial use is more than a quarter of a century away. To supplement these major energy sources there is geothermal energy, trash energy, tidal energy, hydroenergy, magnetohydrodynamics, wind energy, biomass energy, and hydrogen energy, all of which must be evaluated. All aspects of this wide variety of energy sources involve important decisions—decisions that not only involve all of us but often require our input. In the case of nuclear power, citizens are being asked to cast a vote for or against its use in their states. This text is designed to help you make a meaningful energy decision whether in the public poll booth or, for example, in deciding on a solar heating system for a residence.

Just how to study the energy problems of society is debatable. In this text, physical principles are central to the teaching philosophy. At a fundamental level physical principles allow you to understand energy technologies and their inherent benefits, risks, and problems. But there is more to the principles than this, because they apply to much more than energy. Their understanding opens up a whole new world of adventure. Adventures into the world of physics are a key feature of this text.

Like it or not, numbers are a part of our life. We make change. We fill out income tax returns. We are conscious of the speed registered by a speedometer. We evaluate our electricity bills and pay them. The examples are endless. Although we may not be consciously aware of it, many of these transactions are algebraic. Much can be achieved in a study of physical principles and energy without resorting to numbers and elementary algebra, and the thrust of this book favors this emphasis. But in a discussion of physical principles and energy we always arrive at a point where we must ask "how much" and "why." This requires numbers and reasons. Algebra facilitates reasoning. Numbers and elementary algebra are a part of this book. Although you will be dealing with physical quantities like kilowatt-hours and British thermal units, the manipulations are not a lot unlike those you do on an everyday basis. It is my wish that you would adopt this attitude and seek out the analogies with everyday situations. For a small investment, the rewards are big.

While the principles of physics that you absorb from this book will not change, much of the energy information will. Government regulations concerning energy will change. Emphasis on particular energy technologies will change. Some of these changes will occur in short periods of time. We can keep abreast of these changes only by perusing current literature. At the end of each chapter you will find selected references. These particular references are not intended to be the last word. You should look to the current editions of the journals referenced for more current energy information.

Much of the drudgery and fear of numerical manipulations has been removed with the advent of hand-held calculators. With impressive speed and accuracy, they perform routine numerical manipulations involving numbers of any size. These instruments are making an enormous impact on our lives. Take advantage of them! They will help you evaluate physics and energy as well as save you money at the grocery. An appendix has been prepared to help you to be comfortable with the wide range of numbers encountered in the book.

Societal problems with energy are very serious and require careful consideration by all of us. Our life-styles may very well change from coping with these problems. Physics can also be serious business. But physics can also be interesting. I hope that your comprehension of energy problems and physics grow as a result of studying this book. Above all, I hope that physics is made more meaningful to you.

ACKNOWLEDGMENTS

Student interest in physics and energy was the motivation for both editions of this book. I appreciate that interest very much. Comments from instructors using the first edition were enormously useful. I am indebted to them. I am grateful to Professors W. Louis Barrett, William M. Duxler, Herbert Schlosser, and David Soule for their reviews and criticisms. Professor David Uhrich provided detailed comments and suggestions during all phases of the writing. I am especially grateful to him. To the physics department, to the library and audiovisual staffs at Miami University who cooperated in numerous ways, to Juanita Killough who saw to it that the manuscript got typed, and to Melinde Hatfield who typed and edited so capably, thank you very much.

Dear Mary. What can I say. I will now fix the dryer, the porch, the washer, . . .

J. P.

CONTENTS

ENERGY AND THE TECHNOLOGICAL SOCIETY

1

Metropolitan centers epitomize our technological society and energy use. Every element either requires energy for its function or required energy for its construction. An automobile needs gasoline (energy) to make it go. Energy in several forms— human, mechanical, heat, chemical—was needed to make the automobile. Energy was required to make the superhighway on which the automobile travels. Pick out any other component and you can identify a similar energy pattern. Photograph by Bob Smith. (Courtesy of the EPA.)

1.1 ENERGY: A BYWORD FOR THE TIMES

Nothing influences our living style more than the availability and utilization of energy. Yet until the 1970s few of us really thought much about this crucial societal role even though energy, as a word, very likely entered our vocabulary in elementary school. At some time, all of us "have energy to do homework" or profess to "not being very energetic." We know that energy is fed into our homes to run electrical appliances and that the purpose of gasoline is to furnish energy to run automobiles. Still it took a sudden reduction in energy supplies, an "overnight" increase in gasoline prices, and an intolerably putrid atmosphere to bring us to a confrontation with energy and the associated problems that have been developing for years. Now we speak candidly of "energy shortages" and "energy crises." The news media teem with energy information. Charges of contrivance are leveled against the energy industries. The industries counter with full-page newspaper items and lengthy television statements professing their innocence. In the middle is the bewildered public not knowing who or what to believe but, nevertheless, paying high prices for the energy content of gasoline, oil, natural gas, and coal. The public bewilderment is understandable because the total energy picture is incredibly complex. The dilemma is due partly to poor planning, contrivances, environmental restrictions, and political indifference. Apart from this there is the reality of depleting limited energy resources and dealing with the by-products that are generated when energy is utilized. Whether or not we have the energy to operate our technological society will involve important decisions and trade-offs by all elements of our society. It is hoped that this study of energy and society will encourage you to get involved in this decision making at both personal and governmental levels and that it will help you to arrive at acceptable alternatives.

1.2 ENERGY AND SOCIETY

As we delve into energy problems a precise meaning will be given to energy. For the present, let us think of energy as being an essential ingredient for our everyday functions. For example, energy is required to heat a home. Heat might be derived by burning a common fuel like oil, natural gas, or coal. Energy is required on your part to actuate a switch for a reading lamp in your home. This involves human energy that was derived from food. Energy was utilized by some industry to make the switch and energy was used to transport the switch from the factory to the store where purchased. Once the light is on, energy is needed to keep it operating. In operation, it produces light, which is energy. We could go on and on with this exercise. The points being made are (a) energy is involved in all our activities and (b) energy has many forms. Some forms easily identified in this example are heat, light, electric, chemical (fuels), and mechanical (motion). Others will emerge as we proceed.

The major sectors of our society that use energy are usually categorized as industrial, transportation, and residential and commercial. Each of these is readily identified in the previous illustration. Some industry used energy to make the light switch, which was probably transported to a commercial store by truck. The store requires air conditioning and heating, for example, and the switch controlled a lamp in a residence that uses energy for heating, cooking, etc. Figure 1.1 characterizes how the United States converts energy for use in the residential and commercial, industrial, and transportation sectors. About 93 percent of the energy feeding this huge energy-converting system is derived from natural gas, coal, and oil, which are remnants of prehistoric living systems. By their very nature, these fuels are finite and only coal is plentiful in the United States. Of the 16.8 units of oil used daily in 1976, 7.3 were imported—a situation of grave concern. As seen in the oil pathways in Fig. 1.1, over one half of the oil is used in the transportation sector. Of the 9.1 units of energy used for transportation, only 24 percent actually gets into energy of motion because of the inefficiency of the transportation systems. This means that for every unit of energy saved by not driving or driving slower, over four units are saved at the input. Similarly in the production of electric energy, it takes about three units of input energy to generate one unit of electric energy. Saving one unit by turning off unused lights, for example, means saving three units of input energy. This is why energy conservation is so very important and is inevitably an essential attribute of any energy program concerned about energy supplies.

Figure 1.1 The sources, uses, and flow pattern of energy in the American society in 1976. The units are in terms of the equivalent of millions of barrels of oil used per day. Coal, for example, contributed the equivalent of 7.2 million barrels of oil each day; 4.7 of these units were used to make electricity, 0.6 units were exported, 1.8 units went for industrial purposes, and 0.1 units were used for residential and commercial purposes. This manner of presentation is common in energy discussions. The numbers may vary because of different data sources and assumptions. The data for this diagram were taken primarily from "Energy in Focus: Basic Data," published by the Department of Energy.

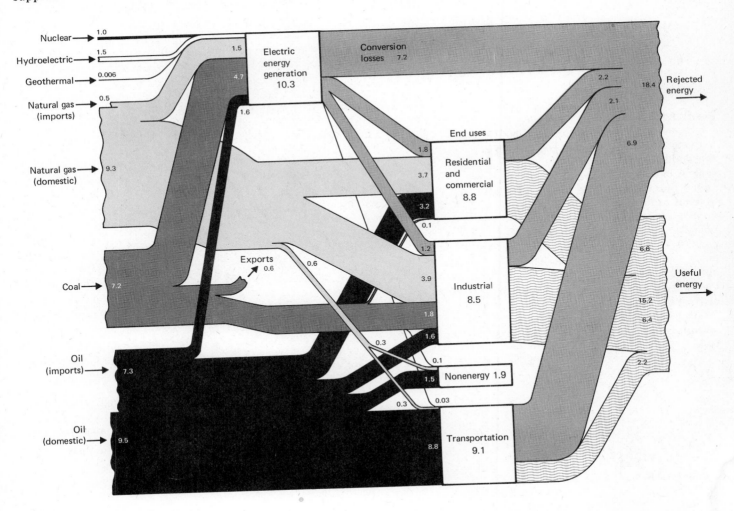

An analysis of energy use in a home would be very different for a society less technologically oriented than ours. That home may not have the use of electric energy, for example. Regardless of the country chosen, though, you would find that no country uses more energy per person than the United States. With only 6 percent of the world's population, the United States consumes 35 percent of the total energy produced in the world. The rationale for this enormous use of energy is that it "buys" a higher standard of living. A suggested indicator of standard of living is the gross national product (GNP). The GNP is defined as the total market value of all the goods and services produced by a nation during a specified period. Normally, the time period is a year. If the GNP per person is compared with energy usage per person, as in Fig. 1.2, a noticeable correlation is observed. Those countries with high per capita GNP and energy consumption are also the countries that are technologically inclined. The less technologically oriented countries are eager to improve their standard of living and they recognize energy as being a necessary ingredient. This is the motivation for the Aswan Dam project in Egypt and the development of nuclear energy in India and Brazil, for example. This natural desire for energy puts added strains on the world's available resources and

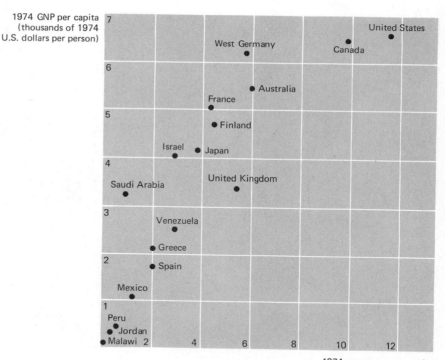

Figure 1.2 Comparison of energy use per capita and gross national product (GNP) per capita for some selected countries in the world for the year 1974. The GNP scale is in terms of the value of 1974 United States dollars. The energy scale is in terms of the energy derivable from burning a thousand kilograms (2200 pounds) of coal. Note that the GNP per capita is roughly the same for West Germany, Canada, and the United States, but that the energy use per capita for the United States is more than twice that of West Germany. The data are from "Statistical Yearbook 1975, Twenty-seventh Issue," United Nations, New York, N.Y. 10017.

makes worldwide the pollution problems characteristic of the high technology countries. It is important to note that while West Germany and the United States have essentially the same GNP per capita energy use differs by more than a factor of two. This suggests that through energy conservation measures and the use of more energy efficient converters a reduction of American per capita energy use could be accomplished without compromising the standard of living.

1.3 A CRISIS IN ENERGY SUPPLY

"The Energy Crisis" is a popular title for a newspaper article, a magazine article or a television program. Both "energy" and "crisis" are words frequently used in our everyday lives. Crises develop when one performs poorly on an examination or when the dam bursts. Although both of these occurrences have an element of surprise, they probably have resulted from developments taking place over a period of time. Doing poorly on an examination may be simply the result of studying the wrong things. Indeed, an energy crisis exists in the sense of sudden price increases and shortages. But the stage for the crisis of the 1970s was set some time before the actual event. Until about 1955 the United States produced the energy it consumed. Thereafter, consumption exceeded production and the United States began relying on foreign energy sources to meet its needs. The trend has persisted and the gap between production and consumption widened noticeably about 1970 (Fig. 1.3). This gap necessitated an increased reliance on foreign sources for petroleum. If for some reason a

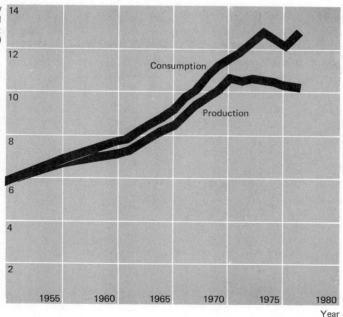

U.S. annual total energy production and consumption (equivalent of billions of barrels of oil)

Consumption

Production

Year

Figure 1.3 Yearly production and consumption of energy of all forms in the United States. The energy units are expressed in terms of the energy derivable from burning a barrel of oil. The difference between consumption and production is made up by imports of oil and natural gas. Note that the imported amount has steadily increased since about 1955. The data are from the Department of Energy.

foreign source decides to curtail its contribution or elects to escalate the price, a crisis results. This is the situation that evolved in 1973 when the Arab sources curtailed their oil shipments and effected a worldwide escalation of petroleum prices. Because the Arab countries control 56 percent of the world's known oil reserves, this cost aspect of the crisis is not likely to change even though the Arabs may choose to contribute to the world's petroleum markets. It is important to realize that the premise for this action was set in 1955 and the crisis atmosphere in the United States remains as long as reliance on foreign energy persists. Some foresighted people recognized this possibility from the outset and gave repeated warnings. In 1970 the Federal Power Commission stated, "A crisis exists right now. For the next three decades we will be in a race for our lives to meet our energy needs." Yet it took a sudden turn of events to awaken the American public.

It is easy to look at production and consumption data and identify the crux of our energy dilemma. It is not as easy to say why the production–consumption gap developed and widened. Consumption increased for many reasons, mostly political and economic. Price structures and advertising encouraged energy usage. The development of the Interstate Highway System and the population shift to suburbia fostered travel by automobile. The widespread and expensive use of energy in air travel added significantly to energy consumption. Production lagged for many complicated reasons—political, economic, and technological. Through favorable tax structures, exploration for and production of foreign oil was encouraged. The development of offshore oil wells in the continental shelf was curtailed for one-and-a-half years following the massive oil spill in California's Santa Barbara channel in 1969. The desire to protect the environment had its impact. The National Environment Policy Act (NEPA) of 1969 demands detailed assessments on environmental impact before energy sources may be developed. These demands led to delays in the development of the Alaskan pipeline and in the construction of nuclear power plants. In order to comply with the Clean Air Act of 1970, many industries shifted from burning coal to burning less-polluting natural gas and oil. A complicated picture with challenging problems emerges. Solving the problems involves environmental, economic, and social trade-offs. Hopefully, these trade-offs will be beneficial. Apart from these rather short-term problems, which can be solved by better planning and conservation, there remains the fact that we are beginning to see the end of cheap energy resources. We are depleting these resources on a time scale which is staggeringly short. Although other energy resources exist, they present their own economic, political, environmental, and technical problems and they, too, involve inevitable trade-offs. Above all, their development takes time and support. It is the acceptable development of these alternate sources that constitutes our long-range solution to the energy problem and that is the essence of our study.

1.4 A CRISIS IN POLLUTION

There is a certain irony in the energy crisis because even if unlimited fossil fuel reserves existed, burning them would still cause problems. In many localities, the effluents from burning gasoline in cars, trucks, and buses—and coal in

Figure 1.4 Donora, Pennsylvania, as it appeared on a polluted day in 1949. Air pollution was the cause of a major epidemic in this city in 1948. (Photograph courtesy of the Environmental Protection Agency.)

industries and electric power plants—are threatening the health and well-being of the public. Additionally, the annual cost of pollution damage to property, materials, and vegetation amounts to about $100 per person.*

Pollution problems have not befallen us suddenly. Isolated incidences of pollution caused by burning coal in this country have been recorded for several decades. The case most often cited occurred in the industrial city of Donora, Pennsylvania, in 1948 (see Fig. 1.4). A combination of stagnant air conditions and effluents from burning coal produced a polluted atmosphere in which over 40 percent of the population was affected to some extent. Problems caused by burning gasoline in cars were recognized in Los Angeles following World War II. The nation, and especially California, has not been completely negligent in trying to head off widespread problems. But with the growing use of energy and the tremendous growth in the number of automobiles, little nationwide headway was made until the Clean Air Act of 1970 was passed. Stringent maximum allowable levels for pollutants were invoked and emission limits for automobiles and industrial sources were set as a result of the act. Some advances toward bringing pollution under control have followed. But these gains have not come without opposition. Industries have insisted that the regulations are too severe and that technology for meeting the standards is not at hand. As energy resources become premium the attraction for burning dirtier fuels blossoms and the pressures for relaxing the standards mount.

1.5 WHAT DOES THE FUTURE HOLD?

At present, the decision to have enough somewhat dirty energy or not enough fairly clean energy is difficult and requires an assessment of the total scene. The prospect of having abundant clean energy is exciting! From the time of the successful operation of the first nuclear reactor at the University of Chicago in 1942, expectations for nuclear fission energy were high. However, nuclear fission energy involves some unique pollution and security problems that makes

---

* "Cost of Air Pollution Damage: A Status Report," Environmental Protection Agency, Publication AP-85 (1973).

its use highly controversial. Although it is presumptuous to hope to develop a perfectly nonpolluting energy source, there are some exciting potential sources that can produce abundant energy and appear to have minimal side effects. The sun, which provides energy to grow our food, is an enormous potential source of energy for warming (and cooling) our homes and factories. The nuclear fusion process, which is responsible for the sun's energy and which requires temperatures of millions of degrees to function, looms on the horizon as a practical energy source for producing electricity or, perhaps, disposing of trash (a huge problem in itself). The geothermal energy that heats geysers and hot springs has been used as a clean energy source in isolated areas for over half a century. Geothermal energy is now being evaluated for large-scale use, and many scientists predict a bright future for it. The wind, which once propelled sailing vessels all over the world and still powers water pumps and electric generators for farms and other remote areas, is now being investigated seriously for producing electricity on a scale competitive with contemporary electric power plants. Heat energy from the oceans and energy from the ocean tides also offer promise. An environmentally acceptable method of trash and garbage disposal has boggled the minds of women and men for generations. Now the feasibility of burning trash and garbage to produce both electricity and heat for homes has been demonstrated. The development of these alternative energy sources is, perhaps, not as glamorous as putting a man on the moon. It is, however, probably no less difficult. Certainly their successful development will reward mankind to a degree that is hard to comprehend. The prospect of developing acceptable energy sources is what motivates this study of energy and its blessings and problems in our technological society. Many of the chapters that follow examine the energy sources mentioned above in detail with the aim of giving the reader a glimpse at both their potential problems and their potential successes.

REFERENCES   **Energy and Society**

1. *Energy and Society* by Fred Cottrell, McGraw-Hill, New York (1955). An extremely interesting and prophetic book. Cottrell's discussions of the relations between energy, social change, and economic development apply to today's problems.
2. *Man, Energy, Society* by Earl Cook, W. H. Freeman and Co., San Francisco (1976). A current and detailed treatment of the societal role of energy.

How Does an Energy Crisis Arise?

1. The Ford Foundation sponsored a comprehensive "Energy Policy Project." The final detailed report is published as *"A Time to Choose: America's Energy Future,"* Ballinger, Cambridge, Mass. (1974).
2. *Time* and *Newsweek* magazines' energy sections do a good job of reporting the energy situation.

3. "The Energy 'Joyride' Is Over," by Edmund Faltermayer, *Fortune* **86**, No. 3, 99 (September 1972) is worthwhile reading.

1.1 Energy: A Byword for the Times

1. Scan a daily newspaper for references to energy. Usually you will find energy-related articles in the editorial, business, and general news sections. Sometimes you will find references to energy in the comics and crossword puzzles.

2. Presently the United States with 6 percent of the world's population uses about 35 percent of all the energy used in the world. Forty years ago the United States had considerably fewer people and used substantially less energy. Recognizing the trend toward worldwide industrialization, speculate on whether the United States used more or less than 35 percent of the world's energy 40 years ago. Try to find a source of information to check your speculation.

1.2 Energy and Society

3. Figure 1.1 shows that the residential and commercial, industrial, and transportation sectors use essentially the same amount of energy. Determine the fraction of energy used by each of these sectors that gets into the useful form.

4. What fuel does the United States export in significant amounts?

5. The category labeled nonenergy in Fig. 1.1 represents oil, coal, and natural gas used by the petrochemical industry. What fraction of the total amount of oil consumed is used by the petrochemical industry?

6. How would you define energy for a concerned citizen?

7. To see how energy is used in your daily life, trace your "steps" for a few hours starting from the time you arise in the morning. Note how you use energy and how it was involved in the items you use.

8. Having thought about how energy is involved in your own life, consider the energy differences between your life and the lives of your parents and grandparents when they were your age. One factor to consider is the difference in transportation.

9. The development of a technological society can be characterized by the ways that its members have learned to increase the rate of using energy. Apply this idea to the development of transportation in the United States.

10. Choose a few countries of interest to you and think about how energy is used and how energy affects the life-style of the people. As a starter, consider Japan. Why is foreign petroleum so important to Japan's economy?

11. How has the availability of energy influenced suburban growth in the United States?

12. Figure 1.2 shows that Saudi Arabia and the United Kingdom have essentially the same GNP per capita. However, the per capita energy use in the United Kingdom is more than five times larger than that in Saudi Arabia. What is the origin of this huge difference in energy use?

13. France, which ranks fairly high in both per capita energy use and per capita GNP (Fig. 1.2), has made a decision to derive much more energy from nuclear sources. What has prompted this decision by the French?

14. Determine from Fig. 1.3 the percentage of total energy used in the United States that is obtained from foreign sources.

1.3 A Crisis in Energy Supply

15. The possibility of the energy dilemma seen in the 1970s was recognized at the time the United States began its program to put a man on the moon. If the American public had been given the choice of spending money to develop new energy sources or putting a man on the moon, what do you think the choice would have been?

16. For the particular part of the country in which you live, speculate on some possible alternative energy sources which could either supplement or replace the energy presently obtained from fossil fuel sources.

17. Look up the definition of crisis and judge whether the 1973 energy situation can be categorized as a crisis.

18. "Contrive," like "crisis," is a word with varied meanings. The oil interests have been accused of contriving the high gasoline prices in order to derive huge profits. They (the industries) say that the profits derived are needed to develop new energy sources. What is your interpretation of "contrive" when used this way? For further insight read "How to Think about Oil Prices," by Carol J. Loomis, *Fortune* **89**, No. 4, 98 (April 1974).

19. The following quote appeared in an energy publication in 1924:

In 1920 it was estimated that one half of the oil resources of the United States had been used. This estimate probably was too large, for new fields will be discovered; yet the country has been so thoroughly explored for evidences of oil that it is not probable that many important fields have been overlooked. We are consuming our petroleum with great rapidity, and its exhaustion seems only a few decades off, unless experiments with compressed air and water, forced into the oil-bearing rocks, shall result in the recovery of much oil now left in the ground.

How do forecasts like this one influence public opinion about the current energy problems?

1.4 A Crisis in Pollution

20. As more and more oil is shipped across the oceans, concerns for oil spills increase. To what extent are these concerns realized? (Check, for example, newspapers and news magazines for the week of March 19, 1978).

21. Pollution is a social cost for energy. What are some other social costs?

FOSSIL FUELS AND THEIR ROLE IN THE ENERGY PICTURE

2

Massive dragline shovels strip away the overburden of coal deposits. The ugly scars of the aftermath are becoming rarer as legislation requiring restoration becomes increasingly effective. Photograph by Lyntha Scott Eiler. (Courtesy of the EPA.)

2.1 USE AND FORMATION OF FOSSIL FUELS

Although you may have been unaware of the role of energy in our lives, you certainly would not question the utility of money. Interestingly, the use of energy and money are remarkably similar. A person, family, business, or government draws money (energy) for its operation from some reserve. This reserve is referred to as capital money (energy). These systems rely on income (money, energy) both for their operation and to replace or increment their reserves. Income energy for our society is provided by the sun. Historically, reserve energy has been provided by several sources but in 1976, 93 percent of the reserve energy came from burning fossil fuels (Fig. 2.1).

Fossil fuels are remnants of plants and animals which died millions of years ago. Depending on the formation conditions, the fuel can be liquid (crude oil), gaseous (natural gas), or solid (coal, peat, lignite). The fuels are produced by very slight imbalances in the growth and decay processes of plant and animal life. For illustration, consider the formation of a fossil fuel from plants. A living plant is a complex system of carbohydrates. Sugar, starch, and cellulose (the principle building material of wood) are common carbohydrates. The carbohydrates are produced by the process of photosynthesis. In photosynthesis, water, carbon dioxide,* and energy from the sun combine to form carbohydrates. Oxygen is released in the process. When a plant dies and decays, the formation process is reversed. Carbohydrates and oxygen combine to form carbon dioxide and water, and energy is released in the process. In the long term there is *almost* perfect balance between formation and decomposition of all plant life. *But* a very slight amount of the carbohydrates formed does not decay because conditions, such as in a swamp, limit the oxygen supply. These carbohydrates become the fossil fuels (petroleum† and coal). Fossil fuels have stored the energy that was used in their formation. This energy is then released when oxygen is supplied and the fuel is burned.

---

* Carbon dioxide is a common gas that we will learn more about later. It is released when a person exhales, for example.

† Petroleum includes natural gas, crude oil, and a number of waxy or asphaltic materials.

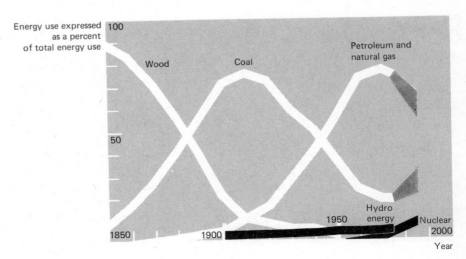

Figure 2.1 Historical dependence of the United States on various sources of energy. While shifts from one form of energy to another are evident, the country has relied almost totally on fossil fuels (coal, petroleum, and natural gas) since the turn of the century. In 1976, petroleum, natural gas, and coal provided 47.3, 27.4 and 18.4 percent, respectively, of the total energy. Note the onset of nuclear energy in 1960. The shaded portions ending at 1985 represent low and high estimates. The data are from the Department of Energy.

Nearly all the fossil fuels used today were produced in the last 100 million years. However, we are depleting these reserves on a time scale that is of the order of the average lifetime of a human being. This is a difficult proposition to accept but it can be understood from elementary considerations.

We read often that the United States oil and natural gas reserves will be exhausted by the year 2000. Such a statement is plausible, but it is extremely important for us to understand what is meant by *reserves* and depletion. A personal money reserve is a fund set aside and available for use as the person sees fit. A reserve may exist in the form of money from a parent but often certain conditions must be met before it becomes available. A *proven oil reserve* is not necessarily one in which the oil has been pumped from the earth and set aside in tanks for use as the consumer sees fit. Usually a proven oil reserve consists of petroleum in the ground ready for extraction. Detailed geological analyses, such as borings to see if the oil is present, have already been done at the proven reserve sites. This oil is available at current prices using current extraction technology.

Apart from these known reserves there are less certain, *unproven oil reserves* of several different kinds. Some unproven reserves are inferred from geological studies and oil exploration experience. Other unproven reserves may exist in areas where the technology for extracting oil has not been developed. Oil such as the residual oil that remains in conventional oil wells may also exist in quantities too small to permit its economic extraction. Conventional pumping technology is capable of extracting only 30 percent of the oil from an established well.* More advanced technology is available for removing some of this residual oil, but if the cost of extracting it is excessive, it is not marketable. As oil prices escalate, its extraction could become economically feasible.

Still another unproven reserve is the huge amount of oil that might be gleaned from the shale deposits in the Western United States if the technology for removing it can be developed in an economically and environmentally acceptable manner. Having given these definitions of proven and unproven oil reserves, let us proceed with estimating their lifetimes.

Determining the lifetime of a reserve appears to be analogous to asking, "How long will $100 last if you spend it at the rate of $10 a day?" Clearly, the answer is ten days. Although you may not have arrived at this answer by using a formal method, the solution simply involves dividing the amount ($100) by the rate ($10 a day):

$$\text{time} = \frac{\text{amount}}{\text{rate}} = \frac{\$100}{\$10/\text{day}} = 10 \text{ days}. \qquad (2.1)$$

So, seemingly, to determine the lifetime of proven reserves of a fossil fuel one needs to know the amount of the reserves and the production rate. But some hitches arise. The first is the inherent difficulty in estimating the quantity of the reserves. It is particularly difficult to estimate the amount of oil and

_____
* John C. Fisher, "Energy Crisis in Perspective," *Physics Today,* vol. 26, no. 12 (December 1973).

2.2 LIFETIME ESTIMATES FOR THE FOSSIL FUELS

natural gas reserves in deposits formed in geological traps several thousand feet below the surface of the earth. For example, the average depth of wells in the United States in 1971 was 5924 feet. A variety of techniques is used to locate the traps. One technique is to measure any characteristic of the earth's subsurface, such as its magnetism or density, that might be altered by the presence of oil. In addition, geological mapping is made of the first 30,000 feet of the subsurface. But the foremost technique used for locating oil is seismic. An underground explosion is set off in an area which might contain oil. This explosion creates a disturbance which travels through the ground somewhat like a water wave moving across the surface of a pond. The characteristics of the seismic waves are measured after they have passed through and bounced from constituents of the subsurface. Of the 100 wells located by this sophisticated technique in 1970, only nine produced oil or natural gas and only two of those nine were commercially significant. Coal reserves are somewhat easier to estimate because coal occurs in broad, thick, stratified deposits near the surface of the earth. Drilling into the ground is sufficient to determine the presence and quantity of coal in a deposit.

A second difficulty in estimating the lifetime of the reserves is caused by the fact that the production rate changes with time (Fig. 1.3). For example, the production rate in 1976 differed from that in 1970. If one assumes a constant rate in order to use Equation 2.1, a precise estimate will not result. However, it will give a rough assessment of the depletion times. Table 2.1 gives the proven and unproven reserves and 1976 production rates for oil, natural gas, and coal in the United States.

Table 2.1 United States proven and unproven reserves and 1976 production rates for oil, natural gas, and coal. Data for proven reserves are from "Energy in Focus: Basic Data," Department of Energy (1977). Data for unproven reserves were determined from Tables 3 and 4, Appendix C of "Exploring Energy Choices," Ford Foundation Energy Project (1974). The numbers in parentheses are from the Department of Energy source for the year 1973. These numbers are controversial and may not agree with those provided by a different source. For example, the 1973 proven reserves quoted in the first edition of this text were somewhat higher than the values quoted here. The fact that the reserves are declining is significant and is likely characteristic of the estimates provided by any source. There is very little variation in the estimate of the unproven reserves for any of the three fossil fuels. However, the unproven reserves are sometimes broken down into two categories called recoverable resources and remaining resource base and often there will be disagreement on estimates of the amounts in these categories.

	Proven reserves	Unproven reserves	1976 production rate
Oil (in billion barrels)	30.9 (35.3)	2890	2.96 billion barrels per year
Natural gas (in trillion cubic feet)	216 (250)	6580	19.9 trillion cubic feet per year
Coal (in billion tons)	164	2560	0.631 billion tons per year

The data in Table 2.1 show the lifetimes for depletion of the proven reserves are 10, 11, and 260 years for oil, natural gas, and coal, respectively. Thus the lifetimes for oil and natural gas are shockingly small. The corresponding calculation for coal shows that there is a significant amount of this resource remaining. The same calculations for the unproven reserves yield 976, 331, and 4060 years for oil, natural gas, and coal, respectively. Granting that these latter results are subject to large errors because of the uncertainty of the unproven reserves, they still indicate that significant amounts of fossil fuels are available. But the development of these unproven reserves is difficult and expensive. The average cost of a well in the contiguous United States in 1969 was $68,736. The corresponding costs for offshore wells in the continental shelf and in the recently developed oil fields of Alaska were, respectively, $559,309 and $2,087,471. Thus we conclude that the proven reserves of oil and natural gas are short-lived. Although there are significant amounts of all fossil fuels in existence, they will be expensive, and difficult to extract.

2.3 OIL SHALE AND TAR SANDS

Petroleum from which fuel oil and gasoline are refined is pumped from natural reservoirs within the earth. Coal is mined from areas within the earth where it was formed and is ready to be used as a fuel as soon as it is mined and broken into burnable pieces. In certain regions of the earth, petroleum products exist which are not trapped in concentrated pockets but are distributed throughout some containing medium. Oil shale and tar sands are two media of potentially great significance. Both materials must be processed to obtain the petroleum products they contain.

Shale is a layered, rock structure composed of fine-grained sediments. The name oil shale suggests that liquid oil is distributed throughout the shale. The name is deceiving, though. Oil shale is a brownish rock with no oily characteristics. The petroleum product in shale is a solid material called kerogen. Like conventional petroleum, kerogen is a fossil remnant. Petroleum products are extracted from the shale by chemical methods. A high-grade oil shale deposit will produce at least 0.6 barrels of oil per ton of shale. The primary oil shale deposits in the United States are located in the Green River Formation, which extends into Colorado, Utah, and Wyoming. It is believed that 600 billion barrels of oil could be extracted from this formation alone. Clearly, such a quantity is attractive. However, the technology for recovering it on a commercial basis is not at hand. The magnitude of the technological difficulty involved in extracting the oil can be appreciated by considering the amount of shale that would have to be processed in a commercially useful source. A useful source would produce, say, a million barrels of oil a day. At a production rate of 0.6 barrels per ton of shale this means that nearly two million tons of shale would have to be processed each day. Processing such a quantity would have enormous environmental impact on the areas containing the shale. The fact that the shale actually increases in volume during the processing magnifies the problem. Some hope exists for extracting the oil without first digging the shale. Although the shale deposits have not been exploited, they are extremely attractive and developmental work toward using them is proceeding.

Tar sands mean what the words imply. Tar, a viscous, oily, petroleum liquid, is intermixed in the pore spaces of sandstone. The sandstone must be processed in order to obtain the petroleum product. As for shale, the yield per ton is around 0.6 barrels. There are substantial tar sand resources in North America but few in the United States. The most famous of these are in northern Alberta, Canada. Estimates of the amounts of extractable oil are in the range of hundreds of billions of barrels. The Canadian tar sand deposits are already producing oil on a commercial basis and continued expansion is expected.

2.4 EXPONENTIAL GROWTH OF ENERGY USE

Change is an integral part of life. The travel time from New York to San Francisco has changed from a few days by train to a few hours by plane within the last twenty years. Salaries and the cost-of-living escalate. The number of automobiles on the highways more than doubled between 1950 and 1970. Change is a fact of life. Unfortunately, some of the changes are of a type that if they are unchecked, can lead only to disaster. Energy consumption and population growth in some countries are typical of this type of change called exponential. Exponential is a common word and very likely you have encountered it before. Its use in the following quote* is typical in popular writing.

> The art of projecting technological requirements may be likened to a feat of archery. One sets down on a century-wide chart the past 70-year record (logarithmic graph paper is essential because of the exponential character of the 20th century's technology), and the archer then lines up his arrow shaft with the progression of historical points. The arrow is sent upward on a straight line course and some celestial observer notes where it crosses the terminus of the century. If this sounds somewhat zany, it really is not because most of the points on the chart fall on a straight line. They reveal man's monotonously ravenous appetite for power in all forms. The only real variation in the chart is a pronounced dip at the time of the Great Depression in the 1930s. Incidentally, kilowatts and the Gross National Product march in lockstep over the course of the past 70 years.

There are several ways of examining exponential behavior. Dr. Lapp suggests that "logarithmic graph paper is essential." It is not for our purposes. The following description suffices. If a quantity doubles (or halves) in value at regular intervals of time it is said to be increasing (or decreasing) exponentially. To illustrate suppose that you start the first day of a 31-day month with one cent. Then on each successive day you somehow double the amount you had on the previous day. On the second day you have two cents, the third day four cents, the fourth day eight cents, and so on. This doesn't seem particularly exciting or profitable but if you were to continue this through the 31st day, i.e., thirty doublings, then you would have over $10,000,000! And this is exciting! When did you get the last half of your final amount? Answer—on the last day. The point is that when a quantity changes exponentially very significant changes in that quantity occur in very short time periods once a significant amount of that quantity has accumulated.

---

* From Ralph E. Lapp, "Where Will We Get the Energy?" *The New Republic,* July 11, 1970. Reprinted by permission of The New Republic, © 1970, Harrison-Blaine of New Jersey, Inc.

The numerical ramifications of exponential growth are most easily assessed by working with the doubling time. Often, however, the exponential growth rate is given rather than the doubling time. But don't despair. The mathematics of exponential behavior lead to a useful relation between doubling time and exponential growth rate. The doubling time is determined approximately by dividing 69.3, or to a good approximation 70, by the growth rate expressed as a percentage per unit time. For example, if something is growing exponentially at a rate of 3.5 percent per year, the approximate time for that something to double is

$$\text{doubling time} = \frac{70}{3.5} = 20 \text{ years.} \tag{2.2}$$

If the rate had been 3.5 percent per minute, the approximate doubling time would be 20 minutes.

The growth of electric energy production in this country is a classic example of exponential behavior. Since 1935, electric energy production has doubled about every 10 years (Fig. 2.2).* A calculation often done to dramatize the growth of electric energy use is to determine the time required for the United States to be covered by electric power plants if the production continues to double every ten years. In 1976, all of the electric energy in the United States could have been produced by 300 large plants. Each plant might occupy a square of land about 1000 feet (about 0.2 miles) on a side so that all 300 plants would occupy about 12 square miles, which represents the area taken up by an average size city. The contiguous United States is about 3000 miles long and 1000 miles wide so that its total area is about 3 million square miles. But if

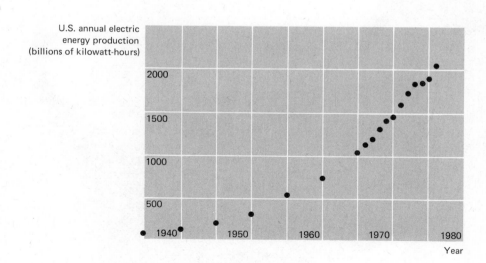

U.S. annual electric
energy production
(billions of kilowatt-hours)

Year

Figure 2.2 United States annual production of electric energy. Each successive year has seen an increase in electric energy production. The energy units are in billions of kilowatt-hours. For comparison, one billion kilowatt-hours is the energy equivalent of 588,000 barrels of oil or 118 million kilograms (130,000 tons) of coal. The data are from the Department of Energy.

* The energy units in the graphs in this chapter may not be familiar. This is not important at this time because the graphs are intended only to show the trends over the years. The energy units are defined in the next chapter.

the present exponential growth of electric energy continues it would take only 18 doublings or 180 years for electric power plants to completely cover the United States. This cannot happen, but that the trend is that way is shown dramatically in the distribution of electric generating centers in the United States for 1970 and projected for 1990 (Fig. 2.3).

The amount in any doubling period of an exponential process is more than the accumulated amount in the entire time period up to the start of the doubling period. This means, for example, that between 1963 and 1973 more electric energy was used than in the entire period up to 1963. If the use of electric energy continues to grow exponentially more will be used between 1973 and 1983 than in the entire time period up to 1973. This is a frightening aspect and points to the burden placed on the electric energy industry to supply the energy.

Figure 2.3 (a) The major steam generating centers in the United States in 1970. (b) Expected major steam generating centers in the United States in 1990. (From "Power Generation and the Environment," by Rolf Eliassen, *Bulletin of the Atomic Scientists,* September, 1971.)

(a)

(b)

Total energy use has also grown much as electric energy use has but at a lower rate. The doubling period for total energy use is about 20 years. This growth is significant and has important implications for the lifetime of the fossil fuels. To illustrate, consider the following example.

Suppose that the amount of oil in a given reserve were actually twice the original estimate. If the oil is taken at an exponentially increasing rate, does that mean that the lifetime of the reserve is doubled because there was twice the anticipated reserves? No, it does not! To illustrate, consider the time it takes to deplete to 1 billion barrels a well thought to contain 8 billion barrels if the amount taken doubles every five years starting with 1 billion barrels during the first five-year period. After 15 years, 7 billion barrels would be taken and 1 billion barrels remain. If the well contained 16 billion barrels rather than 8 billion barrels, it would take 20 years to reduce the reserve to 1 billion barrels. Hence twice the amount of oil in the well increases its lifetime by only five years or one doubling period. This point is an extremely important aspect of exponential behavior. Even though there may be large uncertainties in estimates of the reserves, it makes only small variations in the lifetime if they are being depleted exponentially. It also means that although the Alaskan oil discovery may yield 50 billion barrels, it doesn't add significantly to the lifetime of the reserves if consumption continues to grow exponentially. One of the most difficult decisions facing our society is whether we really need to continue this exploitation of energy resources.

Everyone will not agree on the estimates for lifetimes of the resources, but all must accept the fact that they are finite. When you continually draw from a finite source the end must come eventually. On a smaller scale this is exactly what happens when a gold deposit is worked. When the deposit is first discovered, the gold is easy to mine. More mining equipment and men are brought in and the production rate rises. The local economy flourishes. As the deposit nears depletion, the gold becomes harder to find and the production rate declines. Equipment and men gradually move out tending to further decrease the production. As time goes by the production rate fades away. Perhaps it does not vanish completely because a single miner may stick it out making a livelihood from the scattered gold remnants. Nevertheless, the local economy dies and a ghost town remains. In the same sense oil production will never cease completely. Hence the notion of a lifetime is a little misleading. But as the reserves deplete and it becomes harder and harder to find productive wells the production rate will decline and fade away. M. King Hubbert, a highly regarded geophysicist who has spent a large portion of his life estimating oil reserves, believes that oil production in the United States has peaked.* He speculates that the exponential rise seen in oil production will be followed by an exponential decline. He cannot be refuted at this time because oil production has steadily declined since 1970 (see Fig. 1.3). If he is correct and production declines as anticipated, 80 percent of the United States reserves of oil will be used in a 65-year period, the average lifetime of a person (Fig. 2.4).

The pattern of energy use is symbolic of many other facets of our society that are changing exponentially. Population in many countries grows

_____
* See, for example, *Science*, vol. 185, no. 4146 (July 1974): pp. 127–130.

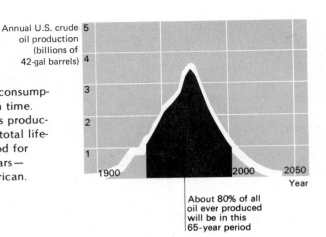

Annual U.S. crude oil production (billions of 42-gal barrels)

Figure 2.4 A possible way that consumption of crude oil will progress in time. The heavy black area represents production of 80% of the anticipated total lifetime production. The time period for 80% production is about 65 years— roughly the lifetime of an American.

About 80% of all oil ever produced will be in this 65-year period

exponentially. The doubling time for world population is about 35 years. There have been periods of exponential population growth in the United States but, while the population is still growing (Fig. 2.5), the rate is less than exponential. In fact, the birth rate in the United States has reached the level needed for zero population growth.* One might suspect that the growth in energy use reflects the growth in population, but that this is not the case is

Figure 2.5 United States population versus time.

* See, for example, "The Populations of the Developed Countries," by Charles F. Westoff, *Scientific American*, vol. 231, no. 3 (September 1974): p. 108.

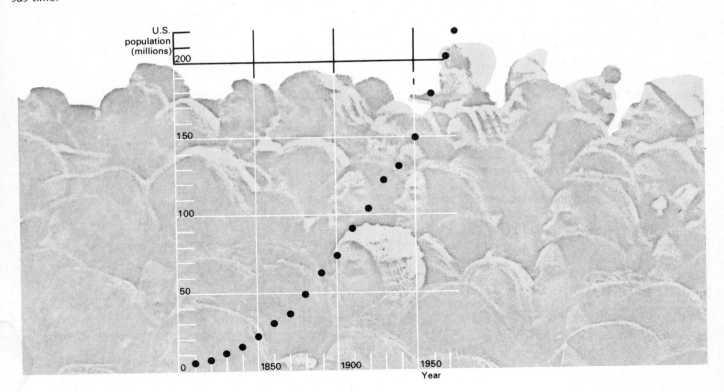

U.S. population (millions)

U.S. annual total energy consumption (equivalent of billions of barrels of oil)

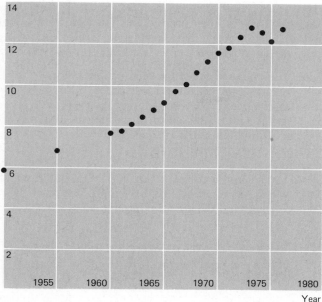

Figure 2.6 United States annual total energy consumption. The energy units are expressed in terms of the energy derivable from burning a barrel of oil. The data are from the Department of Energy.

Year

U.S. annual per capita total energy consumption (equivalent of a barrel of oil per person)

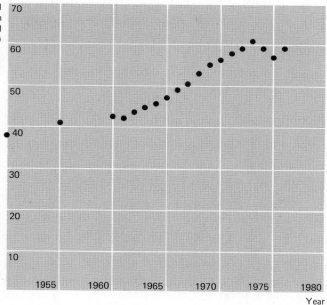

Figure 2.7 United States annual per capita total energy consumption. The leveling off (probably temporary) in per capita energy use around 1973 is produced by an overall decrease in total energy use. The data are from the Department of Energy.

Year

illustrated in Figs. 2.6 and 2.7, which show the growth in total energy and energy use per person. We are demanding more energy to run our automobiles, to air-condition our homes, to manufacture our convenience items, etc. The growth of many of these convenience items, such as aluminum cans, nonreturnable bottles, and plastic containers, is nearly exponential, and this growth has aggravated the disposal problems and placed more demands on energy sources for their production. The exponential depletion of minerals and metals, such as lead, zinc, and iron, is also of great concern. Exponential exploitation of finite resources can lead only to disaster.

2.5 SUMMARY

Fossil fuels have supplied the bulk of our energy for over a century. Their use will escalate for at least several decades although nuclear energy will tend to contribute more to the total energy budget (Fig. 2.1). Apart from the supply problems already discussed there are enormous difficulties associated with the control of pollutants that are generated when the fuels are burned. The effects and control of these pollutants and the alternatives to using fossil fuels provide the theme for our remaining study. We have been able to accomplish our initial tasks without formal study of energy, but physical concepts will need to be developed as we continue. Let us proceed, then, with the development of these concepts and the projected study.

TOPICAL REVIEW

1. Distinguish between income and capital energy.
2. Outline the various transformations of energy in the process of photosynthesis.
3. What are the current major sources of energy in the United States?
4. Distinguish between proven and unproven reserves.
5. Which is the most plentiful of the fossil fuels?
6. What are some of the difficulties involved in estimating the lifetime of the fossil fuel reserves?
7. How are amount, rate, and time related?
8. What is meant by exponential growth? Why is it so important in many environmental considerations?
9. Define doubling time.
10. What are tar sands and oil shale? Why are they of interest?

REFERENCES

Fossil Fuels and Their Role in the Energy Picture

1. The origin of fossil fuels and extraction methods are discussed in most elementary geology books. A good reference is *Man and His Geologic Environment,* David N. Cargo and Bob F. Mallory, Addison-Wesley, Reading, Mass. (1974).
2. The formation of fossil fuel is discussed in the article "The Carbon Cycle" by Bert Bolin in *Scientific American,* September 1970. The same article appears in *The Biosphere,* W. H. Freeman, San Francisco, (1970).

Lifetime Estimates for the Fossil Fuels

1. The article "The Energy Resources of the Earth" by M. King Hubbert in *Scientific American,* September 1971, is a good account of all energy resources.

2. The article "How Much in Reserve?" by F. Case Whittemore in *Environment* **15**, No. 7, September 1973, clarifies the different meanings of "reserves" and discusses many of the problems of estimating fuel resources.

3. The article "World Oil Production" by Andrew R. Flower in *Scientific American,* March 1978, shows, in detail, how the world's oil reserves are being depleted.

Exponential Growth of Energy Use

1. An extended, but still elementary, treatment of exponential behavior is given in Chapter 4 (Exponential Processes in Nature) of *My Father's Watch: Aspects of the Physical World,* by Donald F. Holcomb and Philip Morrison, Prentice-Hall, Englewood Cliffs, N.J. (1974).

2. The article "The Effect of Growth Rate on Conservation of a Resource" by R. T. Robiscoe, *American Journal of Physics* **41**, No. 5, 719 (1973) is enlightening.

Oil Shale and Tar Sands

1. A discussion of oil shale and tar sands is given in Chapter 7 of *Man and His Geologic Environment*, by David N. Cargo and Bob F. Mallory, Addison-Wesley, Reading, Mass. (1974).

2. "Oil Shale: Prospects on the Upswing . . . Again," *Science* **198**, 1023 (9 December 1977).

2.1 Use and Formation of Fossil Fuels

1. Figure 2.1 shows historically the dependence of the United States on various sources of energy. Wood, coal, and petroleum and natural gas exhibit similar historical use patterns and nuclear energy appears to be starting off the same way. Why is the use pattern of hydroenergy distinctly different?

2. The sources of energy listed below are used presently to some extent in the United States. Which are income and which are capital?

> solar
>
> nuclear
>
> geothermal
>
> trash and refuse
>
> wind

3. What would constitute income energy and capital energy for a squirrel? a bear?

4. Discuss some of the problems that arise in trying to estimate fossil fuel reserves.

5. A significant amount of oil can be derived from the shale formations in Western United States. At the present time, why aren't these shale deposits considered to be a proven reserve?

6. The first drilling for oil in the continental shelf off the coast of Eastern United States began in 1978. Does this area constitute a proven oil reserve?

7. When the price of a fuel increases how does this tend to increase the proven reserves of the fuel?

8. Figure 2.1 shows that, on a percentage basis, coal use peaked in 1910. During what two years was the coal use half the 1910 amount? If petroleum and natural gas use follows the coal use pattern, when will the petroleum and natural gas use be one half the maximum use?

9. Using the format shown below, fill in the blanks with words appropriate to the process of photosynthesis.

_____ + _____ + _____ = _____ + _____

2.2 Lifetime Estimates for the Fossil Fuels

10. Using data from Table 2.1 verify the fossil fuel lifetime estimates quoted in Section 2.2.

11. Suppose that you had a fixed amount of money, say $1 million, and you started spending this at an increasing rate. Why must this rate eventually peak and about how much money would you expect to have left when the rate peaks? What is the relation between this example and the depletion of a finite resource?

12. Estimates of the recoverable oil in the Georges Bank commercial fishing region 100 miles offshore from Nantucket Island vary from 180 million barrels to 650 million barrels. At the 1976 consumption rate of 6.0 billion barrels per year, about how many years would this offshore oil last?

2.3 Oil Shale and Tar Sands

13. The recovery of oil from shale requires considerable amounts of water. Why might this be a deterrent to the exploitation of the shale deposits in Western United States?

14. Suppose that a shale desposit yields 1 barrel of oil per ton of material processed. If the deposit yields 500,000 barrels of oil per day, what is the daily amount of shale processed? If this shale is transported by railroad cars each carrying 50 tons, how many carloads are required each day?

15. What are some trade-offs to be made if the oil shale deposits in the Western United States are developed?

16. A cube of shale 10 feet on a side weighs about 70 tons. How many cubes could be formed from 350,000 tons of shale? If a system recovering oil from shale processed 350,000 tons of shale a day and the shale was laid end to énd in cubes 10 feet on a side, how far would the blocks extend?

2.4 Exponential Growth of Energy Use

17. The exponential growth rate of electric energy in the United States is 7 percent per year. What is the doubling time? Does the doubling time agree with that quoted in the text? The doubling time for oil consumption is 20 years. What is the growth rate?

18. What are some rates that have a direct bearing on our environment?

19. The sizes of some of the numbers encountered are difficult to comprehend. For example, about 8 million barrels of oil were produced domestically each day in 1976. A barrel is roughly half the height of a person, say three feet. If 8 million barrels were laid end to end, how would the total length compare with the distance from New York to San Francisco (about 3000 miles)? This should give you an idea of the magnitude of the problem of developing a source of oil that produces a million barrels a day.

20. The following experiment is one that you can do to illustrate exponential growth. Take a sheet of 8½" × 11" paper, typically 0.003 inches thick, and cut it into two equal pieces. Then take the two pieces and cut them into four equal pieces. Continue this process of doubling the number of pieces a total of eight times. Are you surprised that the final stack is about one inch thick? Calculate how thick the stack would be if you continued for a total of 25 steps. This is a good example of how a system grows if it doubles at regular intervals.

 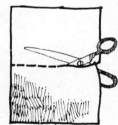 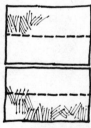 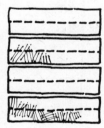

    Start   First cut   Second cut   Third cut

21. A familiar quote states that "Half of *all* the energy consumed by man since the birth of Jesus has been consumed within the last century." Knowing that worldwide energy consumption increases at an exponential rate of about 3%/year, does this quote make sense?

22. In two days a moth caterpillar can consume an amount equal to 70,000 times its birth weight. Assuming that the caterpillar consumes at an exponential rate, show that about 16 doubling periods are required. What would be the doubling time?

23. Presently the rate of electric energy production in the United States is only a millionth of the rate of solar energy flow to the earth. If the rate of electric energy production continues to double every ten years, how long will it be until the rate equals the rate of solar energy flow to the earth?

PHYSICAL BASIS
FOR ENERGY

3

Energy pervades all aspects of our lives. Converting human energy to mechanical energy with a bicycle is one of the more pleasant occasions. Photograph by Jim Horner, Miami University, Audiovisual Services.

3.1 MOTIVATION

The measuring units of tons, barrels, and cubic feet associated with coal, oil, and natural gas are fairly common and present no conceptual difficulty. Likewise, we all know that gasoline for our automobiles is purchased by the gallon and that in the operation of the automobile it is depleted at some rate— 5 gallons per day, for example. It would probably be more meaningful to purchase gasoline in terms of its energy content because that is the usage for which it is intended. Although we have some intuitive feeling for what energy is, the concept involves ideas and measuring units unfamiliar to many of us. Yet if we want to understand more fully the environmental, social, and economic impact of the utilization of our energy resources, it is important that we understand some of the physical principles and measuring units involved.

Some compromise has to be made on the depth of the treatment because complete texts and courses can be built around an energy theme. If we start with the ideas of a force (push or pull) doing work on some object by moving it some distance and energy as a capacity for doing work, it is logical to consider first the basic notions of speed, velocity, and acceleration that form a background for the concepts of forces and their action. These concepts are fundamental for several discussions in later chapters. Our treatment emphasizes those *principles* and *units* which are useful in the study of energy and our society. Much supplemental reading material is available and you may want to delve into it. Several appropriate references are given at the end of this chapter.

3.2 COMMENT ON MEASURING UNITS

Learning a given system of units is not particularly difficult. We master our monetary system at an early age. You may not realize it, but it is a good system in the sense that it is easy to convert from one unit to another, say cents to dollars. Conversion involves only a shift of a decimal point. We are not deterred from taking a trip to another country because that country has a different monetary system. We simply learn how to convert between the two systems. It may be awkward, but we do it. As children we learn our system for measuring lengths. Some of us accept the fact that it is an irrational system. You cannot convert inches to feet by moving a decimal point. The metric system, using meters as the basic unit of length, is much more sensible. You realize this fact when you travel in countries using the metric system. In the discussions that follow the unit of length will be the meter. Time will be measured in seconds. We call this system the meter-kilogram-second or mks system.* When we discuss mass (in Sec. 3.4) we will see that a kilogram is a unit of mass. Energy and power units will be expressed in these basic units. When traveling in other "energy and power" worlds using different units we will make conversions, perhaps reluctantly, but we must make them.

---

* This system of units is accepted internationally in the science community where it is designated SI, meaning Systeme Internationale.

Speed, velocity, and acceleration can all be categorized as rates (Sec. 2.2). Speed is the most common of the three. Here the amount is the distance traveled in some time period so that

$$\text{speed} = \frac{\text{distance traveled}}{\text{time period}},$$

$$s = \frac{d}{t}.$$

If a runner completes a 100-meter dash in 10 seconds, his speed would be

$$s = \frac{100 \text{ m}}{10 \text{ sec}} = 10 \text{ m/sec}.$$

Every automobile is equipped with a speedometer which continuously displays the speed of the vehicle. The speedometer in most contemporary American automobiles denotes the speed in both miles/hour and kilometers/hour. If the speedometer of a car reads 50 kilometers/hour and if this speed is maintained, then in a time period of three hours the car will travel a distance

$$d = s \cdot t$$
$$= \frac{50 \text{ km}}{\text{hr}} \cdot 3 \text{ hr}$$
$$= 150 \text{ km}.$$

It is important to note that the concept of speed avoids all reference to direction. It is easy to imagine situations where the direction in which an object travels is just as important as the distance traveled. For example, a pilot who wants to know how long it will take to fly between New York and Chicago needs to know both the speed and direction of the plane. Velocity provides the means by which the direction of travel is included in the speed. A car traveling due north at 50 miles/hour would have the same speed as one traveling due east at 50 miles/hour. Their velocities, though, would be different. Although there is this distinct difference between speed and velocity, they are often used interchangeably. As a working rule it suffices to say that if no direction is involved then you may use speed and velocity interchangeably. If a direction is implied, then velocity is the only appropriate term.

Acceleration, like speed, finds everyday usage. Alvin Toffler in his best-selling book *Future Shock* talks about the "acceleration of change in our time." Every driver knows that he must depress the accelerator pedal to increase the speed of his automobile. All automobiles do not, however, respond in the same way to a depression of the accelerator. Some will change speed much faster than others. The acceleration response of a car is often characterized by the time it takes to achieve a certain speed, say 60 miles/hour, starting from rest. In the comparative evaluation of two cars, the car achieving the designated speed in the *least* time would have the *greater* acceleration. All these everyday observations and ideas are incorporated in a simple rate expression:

$$\text{acceleration} = \frac{\text{change in speed}}{\text{time required for this change to occur}}.$$

To be a little more rigorous, direction should be included and acceleration defined in terms of a change in velocity rather than change in speed. In other words, an object can be accelerated by changing either (or both) the speed and direction. For example, a car traveling at a constant speed of 50 kilometers/hour on a circular race track is accelerating because its direction is continually changing (see Fig. 3.1).

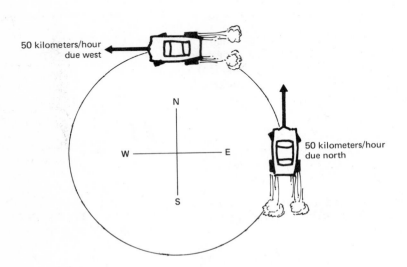

Figure 3.1 A car traveling at a constant speed of 50 kilometers per hour on a circular racetrack is accelerating because its direction is changing continually.

3.4 NEWTON'S THREE LAWS OF MOTION

Although force is a word used in a variety of contexts, it always implies a stimulus for some kind of motion. As examples: we are *forced* to get up for a 7:30 A.M. class, or she was *forced* off the road by another car. Likely, a push or pull is the simplest physical illustration of a force. Newton's three laws of motion provide both a verbal and a quantitative base for discussion of forces acting on some object. Newton's statement of the first law reads, "Every body continues in its state of rest, or of *uniform motion* in a straight line, unless it is *compelled* to change that state by *forces* impressed on it." Uniform in this context means constant and the motion refers to velocity. So if the velocity is constant (both in magnitude *and* direction), the acceleration is zero. The velocity can be changed only by applying a force. Whenever an object is observed to be accelerating, Newton's first law suggests looking for the force responsible for the acceleration. Often this force which produces the acceleration is obvious; other times it is not. A batted baseball is accelerated by the force that the bat exerts on the ball. A ball dropped from the window of a building accel-

erates toward the ground. But what is the force that causes it to accelerate? It turns out that the earth, though not in contact with the ball, pulls on the ball and accelerates it. This force is commonly called gravity.

Suppose that an experiment is conducted in which a moving automobile makes a collision with an automobile which is initially at rest. The struck automobile is accelerated by virtue of a force exerted on it by the moving automobile. We should expect that if the force increases, the acceleration also increases. If the moving automobile always exerts the same force, then we should expect that the struck automobile always achieves the same acceleration. For the sake of argument, suppose that a force of 100 units produces an acceleration of one unit when there are no occupants in the struck automobile. Now suppose that we loaded the stationary automobile with passengers. Would we expect a force of 100 units to produce an acceleration of one unit? Surely we would not. We would probably all agree that the acceleration of the loaded automobile would be less. You need not do this experiment if you remain unconvinced. You might prefer kicking different objects on the floor with the same force. Every object offers some resistance to being put into motion. This resistance is due to the property of mass.

You are probably more accustomed to thinking of mass as a measure of the weight of an object or the amount of matter contained by the object. The ideas of resistance to being put into motion and quantity of matter are equivalent, however. As a way of assigning a number to measure mass we might use a simple balance (similar to that used in a post office to "weigh" a letter) that compares the unknown mass with a known mass (Fig. 3.2). The units used will be kilograms symbolized as kg.

Newton's second law of motion quantifies all these ideas by stating that the acceleration experienced by an object is proportional to the *net* force acting on it and is inversely proportional to its mass:

$$\text{acceleration} = \frac{\text{net force}}{\text{mass}},$$

$$a = \frac{F}{m}, \quad \text{or} \quad F = ma.$$

Using Newton's second law, we can determine the unit of force:

$$F = ma,$$

$$= \frac{\text{kg} \cdot \text{m}}{\text{sec}^2}.$$

One kg · m/sec² is called a newton (N). If a 1-kg mass experiences an acceleration of 1 m/sec², then the force on the mass is 1 N. There are many secondary methods of measuring forces such as with a spring balance, but they are all consistent with this basic scheme.

Newton's third law of motion states that if one object (call it *A*) exerts a force on another object (call it *B*), then *B* exerts an equal but oppositely directed force on *A*. To illustrate, consider the colliding automobile example dis-

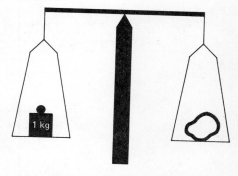

Figure 3.2 A simple balance used to determine the mass of an object. The pans on which the objects rest achieve the same level when the masses are equal.

Figure 3.3 Illustration of the forces on two colliding cars. Newton's third law of motion requires that the two colliding autos experience equal, but oppositely directed, forces.

cussed in connection with Newton's second law. If the moving automobile exerts a force on the struck automobile that pushes it forward, the struck automobile exerts an equal but oppositely directed force on the moving automobile that tends to push it backwards (see Fig. 3.3). Newton's third law does not seem like a particularly useful concept in terms of calculating the motion of objects or systems. It is, however, an extremely important concept and is particularly useful when trying to understand how atoms and molecules are held together.

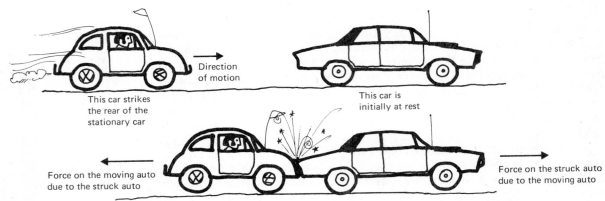

Direction of motion

This car strikes the rear of the stationary car

This car is initially at rest

Force on the moving auto due to the struck auto

Force on the struck auto due to the moving auto

Mass

Arrow denotes gravitational force acting on the mass

Earth

Figure 3.4 A mass is pulled toward the earth by a gravitational force. The arrow denotes the direction of the force acting on the mass.

As an illustration of these ideas, let us verbally and analytically examine the motion of an object falling vertically toward the surface of the earth. A practical example of this is water falling vertically from the top of a dam in a power plant producing electricity.

Stop for a moment and perform this experiment. Hold two objects, say a pencil and a coin, the same distance above the floor and release them at the same time to the best of your ability. You probably conclude that they struck the floor at the same time. On the other hand, if you did the same experiment using a coin and a small scrap of paper, the coin would surely strike the floor before the paper. You might conclude that the coin falls faster because it is more massive. But if you were to use a coin and a pad of paper (which obviously is more massive) dropped with its surface parallel to the floor, you would see that the coin wins again. Observation of the flights of the dropped objects reveals that the air affects the motion. It is not easy to contrive a set of experimental conditions to remove the air surrounding the objects being dropped. In a moderately sophisticated laboratory it is possible to evacuate a reasonably large container; on the moon which has no atmosphere, the Apollo astronauts could drop objects in an air-free environment. Under these conditions two objects dropped simultaneously will take the same time to travel a given distance. In many situations even though the air is present its effect is minimal and can be neglected. Under this assumption gravity is the only force acting on an object. When gravity is the only force acting, experiment demonstrates that the acceleration is the same for all masses. This acceleration is given the symbol g

and is numerically equal to 9.8 meters per second per second. That is to say, the velocity changes by 9.8 meters per second for every second that the object falls. The force of gravity on an object is called its weight. Thus, using Newton's second law, the weight of an object is simply its mass multiplied by the acceleration due to gravity. In symbols, $W = mg$. The connection between mass and weight is now clear. They are proportional but not equal. Weight is a force and in the mks system of units is expressed in newtons. In our everyday world, the unit of weight is the pound. For comparison, one pound is equivalent to 4.45 newtons.

In everyday language, the word "work" is associated with labor or toil. In this sense, it can be thought of as "a physical or mental effort directed *toward* the production of something." From the physics standpoint, the effort is supplied by a force, and if a force acts on a mass while it moves through some distance (d), work is said to be done. The amount of work is defined as the product of force and distance:

work = force × distance

$W = Fd$.

Work has the units of force times distance, that is, kg-m²/sec². One kg-m²/sec² is called a *joule* (J). Although the physical definition of work retains much of the popular notion, there is a subtle but important difference. No matter how much force (or effort) is exerted, no work is done in the physics sense if the mass does not move. It is important also to recognize that a force can act either in the same direction as the mass is moving or opposite to the direction in which the mass is moving (Fig. 3.5). We distinguish between these two physically different situations by calling the work positive if the force is in the direction of motion, and negative if the force is opposite to the direction of motion. The *net* work on an object is the algebraic sum of the works done by each force acting on the object. For example, suppose that two forces of +100 N and −50 N caused an object to be moved a distance of +2 meters. Then the work due to the two forces, respectively, would be (+100) (+2) = +200 J and (−50) (+2) = −100 J. The net work would be +200 − 100 = +100 J. The reason for making such a distinction will become clearer in the next section.

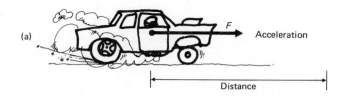
(a)

F

Acceleration

Distance

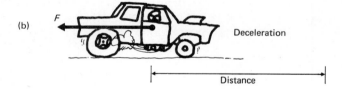
(b)

F

Deceleration

Distance

Figure 3.5 (a) Force acts in the same direction that the car moves tending to increase its speed. This is a case of positive work. (b) Force acts in a direction opposite to that in which the car moves tending to decrease its speed. This is a case of negative work.

3.6 ENERGY AND WORK

Energy in almost any context can be thought of as the capacity for doing work. Thinking in terms of physics we might ask, "Under what conditions does something have a capacity for moving an object through some distance?" A car in motion has this capacity because if it rams into the rear of a car stopped at a red light, the struck car will surely move some distance. The energy associated with motion is called *kinetic* energy. Experience tells us that a big, massive car colliding with a stopped car will "do more work" on the car. Similarly, a fast moving car will "do more work" on a stopped car than a slowly moving car of the same type. Thus a formal mathematical definition of kinetic energy should reflect these observations. Formally, the kinetic energy of a mass m moving with speed v is defined as one half of the product of mass and square of velocity:

$$E_k = \frac{1}{2} m v^2.$$

There are some subtle reasons for choosing this particular form, but they are not essential for our discussion. Nevertheless, as required, it shows that kinetic energy increases if either or both the mass and speed increase. The units are kg-m²/sec², which are precisely those of work. If we retain the concept of energy as the capacity for doing work, it would have probably been more surprising had work and energy not had the same units. The work–energy principle ties these concepts together quantitatively by stating that the *net* work done on a mass is equal to the *change* in kinetic energy of the mass:

net work = change in kinetic energy

= final kinetic energy − initial kinetic energy

$$W_{net} = E_k \text{ (final)} - E_k \text{ (initial)}$$

$$= \frac{1}{2} m v_f^2 - \frac{1}{2} m v_i^2 .$$

where v_f and v_i are the final and initial speeds, respectively. The reason for distinguishing positive and negative work is now clearer. If W is positive, the final kinetic energy is greater than the initial kinetic energy. The converse is true if W is negative.

You see this principle in operation in many processes of interest to us. The kinetic energy of water flowing over a dam increases if its velocity increases. This increase in kinetic energy is due to work done on the water by the force of gravity. When a car is set into motion starting from rest, its kinetic energy increases. This is due to work done by a force in the direction of motion of the car. Likewise, when the car slows down, its kinetic energy decreases. This is due to (negative) work done on the car by a force acting in a direction opposite to the direction of motion.

A car at rest has no kinetic energy, yet it is *potentially* capable of moving another object *if* it can be put into motion. The potentiality arises from the intrinsic energy of the gasoline that, when burned in the automobile engine, can be *converted* to energy of motion. Water at the top of a hydroelectric dam is capable of exerting a force on the rotor of a turbine *if* it is released. Before release, the energy is potential. Once released and set into motion, the water possesses kinetic energy. There is potential energy in coal, oil, and gas because *if* ignited, each would release heat energy. There is also potential energy in the mass of the nucleus of the atom which can be released *if* an appropriate reaction is initiated.

The flow of water over a dam and onto the blades of a rotating paddle wheel provides an ideal illustration of the connections between kinetic energy, potential energy, and work. Because it is easy to visualize the motion of an object that has some boundaries, let us follow the motion of a plastic bag of

3.7 POTENTIAL ENERGY

Figure 3.6 The potential energy of dammed water is converted to electric energy at the Fontana hydroelectric station located on the Little Tennessee River in North Carolina. This 480-foot high dam is typical of the many dams in the Tennessee Valley Authority. (Photograph courtesy of the Tennessee Valley Authority.)

Figure 3.7 Energy transformations involved in water flowing over a dam.

Position 1. At the top of the dam, a bag of water has kinetic energy ($\frac{1}{2}mv^2_{top}$) due to its motion, and potential energy (mgh) due to its position relative to the bottom of the dam.

Position 2. At the bottom of the dam, all of the potential energy has been converted to kinetic energy. The energy of the bag of water is all kinetic and includes the kinetic energy it had atop the dam plus the kinetic energy acquired by falling over the dam.

Position 3. Passing by the paddle wheel, the bag of water transfers energy (W) to the wheel. Its energy after passing the wheel is still all kinetic and is equal to its energy before impinging on the wheel minus the energy imparted to the wheel.

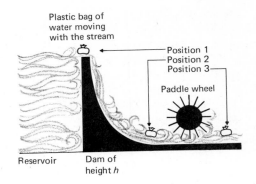

water moving along with, and with the same speed, as the water (Fig. 3.7). At the top of the dam, the bag of water has kinetic energy ($\frac{1}{2}mv^2$) by virtue of its mass and speed, and potential energy (PE) by virtue of its position at the top of the dam. As the bag flows from the top of the dam to the bottom, work is done on it by the gravitational force. This work is just the product of the weight (mg) and the distance (h) through which it falls. Its total energy at the bottom of the dam includes the energy it had at the top plus the energy acquired through falling. Potential energy atop the dam + kinetic energy atop the dam = kinetic energy at the bottom of the dam:

$$mgh + \frac{1}{2}mv^2_{top} = \frac{1}{2}mv^2_{bottom}.$$

The bag of water arrives at the paddle wheel with the speed it has at the bottom of the dam. It exerts a force on the paddle wheel and does work (W) on it producing rotational motion. After it has passed the paddle wheel, its speed decreases because it has lost energy to the paddle wheel. Kinetic energy beyond the paddle wheel = kinetic energy at the bottom of the dam − work done on the paddle wheel.

$$\frac{1}{2}mv^2_{beyond\ wheel} = \frac{1}{2}mv^2_{bottom} - W.$$

It is somewhat more difficult to visualize the energy transformations in an automobile engine, for example, by visualizing moving particles. The idea, though, is the same, and it is always revealing to follow the energy flow and transformations in any energy system.

3.8 ENERGY CONVERSION, CONSERVATION OF ENERGY, AND EFFICIENCY

We have dwelled at some length on the usefulness of energy in many facets of our society. When put to use, energy is kinetic. An automobile is useful when it is in motion. Heat energy involves the kinetic energy of atoms and molecules. Electricity, probably the most useful form of energy, results from the motion of electric charges. Useful energy is always derived by converting some other form of energy into the useful form through the use of some appropriate converting device. For example, heat energy is useful for heating a room in a home and can be derived by converting the intrinsic chemical potential energy in a fossil fuel through burning in a furnace. An idealist might expect to get all the energy converted into the useful form. A realist suspects from the way things work in nature that you cannot expect to get all the energy converted into the useful form. Some of the energy will be diverted to other areas or uses. In physics, this idea is contained in the principle of conservation of energy or equivalently, the first law of thermodynamics, that reads—there is no loss of energy in any energy transformation. If 1000 joules of energy are derived from burning a certain amount of gasoline (the potential energy) in an engine, then there will still be 1000 joules after the burning. Perhaps only 300 joules were transformed to the useful form, i.e., mechanical energy of motion. The other 700 joules were diverted to other forms.

The conservation-of-energy principle provides an indispensable book-keeping procedure. It is, in a sense, a budget where energy rather than dollars is being dispersed. Like any budget, the numbers must total correctly.

Your first thought may be that this concept is inconsistent with concerns for energy supplies. After all, if there is no change in the total amount of energy, why worry about energy shortages? Just because there is energy present does not mean that it can be used or converted to some usable form. The basic conversion mechanisms divert energy to areas where it cannot be recovered. In particular, energy that is diverted to the environment as heat cannot be recovered, and the ultimate disposition of all energy is as heat. Even though the total amount of energy remains unchanged in any energy conversion, the energy ultimately ends up as heat which cannot be recovered.

As a practical matter we would like to know what fraction of the energy converted in any process gets into the useful form. This fraction is designated as the efficiency:

$$\text{efficiency} = \frac{\text{useful energy out}}{\text{total energy converted}},$$

$$\epsilon = \frac{E_{\text{out}}}{E_{\text{converted}}}.$$

Because the useful energy is always less than the total energy converted, the efficiency is always less than unity.* For example, if 300 joules of useful mechanical energy are derived from the conversion of 1000 joules of energy in gasoline, then the efficiency of the engine is

$$\epsilon = \frac{300}{1000} = 0.3.$$

The efficiency is expressed often as a percent. This simply means that you multiply the efficiency by 100. The engine in the previous example has an efficiency of 30 percent. Remember, that in any numerical manipulation, the efficiency has to be expressed as a decimal or fraction. For example, the efficiency of an ordinary light bulb is five percent. The useful output from a light bulb is light energy. The energy converted is electric energy. Thus if 5,000,000 joules

---

* This definition, which incorporates the first law of thermodynamics, is often designated "first law efficiency." In Chapter 6 we shall discuss an efficiency based on the second law of thermodynamics and to avoid confusion we shall call it "second law efficiency."

Table 3.1 Efficiencies of a few important energy conversion devices. A more complete list can be found in "The Conversion of Energy" by Claude M. Summers, *Scientific American* **224**, No. 3, September 1971.

Device	Energy conversion function	Efficiency
Fluorescent lamp	electrical to light	20%
Automobile engine	chemical to thermal to mechanical	25%
Steam turbine	thermal to mechanical	47%
Home oil furnace	chemical to thermal	65%
Electric generator	mechanical to electrical	99%

of electric energy are converted by a light bulb, the amount converted to light energy is

$$\text{light energy} = \text{total energy converted} \times \text{efficiency}$$
$$= (5,000,000 \text{ joules}) \times 0.05$$
$$= 250,000 \text{ joules.}$$

The remainder of the energy that was converted, 4,750,000 joules, emerges as heat energy. The reason that a light bulb tends to get very hot is because most of the energy converted by the bulb goes into heat.

Several efficiencies for some important energy conversion processes are given in Table 3.1. Note that most of these are quite low. We will see that these low efficiencies have contributed to many of our problems resulting from conversion of energy.

The energy for our society, 93 percent coming from fossil fuels, is converted to useful forms at an overall efficiency of about 50 percent. The diverted energy enters the environment as heat. For every unit of useful energy, nearly one unit is wasted as heat. This means that if you can save one unit of useful energy, you then save nearly two units of capital energy. Thus the potential for energy conservation is significant.

While the principle of conservation of energy allows you to account for each joule of energy converted, it says nothing about having converted the energy in the most optimum way. If, for example, the useful energy is needed to heat a home, the principle of conservation of energy does not help you select the best method of conversion. Nor does it tell you the optimum use of a given source of energy. For example, the energy derived from burning natural gas can be used in certain types of engines as well as for heating water and providing household heat. While the engine *must* have the high temperature provided by the burning gas, household heating does not. Thus a case may be made that the most efficient use of natural gas is not in household heating. We shall find that the second law of thermodynamics (to be discussed in Chapter 6) often provides a more realistic evaluation of energy use and energy converters.

3.9 POWER

It is common for a person to have to push a stalled automobile. So if a friend tells you that he did a certain amount of work in pushing his automobile a distance, say 10 meters, you would not be particularly impressed. On the other hand, if he told you he did it in one second you would probably be very impressed because he had done work at an impressive rate. The rate of doing work is a significant quantity. There may exist an enormous energy resource but if it cannot be extracted at a rate appropriate to society's needs, it is of little use. This rate of doing work is called power.

$$\text{Power} = \frac{\text{amount of work done}}{\text{time required to do work}},$$

$$P = \frac{W}{t}.$$

The units of power are joules per second. One joule per second is called a watt (W). An operating 100 W light bulb uses energy at a rate of 100 joules per second. The horsepower (hp) rating of an engine is a measure of the rate at which the engine does work. One horsepower is equivalent to 746 watts. If an engine develops 200 hp, it is doing work at the rate of

$$200 \text{ hp} \times 746 \text{ W/hp} = 149,200 \text{ W} = 149,200 \text{ J/sec}.$$

An electric motor of a small household appliance like a food mixer develops only a fraction of a horsepower, say 1/6 hp. This equals about 120 W.

If the mks system of units were universal, then the units of joules for energy and watts for power would suffice. But as stated at the outset we do not operate in a world where joules and watts are the last word. Thus to travel in these other "worlds" we must conform and convert units as required. Table 3.2 is provided to help in this chore. For example, units in column 1 are convertible

3.10 OTHER COMMON ENERGY AND POWER UNITS

Table 3.2 Some useful energy and power units.

Special energy unit	Study area of main use	Symbol	Equivalent in joules	Other useful equivalents
kilowatt-hour	electricity	kW-hr	3,600,000	3413 Btu, 860 kcal
calorie	heat	cal	4.186	
kilocalorie (food calorie)	heat	kcal	4186	1000 cal
British thermal unit	heat	Btu	1055	252 cal
electron volt	atoms, molecules	eV	1.602×10^{-19}	
kiloelectron volts	X-rays	keV	1.602×10^{-16}	1000 eV
million electron volts	nuclei, nuclear radiation	MeV	1.602×10^{-13}	1,000,000 eV
quintillion	energy reserves	Q	1.055×10^{21}	10^{18} Btu or a billion-billion Btu
quadrillion	energy reserves	quad	1.055×10^{18}	10^{15} Btu or a million-billion Btu

Continued

Table 3.2— *Continued*

Special power unit	Study area of main use	Symbol	Equivalent in watts
British thermal unit per hour	heat	Btu/hr	0.293
kilowatt	household electricity	kW	1000
megawatt	electric power plants	MW	1,000,000
gigawatt	total U.S. or world power production	GW	1,000,000,000
horsepower	automobile engines, motors	hp	746

Energy Equivalents

1 gallon of gasoline = 126,000 Btu   1 cubic foot natural gas = 1030 Btu
1 pound bituminous coal = 13,100 Btu   1 42-gallon barrel of oil = 5,800,000 Btu

Variations of these energy equivalent values will appear in the literature. The values quoted here are typical.

to equivalent units by multiplying by the numbers given in columns 4 and 5. Thus, in joules, 2000 Btu are equivalent to

$$2000 \text{ Btu} \cdot 1055 \frac{\text{joules}}{\text{Btu}} = 2,110,000 \text{ joules}.$$

The origin of many of these units will be explained as we proceed.

3.11 ENERGY CONVERSION SYMBOLISM

The analysis of the wide variety of energy conversion processes using the ideas put forth in this section is the essence of this book. It will be useful to characterize an energy conversion process with the diagram in Fig. 3.8. The principle of conservation of energy allows us to write

total energy converted = useful energy out + energy diverted.

The concept of efficiency characterizes the ability of the system to convert energy to the useful form.

The diagram is not intended to picture the actual converting device. It is used to show the energy pathways and help with the energy bookkeeping. Although these diagrams are extremely useful and essential for our study, one should never lose sight of the complexity of an energy converting system like an electric power plant and the engineering knowledge involved in producing energy at the enormous rates required by our technological society.

A seemingly inescapable feature of energy conversion is the generation of by-products which produce troublesome environmental effects. The automobile engine in its capacity as a converter of chemical energy to mechanical

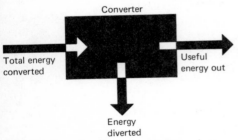

Figure 3.8 Schematic description of an energy conversion system. Regardless of the nature of the converter (engine, person, battery, etc.) the sum of the useful and diverted energies equals the total energy fed into the converter.

energy produces carbon monoxide, hydrocarbons, and nitrogen oxides that in turn have produced undesirable effects in our cities. Coal-burning electric power plants produce sulfur oxides and fine particulate matter that bother persons having respiratory ailments. The control of these emissions involves money and energy. Above all, incentives must exist for wanting to control these emissions. Chapter 4 is devoted to an assessment of these environmental effects.

1. Distinguish between speed and velocity.

. What is acceleration? What are some units of acceleration?

. Define force in both the physics and the everyday senses.

. Formulate Newton's three laws of motion in words. Give physical and nonphysical (if possible) examples of the application of the laws.

. Distinguish between energy and work. What are their units?

. Distinguish between kinetic and potential energy. What are their units?

. What is the physics meaning of "conservation of energy"?

. What is meant by energy conversion? conversion efficiency?

. Distinguish between energy and power. What are their units?

. What is a kilowatt? a megawatt?

Physics of Energy and Power

Any standard physics or physical science text will devote a substantial number of pages to the study of energy and power. Two such texts useful for expansion of the material presented in Chapter 3 are

1. *Physics: A Descriptive Analysis,* by Albert J. Read, Addison-Wesley, Reading, Mass. (1970).
2. *Conceptual Physics,* by Paul G. Hewitt, Little, Brown, Boston (1977).

Conversion of Energy

The conversion of energy is covered in some detail by Claude M. Summers in *Scientific American,* September 1971.

3.1 Motivation

1. Scientifically, energy is thought of as "the capacity for doing work." How is this notion related to the everyday idea of work as applied to a person or factory?

SUGGESTIONS FOR FURTHER UNDERSTANDING OF CHAPTER 3

Further Understanding of Chapter 3   43

3.2 Comment on Measuring Units

2. A gram is a thousandth of a kilogram. How many grams are there in 12 kilograms? Devise a rule involving the decimal point for converting kilograms to grams. An ounce is a sixteenth of a pound. How many ounces are there in 12 pounds? Why can't you devise a rule involving the decimal point for converting pounds to ounces?

3. A consumer certainly would not argue against the importance of price units. It makes a lot of difference whether something costs ten dollars or ten cents per unit. Units for physical quantities are equally important. Using the definitions of kinetic energy and work, show that they have the same scientific units (kg m²/sec²). The work–energy principle equates the work done on an object to the change in kinetic energy. When applying this principle why can't you express the work in joules and kinetic energy in kilowatt-hours even though joules and kilowatt-hours are legitimate units of energy?

3.3 Speed, Velocity, and Acceleration

4. Dividing the distance traveled from your school to your home in kilometers by the time required for the trip yields kilometers per hour. In doing this, have you determined a speed or a velocity?

5. In terms of rates, what does it mean when we say the nationwide use of electric energy is accelerating?

6. When specifying the daily change in the energy output of a power plant, how could you specify mathematically whether the change was an increase or a decrease?

3.4 Newton's Three Laws of Motion

7. Suppose that you had two identically shaped footballs, one filled with air, the other with sand. What differences in motion would you expect when they are kicked from a resting position on the ground with the same force?

8. Alvin Toffler in *Future Shock* writes "The acceleration of change in our time is, itself, an elemental force." What analogies can you draw between this statement and Newton's second law of motion?

9. Comment on the following quotation from *The Quantitative Society* by John Wilkinson. "Massive human institutions, like physical bodies, possess an 'inertia' that keeps them more or less in the motion and trajectory of the present . . . correspondingly massive forces are needed to bring about either a change of direction or a sensible deceleration and . . . such revolutionary forces, when applied, often bring about a state of affairs neither predicted nor predictable."

10. Shortly after a starting signal is given, a runner acquires a velocity. Newton's second law requires that there be a force on the runner to pro-

duce the acceleration. Identify the force, keeping in mind that it must act in the direction of the acceleration.

11. A pound is a measure of force. A gram is a measure of mass. It is customary to compare on the label of a can of food the contents in pounds and in grams. Why is this comparison inconsistent with the notions of force and mass?

12. Imagine two persons tugging at opposite ends of a rope. Pick out some action and reaction forces in this system.

13. A baseball acquires energy from the force of a bat hitting it. What is the reaction force to the force exerted on the ball?

14. Water at the top of a dam is pulled downward by a gravitational force acting on it. What is the reaction force to the force exerted on the water?

3.6 Energy and Work

15. A person tugging feverishly on a rope attached to a car fails to move the car. Has any work been done? Has any energy been expended?

16. In order to maintain a car at constant speed on a straight, level highway it is necessary to depress the accelerator pedal on the car. Still, what is the *net* work done on the car?

17. Give some examples of positive work.

18. Give some examples of negative work.

19. Work must be done on a car to reduce its speed from 50 to 40 miles per hour. Is this work positive or negative? What is the disposition of the kinetic energy when the speed is changed from 50 to 40 miles per hour?

20. Why does a compressed spring have "a capacity for doing work?" Why does it possess energy?

21. If the speed of a car doubles, what change is there in its kinetic energy?

22. Water at the top of a dam has potential energy by virtue of its position. If the height of the dam doubles, the potential energy doubles. Why won't the speed of the water after it flows over the dam also double if the height of the dam doubles?

3.7 Potential Energy

23. An aerosol spray can used to hold household items such as deodorants, is a potential energy source. Think of some other examples of potential energy in a household.

24. A roller coaster is a popular amusement park ride in which an open railroad-type car is released from an elevated position on a curved, undulating track. Discuss the potential and kinetic energy transformations that take place during the travel of a roller coaster. What are some other amusement park rides involving energy transformations?

3.8 Energy Conversion, Conservation of Energy, and Efficiency

25. What energy transformations are involved when a nail is driven into a board with a hammer?

26. There is considerable concern over the depletion of primary metal resources such as copper. But metals like copper are not lost; they are simply redistributed to the extent that it is impractical to recover them. In what ways is this similar to the principle of conservation of energy and the concerns over depletion of energy resources?

27. From the time a bowler picks a bowling ball from a rack beside the bowling lane to the time the ball returns to the rack, the ball is involved in several energy transformations. Trace these transformations from the beginning to the end of the trip.

28. Water impinging on a paddle wheel imparts energy to the wheel. Why is it that all of the energy of the water cannot be transferred to the paddle wheel?

3.9 Power

29. A typical household air conditioner will be rated as 8000 Btu. This rating alludes to the quantity of heat that it is capable of removing from a room. Although time is not mentioned explicitly in this rating why is it important as far as the effectiveness of the air conditioner is concerned?

30. Power and speed are examples of rates. Knowing that distance is equal to speed multiplied by time, what results by multiplying power with time?

31. Why would a megawatt power unit multiplied by a year be a unit of energy?

32. A certain light bulb uses 2 kW-hr of electric energy over a certain time period. Why is it not possible from this information to determine how long the bulb was on or the wattage rating of the bulb?

NUMERICAL PROBLEMS   3.3 Speed, Velocity, and Acceleration

1. Some particles released from a smokestack settle toward the ground at a rate of 0.004 centimeters/second. How long does it take these particles to settle a distance of 1 kilometer?

2. A modern coal-fueled electric power plant uses coal at a rate of about 4000 tons/day. Anticipating a curtailment of supplies, the company desires to have a two-month supply on hand. How much coal will be required?

3. Suppose that a car obtains 15 miles per gallon of gasoline when it averages 45 miles per hour. One might also want to know how many gallons per hour the car uses at this speed. Determine this by combining the units, mi/gal and mi/hr, in a way that gives units of gal/hr.

4. An object falling freely in the atmosphere accelerates at a rate of 32 feet /second per second.

a) Show that this corresponds to an acceleration of 35 kilometers/hour per second.

b) A person stepping off a diving board 5 meters high takes about 1 second to reach the water. What is the person's speed in miles per hour when the water is reached? (You may want to use this information: 39.37 inches = 1 meter and 1 kilometer = 0.62 miles.)

5. A train on the Bay Area Rapid Transit (BART) system has the ability to accelerate to 80 miles/hour in half a minute.

a) Express the acceleration in miles per hour per minute.

b) Starting from rest, how fast will one of these trains be moving after 20 seconds?

6. Velocity, acceleration, force, and inertia, for example, are physics terms that are often used in contexts that have no scientific connotations. Arthur Schlesinger, an adviser to President Kennedy, in a public address spoke of the "velocity of development of technology." As an example, he considered the distance that man could travel in 3 hours at various times in history. Limited to walking he could go about 10 miles, and riding a horse, about 25 miles. Consider the effect of the railroad and piston and jet airplanes, and estimate these distances as a function of the time they appeared in history. If this trend continues, estimate the distance man might travel in 3 hours in the year 2000. Do you see any form of commercial transportation being developed that will allow you to travel this estimated distance in the year 2000?

3.4 Newton's Three Laws of Motion

7. Knowing that one pound is equivalent to 4.45 newtons, express your weight in newtons. Why is it unlikely that the newton, which is the metric unit of weight, will catch on as a popular measure of body weight?

3.6 Energy and Work

8. A person riding a bicycle at 12 miles per hour uses about 25 food calories for each mile of travel. How many miles of cycling at this speed are required to use up 1000 food calories of energy? Make some judgment as to the effectiveness of cycling for losing weight knowing that a person uses about 2500 food calories per day.

3.8 Energy Conversion, Conservation of Energy, and Efficiency

9. At the bottom of a hill, the engine of a car stalls and the car coasts to the top of the hill. At the top, the kinetic energy is one four   kinetic energy at the bottom. In words, write an equation depicti   energy transformations of the car from the bottom to the top of the

10. A car having a mass of 1000 kg starts from rest from the to   hill 100 meters high. At the bottom of the hill the car is moving 30 m   second.

The energies involved in this situation are kinetic $\frac{1}{2} mv^2$, potential (mgh), and friction (call it FR).

a) Write an equation involving the energies involved.

b) Determine the energy due to friction.

11. A car atop a hill has potential energy mgh. Starting from rest at the top of the hill, it acquires kinetic energy $\frac{1}{2} mv^2$ at the bottom. All the initial potential energy does not go into kinetic energy because some energy is lost through frictional forces.

a) Show that the efficiency for converting potential energy to kinetic energy is $v^2/2gh$.

b) Evaluate the efficiency for a hill 50 meters high if the speed is 20 meters/second at the bottom of the hill.

12. An oil furnace for a house has an efficiency of about 65% for converting the chemical energy of oil into heat for the house. How many Btu of heat are delivered to a house by an oil furnace burning 100 gallons of oil? (Refer to Table 3.2 for the energy content of a barrel of oil.)

13. A steam turbine often is used to drive an electric generator that produces electric energy. Although this involves two distinct energy conversion processes, it can be thought of as a single system that converts heat energy to electric energy. If the efficiencies of the engine and generator are 50 percent and 90 percent, respectively, show that the overall efficiency of the combined system is 45 percent. A straightforward way of doing this is to "feed" one unit of energy into the engine and follow it through the system seeing how much electric energy can be derived from it.

14. A steam engine converts heat energy to kinetic energy. Typically, it takes 2 Btu of heat energy to produce 1 Btu of kinetic energy.

a) Draw an energy diagram like that shown in Fig. 3.8 which characterizes this typical steam engine.

b) Assuming that in a given operation the engine produces 1,000,000 Btu of energy, determine the input energy and diverted energy and label these on your diagram.

c) What is the efficiency of this engine?

3.9 Power

15. How many joules of energy are delivered in one hour by an engine developing 10 horsepower?

16. Chauncey Starr in his article in *Scientific American* **224**, 36 (Sept. 1971), uses watt-years as an energy unit. Why is a watt-year a unit of energy? It is stated that 1 watt-year is equivalent to 29.9 thousand Btu. Prove this statement given that 1 watt-hour is equivalent to 3.413 Btu.

17. A person applies a 200 N vertical force to lift a box through a distance of 1 meter.

   a) How many joules of work are done?

   b) How much power is developed if the work is accomplished in 4 seconds?

   c) How does this power compare with the power developed by a motor on a household food mixer? (See Sec. 3.9.)

3.10 Other Common Energy and Power Units

18. Burning a pound of coal produces about 13,000 Btu of energy. How many joules of energy are produced by burning a pound of coal?

19. A person derives energy from food at the rate of about 2500 food calories per day. Show that this is equivalent to a consumption rate of about 100 watts.

20. Heat energy for a steam engine that drives an electric generator is normally expressed in British thermal units (Btu) but the electric energy output is normally expressed in kilowatt-hours (kW-hrs). Knowing that 1 kW-hr equals 3,600,000 joules and 1 Btu equals 1055 joules, show that 1 kW-hr equals 3413 Btu.

21. John C. Fisher in his article "Energy Crisis in Perspective," *Physics Today,* December 1973, uses an energy unit symbolized C. One C is equivalent to 10^{16} Btu (10^{16} is the number 1 followed by 16 zeros). The heat produced by burning 400 million tons of coal is roughly 1 C. Prove this using data from Table 3.2.

ENVIRONMENTAL ASPECTS OF BURNING FOSSIL FUELS

4

Our technological society is extremely dependent on private automobiles. Effluents from the multitude of automobiles have produced serious smog problems in nearly every American city. Photograph by Chester Higgins. (Courtesy of the EPA.)

4.1 MOTIVATION An ideal fossil fuel would be pure carbon and an ideal petroleum or natural gas would be a pure hydrocarbon meaning that it is composed only of hydrogen and carbon. Ideally, when burned to provide energy for an automobile engine or heat for a house, water and carbon dioxide would be produced. The water presents no problems but the carbon dioxide, which is always produced when fossil fuels are burned, may accumulate in the atmosphere and alter the earth's climate (Section 8.8). As you might expect, ideal fossil fuels and ideal burning conditions do not exist in practice and undesirable products are generated in the burning process. Although fossil fuels are composed primarily of hydrocarbons, varying amounts of other ingredients are always present. Sulfur and unburnable ash are particularly noteworthy contaminants. All fossil fuels contain some sulfur but it is most prevalent in coal. The sulfur content in coal varies from about 1 percent to about 5 percent depending on where the coal is mined. Coal from the extensive coalfields of Ohio, Kentucky, and West Virginia typically has 3 percent sulfur content. The coalfields in Wyoming and Montana contain about 1 percent sulfur. The sulfur content of oil is usually very low, 1 percent or less, while that in natural gas is negligible. When fossil fuels are burned, damaging sulfur oxides are produced. As long

Table 4.1 Total air pollutant emissions for the year 1974. The units are millions of tons. To indicate the general decline in emissions since 1970, data for 1970 are shown in parentheses. Note that transportation is primarily responsible for the production of carbon monoxide, nitrogen oxides, and hydrocarbons, while stationary fuel combustion sources and industrial processes produce most of the sulfur oxides and particulate matter. These data are from "National Air Quality and Emissions Trends Report, 1975," U.S. Environmental Protection Agency Report Number 450/1-76-002. They represent the most sophisticated estimational procedures available at the time of the report and supersede any previously published emissions estimates.

	Carbon monoxide	Nitrogen oxides	Hydro-carbons	Sulfur oxides	Particulate matter	Total
Transportation	82.1 (88.0)	10.6 (9.3)	12.5 (14.1)	0.8 (0.7)	1.3 (1.3)	107.3 (113.4)
Fuel combustion (stationary sources)	1.4 (1.5)	13.3 (12.3)	1.7 (1.6)	26.8 (26.6)	7.0 (9.7)	50.2 (51.7)
Industrial processes	11.0 (11.5)	0.7 (0.6)	3.7 (3.6)	6.3 (6.7)	10.6 (13.6)	32.3 (36.0)
Solid waste refuse disposal	3.5 (6.8)	0.2 (0.3)	1.0 (1.9)	0.1 (0.1)	0.6 (1.2)	5.4 (10.3)
Miscellaneous uncontrollable	5.3 (5.9)	0.2 (0.2)	14.0 (12.7)	0.1 (0.1)	0.8 (1.0)	20.4 (19.9)

ago as 1968, it was estimated conservatively that sulfur oxides caused $8 billion of environmental damage.*

Unburnable ash is significant only in coal. Most coal contains about 10 percent unburnable ash. When coal is pulverized and burned, very fine particulates are produced from the unburnable ash that unless collected enter the environment and contribute to environmental damage. Additionally, nitrogen oxides, hydrocarbons, and carbon monoxide also regarded as pollutants by the Environmental Protection Agency are produced in copious amounts. Table 4.1 shows the amounts of these five pollutants produced in 1974. Note that the quantities are in millions of tons and that fuel burned in transportation contributes most of the carbon monoxide, nitrogen oxides, and hydrocarbons while stationary fuel combustion sources (this category includes electric power plants) and industry contribute most of the sulfur oxides and particulates.

Recognizing the necessity of reducing these monetary and human costs even in the face of potential shortages of fossil fuels, the federal government is taking steps to bring air pollution under control. We will examine these measures and look more deeply into the origin and effects of the five air pollutants identified by the Environmental Protection Energy. Our study is divided into two parts. The first covers the primary emissions (sulfur oxides and particulates) from stationary fuel combustion and industrial sources. The second covers the primary emissions (nitrogen oxides, hydrocarbons, and carbon monoxide) from transportation sources.

EMISSIONS FROM STATIONARY FUEL COMBUSTION AND INDUSTRIAL SOURCES

A burning process evokes a variety of responses. A romanticist enjoys the enchanting effect of a crackling fire. Primitive man saw fire as a mechanism for survival. In addition to appreciating fire and burning, all—including the romanticist and primitive man—have probably tried to understand what is involved. There are two formal approaches taken for seeking an understanding of burning (and physical systems in general). One is to devise a scheme for measuring the effluents of interest—heat, water, carbon dioxide, sulfur oxides, particulates, etc. Information derived in this fashion is extremely practical. For example, if you know the amount of sulfur oxides generated per pound of coal burned, then you can determine the sulfur oxides produced by a coal-burning electric power plant. These data would provide input for an assessment of the impact of sulfur oxides on the environment. A second

4.2 FUNDAMENTAL ORIGIN OF SULFUR DIOXIDE, CARBON MONOXIDE, AND NITROGEN OXIDES

---

* "Cost of Air Pollution Damage: A Status Report," Environmental Protection Agency Publication AP–85 (1973).

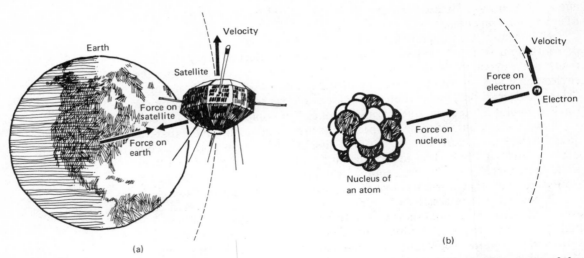

(a)   (b)

Figure 4.1 (a) An orbiting satellite and Earth. Earth and the satellite are bound together by a gravitational force. The force of gravity continually pulls inward on the satellite and keeps the satellite progressing in a circular orbit. (b) A nucleus and an electron. The nucleus and electron are bound together by an electric force. The electric force continually pulls inward on the electron and keeps the electron progressing in a circular orbit.

approach views the burning from the level of the elemental components of the materials involved. In this sense, elemental means atoms and molecules. This latter approach is more fundamental and provides the most insight into the origin of energy and the undesirable by-products that are produced. As we proceed we will capitalize on both the basic and the practical approaches to understanding physical systems.

An atom is one of a large class of physical systems appropriately called bound systems or bound states. These systems are bound in the sense that they are held together by some binding force. Often the binding force behaves like the force of a spring that pulls a door closed. An orbiting satellite and the earth constitute a bound system which is held together by the gravitational force between the two (Fig. 4.1a). At any instant, the satellite tends to shoot off in a straight line but it is continually pulled toward Earth by the force of gravity and it executes circular motion about the earth. Conceptually, an atom is much like a system composed of Earth and satellites revolving about it. In the atom, electrons are satellites and a nucleus plays the role of Earth (Fig. 4.1b). Just as Earth is much more massive than a satellite, a nucleus is much more massive than an electron. The mass of the nucleus is confined to a volume approximating a sphere having a radius of about 0.000000000000001 (1×10^{-15}) meters. The closest electron is about 0.0000000001 (1×10^{-10}) meters from the nucleus. In perspective, if the radius of the nucleus were about the thickness of a dime, the nearest electron would be about the length of a football field away. Like the earth–satellite system, the electrons and nucleus experience the attractive gravitational force that exists between any masses. However, this force is much too weak to hold the atom together. The primary force that binds electrons to nuclei is electric. This force is due to the property of charge rather than the property of mass as in the case of the gravitational force. The existence of the electric force can be demonstrated easily. If two balloons are rubbed on a person's hair and then held on threads alongside each other, they will repel each other (Fig. 4.2). This is because each balloon acquired a net charge as a result of the rubbing process, and they are pushed apart by the electric force between charges on each balloon. Two types of charge exist in nature; they are distinguished by the different direction of the force they produce on each other.Two charges of the same type will repel each other. Two unlike charges will attract each other. The charges could be differentiated by calling one A and the other B, for example, but because a charge of one type can be neutralized by a charge of the other type, it is appropriate to call the charges positive and negative. The unit of charge in the mks system of units is the coulomb (C). Electrons possess a negative charge. There is no confirmation of a unit of charge smaller than that of the electron which is -1.602×10^{-19}C. This means that 6¼ billion billion electrons are needed to make 1 coulomb of charge.

The nucleus of an atom is positively charged. The electrons are bound to the nucleus by the electric force that exists between any oppositely charged objects. In a normal atom the total negative charge of the electrons balances the charge of the nucleus and the atom is said to be electrically neutral. This model of the atom allows us to interpret the experiment with the balloons. The rubbing process initiated a transfer of electrons from the hair to the balloons. The balloons acquire electrons and become negatively charged; the hair loses electrons and is left with a deficiency of negative charge and an excess of positive charge. The balloons repel each other because they both have a net negative charge.

The number of electrons in a neutral atom is called the atomic number. The atomic number distinguishes different atomic species. For example, carbon has six electrons and oxygen has eight electrons. Table 4.2 shows the first 18 atoms in the order of increasing number of electrons. As the atomic number increases, the mass of the atom also increases. This is due both to the additional electrons and to an increase in the mass of the nucleus. There are a total of 92 naturally occurring chemical species, but the air pollutants of interest involve five of the atoms (H, C, N, O, S) shown in Table 4.2. It is possible to have atoms with the same number of electrons, but with nuclei of different masses. These are called isotopes and later we will see the origin of their mass differences. Since isotopes have the same number of electrons, they share the same chemical characteristics.

A molecule is a bound system of atoms which is also held together by electric forces. This seems contradictory because an atom alone is electrically neutral and it would appear that no electric force should exist. The electronic charge, especially the charge farthest from the nucleus, redistributes in such a way that net attractive electric forces come into play. For example, gaseous hydrogen normally exists in a molecular state in which each molecule consists of two hydrogen atoms. The atoms are close enough in the molecule that the electron of one atom may temporarily be transferred to, or shared by, the other. The redistribution of charge results in a net attractive force which holds the atoms together. Because of the motion of the electrons, the distribution of charge changes continually, but, on the average, there is some net attractive

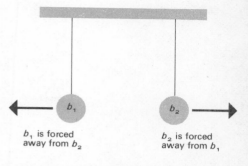

b_1 is forced
away from b_2

b_2 is forced
away from b_1

Figure 4.2 Two balloons possessing a net charge of the same type will repel each other. This is due to a repulsive electric force between the two. If the charges on the balloons were of opposite sign, then the balloons would be drawn together.

Table 4.2 The first 18 atoms in order of increasing number of electrons and mass. The mass is that of the most abundant isotope and is based on a scale of exactly 12 for the most abundant form of carbon.

Atom	Chemical symbol	Number of electrons	Approximate relative mass
Hydrogen	H	1	1
Helium	He	2	4
Lithium	Li	3	7
Beryllium	Be	4	9
Boron	B	5	11
Carbon	C	6	12
Nitrogen	N	7	14
Oxygen	O	8	16
Fluorine	F	9	19
Neon	Ne	10	20
Sodium	Na	11	23
Magnesium	Mg	12	24
Aluminum	Al	13	27
Silicon	Si	14	28
Phosphorus	P	15	31
Sulfur	S	16	32
Chlorine	Cl	17	35
Argon	Ar	18	40

force. This type of binding is called covalent. There are various types of fairly complex binding (or bonding) that in chemistry have names like ionic, metallic, or hydrogen bonds. The details of these bonds are important, but not necessary for this discussion.

Energy must be supplied to break the bonds between atoms of a molecule. For example, the simplest and most stable oxygen molecule consists of two oxygen atoms bound together. We designate the molecule as O_2, meaning a bound system of two oxygen atoms. Energy must be added to the molecule to separate it into two oxygen atoms. A symbolism (or model) for this process is

$$O_2 + energy \longrightarrow O + O.$$

The oxygen atoms are rearranged into different configurations but there is no change in the number of atoms in the rearrangement. This is a conservation-of-atoms law that applies to all reactions involving atoms and molecules. Energy is released when two oxygen atoms join to form an oxygen molecule. A symbolism for this process is

$$O + O \longrightarrow O_2 + energy.$$

Energy is released in chemical reactions when atoms and molecules are rearranged to form more tightly bound configurations. This happens in the combustion of coal. Coal is primarily carbon (C). If the combustion were perfect the carbon would unite with oxygen (O_2) to form carbon dioxide (CO_2) and energy.

$$C + O_2 \longrightarrow CO_2 + \text{energy}.$$

The combustion is not perfect and some carbon monoxide (CO) is formed.

$$2\,C + O_2 \longrightarrow 2\,CO + \text{energy}.$$

As stated earlier, coal contains a certain amount of sulfur (S). When the coal is burned the sulfur combines with oxygen to form sulfur dioxide (SO_2).

$$S + O_2 \longrightarrow SO_2 + \text{energy}.$$

To a lesser extent other sulfur oxides such as SO_3 are formed. Oxygen for the process comes normally from "air" in the environment. About 21 percent of this air is oxygen in the two-atom molecular form. Seventy-eight percent is nitrogen in a two-atom molecular form (N_2). In the presence of oxygen and the high temperature resulting from combustion, nitrogen and oxygen combine to form nitric oxide (NO) and nitrogen dioxide (NO_2).

$$N_2 + O_2 \longrightarrow 2\,NO.$$

$$N_2 + 2\,O_2 \longrightarrow 2\,NO_2.$$

There are many other aspects of physics and chemistry that can be built into these chemical reaction models. This treatment is sufficient to see the fundamental origin of the problematical by-products and heat energy. Let us proceed to examine the environmental and technological consequences of these by-products.

General Properties

The particulate matter (fly ash) emerging from the smokestack of a coal-burning electric power plant (Fig. 4.3) enters the atmosphere in the form of fine, solid particles. These particulates intersperse with those from many other sources of particle-like pollutants. Often pollutants are in the liquid state and it is appropriate to consider these along with the solids. The word particulate is intended to cover both states.* Like a dust particle, particulates will generally have peculiar and poorly defined geometric shapes. Even though particulates are shaped peculiarly, they are classified according to size. The size refers to a length associated with an assumed shape of the particulate. For example, if the particulate is assumed to be spherical, the size would be determined by the diameter (or radius) or if cubic, the length of a side. Regardless of the assumed shape, the size will be roughly the same unless the assumed shape is very odd.

_____

* The word *aerosol* is often used interchangeably with particulate. Some workers, though, prefer to consider aerosols as particles with diameters less than some particular value, for example 0.0001 meters.

Figure 4.3 A fossil-fuel electric power plant converts the chemical energy of coal into heat. Gaseous carbon monoxide, carbon dioxide, sulfur dioxide, nitrogen oxides, and solid particulates are produced in the process. Unless controlled, the gaseous products and the particulates are released to the environment from a smokestack. (J. M. Stuart Generation Station. Photograph courtesy of the Dayton Power and Light Company.)

For simplicity, it is generally assumed that the particulates are spherical. To see how this works, we need the concept of mass density. Objects of the same size do not necessarily have the same mass (or weight)—a quart of air has substantially less mass than a quart of milk. Mass density is a measure of how much mass is contained in a prescribed volume. Formally,

$$\text{mass density} = \frac{\text{mass of an object}}{\text{volume occupied by the mass}}$$

or

$$\rho = \frac{M}{V}.$$

(The Greek letter rho (ρ) is the symbol used normally for mass density.) Mass density has units of mass divided by the units of volume. Typical units are grams per cubic centimeter (g/cm^3) and kilograms per cubic meter (kg/m^3). Water has a mass density of 1 g/cm^3. Regardless of geometric shape, 1 cm^3 of water has a mass of 1 g. If you measure the mass of a certain amount of water to be 10 g, using a balance like that used to "weigh" a letter for mailing, you know that its volume is 10 cm^3. Similarly, if you know that the density of particulate matter is 2 g/cm^3, then, knowing its mass from some measurement, you can calculate its volume from the expression

$$V = \frac{M}{\rho}.$$

For example, if a given particulate has a mass of one millionth of a gram (10^{-6}g) and a density of 1 g/cm^3, it would have a volume of one millionth (10^{-6}) of a cubic centimeter. If this particulate were rolled into a tiny ball, it would have a diameter of 0.012 centimeter. A dime has a diameter of about 1 cm so the diameter of the particulate would be about one hundredth the diameter of a dime. This seems small but actually this would be a fairly large particulate. The sizes of particulates range from about 0.00000002 cm to 0.05 cm. Rather than work with such very small numbers it is common to express these sizes in terms of one millionth of a meter (or one ten-thousandth of a centimeter). One millionth of a meter is called a micrometer and is symbolized μm. On the micrometer scale, the particulate range 0.00000002 cm to 0.05 cm would be 0.0002 μm to 500 μm.

Particulate concentrations in the atmosphere are reported as the total mass contained in a cubic meter of air. Typical units are millionths of grams (micrograms) per cubic meter, symbolized μg/m^3. Mass concentrations range from about 10 μg/m^3 in remote nonurban areas to 2000 μg/m^3 in heavily polluted areas. The average concentration in American cities is about 105 μg/m^3. Even though the concentrations vary substantially throughout the country, the distribution of sizes is essentially the same everywhere. (Fig. 4.4). Note that the bulk of the particulates are less than 0.1 μm in size. However, particulates with size greater than 0.1 μm account for about 95 percent of the total mass. This fact should not be overlooked. Because of health-related effects, the number of the smaller-sized particles may be more important than the fact that the bulk of the total mass is composed of the larger-sized particles.

Particulates evolve from a number of natural and man-made sources. Those smaller than 1 μm come principally from condensation and combustion processes. Those with sizes between 1 μm and 10 μm come from such things as industrial combustion products and sea sprays. Sizes greater than 10 μm result from mechanical processes such as wind erosion and grinding. The sizes and origin of atmospheric particulates are shown in Fig. 4.5.

Of the several environmental concerns over particulates in the air, the following are noteworthy.

1. The dirt they produce after settling to the ground is annoying and requires energy and money for removal.

Figure 4.4 Typical measured size distribution of atmospheric suspended particles. This curve will be higher or lower or shifted left or right depending on the geographic location of the source of the samples used. Its shape, though, will always be very much like that shown. The sizes shown here would be like those found in smoke (Fig. 4.5). This type of graph may be unfamiliar. It is just a way of classifying particles according to their size. You could make a similar graph by taking a sample of dried peas, sorting them according to size, and then plotting the number having a given size on the vertical axis and the corresponding size on the horizontal axis.

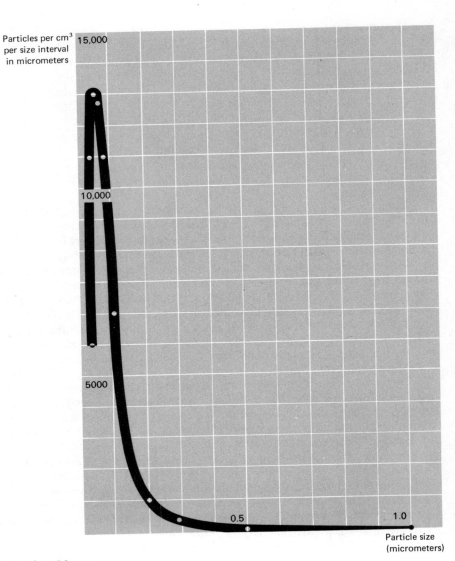

Particulate Matter

2. They produce effects detrimental to materials, plants, and animals, including man.
3. There is concern over their effect on the heat balance of the earth through reflection and absorption of solar radiation.

These effects are complicated by the fact that many of the particulates remain aloft for unusually long periods and in some cases become permanently airborne. Let us examine the reasons for this and then elucidate the environmental effects.

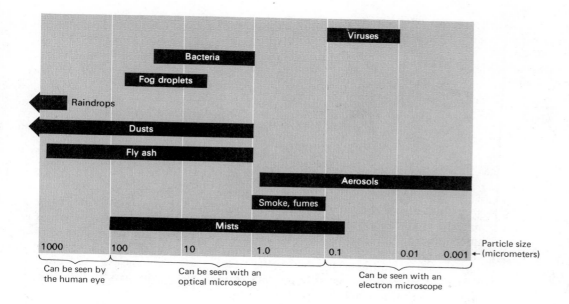

Figure 4.5 The sizes and origin of atmospheric particulates. Note that this source chooses to call aerosols particles with diameters less than one micrometer. Note, also, that larger sizes are toward the left on this figure.

Settling of Particulates in the Atmosphere

A solid plastic sphere dropped in a viscous liquid like oil or shampoo is pulled downward by gravity and held back by (1) a viscous force exerted by the fluid on the sphere, and (2) a buoyant force which tends to float the sphere. The viscous force is like that exerted on your hand when you hold it out the window of a moving car. The buoyant force is like that which causes a child's helium-filled balloon to rise.* As the sphere gains speed the viscous force increases. Ultimately the force of gravity is balanced by the opposing viscous and buoyant forces and the sphere descends with constant speed. While any sphere, regardless of size, will ultimately achieve constant speed, the limiting speed achieved by spheres made of the same material increases as the diameter increases. Particulates in the atmosphere are not plastic spheres and their motion is influenced greatly by winds, but the larger ones do settle faster for exactly the same reason. Some smaller particulates become permanently airborne and do not settle at all. A simple calculation using the rate concept shows why.

Table 4.3 shows the relation between settling velocity and particulate diameter. Let us determine the time it takes a 1-micrometer diameter particulate

---

* Buoyant force is discussed in more detail in Section 8.3.

Table 4.3 Approximate settling velocities in still air at 0°C and 760 mm pressure for particles having a density of 1 g/cm³ (From "Air Quality Criteria for Particulate Matter," National Air Pollution Control Administration Publication No. AP-49.)

Diameter (micrometers)	Settling velocity (cm/sec)
0.1	0.00008
1	0.004
10	0.3
100	25
1000	390

to settle 1 kilometer (0.62 mi). A 1-micrometer diameter particulate has a settling velocity of 0.004 cm/sec. Thus

$$\text{settling time} = \frac{\text{distance traveled}}{\text{settling velocity}}$$

$$= \frac{100,000 \text{ cm}}{0.004 \text{ cm/sec}}$$

$$= 25,000,000 \text{ sec}$$

$$= \frac{25,000,000 \text{ sec}}{3600 \,(24) \text{ sec/day}}$$

$$= 290 \text{ days}.$$

Particulates with diameters less than 1 micrometer settle so slowly that they tend to migrate thousands of miles before settling to the ground. It is not uncommon to find dust particles characteristic of the earth in Arizona to have migrated as far east as New York. Or to find radioactive debris from nuclear bomb tests migrating across the Pacific Ocean to the east coast of the United States. Many particles actually become permanently airborne and it is these which cause concern about long-term weather effects.

Environmental Effects

Health Effects   As mentioned earlier, particulate concentrations average about 105 μg/m³ in American cities. If the concentration were close to this value at all times, then it would be difficult, if not impossible, to assess particulate effects on human health. But if for some reason the concentrations increased to some abnormally high value and were accompanied by an unusual number of health effects, then some association would be possible. Still there is considerable margin for error because high particulate levels are normally accompanied by high sulfur oxide levels and it is difficult to separate the effects of each. In fact, the two pollutants together can produce effects that neither is capable of doing alone. This effect is referred to as synergism. There are several incidents of unusually high pollution levels brought on by peculiar weather conditions that produced stagnation of the air. (The physical origin of these conditions is discussed in Chapter 8.) In some areas, particularly London, the conditions were complicated by heavy fogs. Many of the incidences were accompanied by significant increases in health effects. The most-cited incidents of this type occurred in the Meuse Valley, Belgium (1930), Donora, Pennsylvania (1948), London (1952), and New York City (1953, 1966). The London episode produced some 4000 excess deaths. Some 5910 persons (42.7 percent of the total population) were affected to some degree in the Donora incident. Based on studies such as these, certain conclusions have been reached. They are not absolute but they are helpful as guidelines for setting standards. Two typical fairly reliable conclusions are as follows:

1. AT CONCENTRATIONS of 750 μg/m³ and higher for particulates on a 24-hour average, accompanied by sulfur dioxide concentrations of 715

Figure 4.6 Cincinnati City Hall photographed during cleaning in 1963 shows the accumulation of 34 years of dirt. (Photograph courtesy of the EPA.)

μg/m³ and higher, *excess deaths* and a considerable *increase in illness* may occur.

2. IF CONCENTRATIONS ABOVE 300 μg/m³ for particulates persist on a 24-hour average and are accompanied by sulfur dioxide concentration exceeding 600 μg/m³ over the same period, *chronic bronchitis* patients will likely suffer *acute worsening of symptoms.*

More of the details leading to these and several other similar conclusions are given in "Air Quality Criteria for Particulate Matter," EPA Publication AP-49. The points to be made here about health effects are as follows:

1. There are documented correlations between high particulate concentrations and adverse human health effects.

2. Damage is done primarily to the respiratory system. In particular, those suffering from bronchitis and other respiratory ailments are adversely affected.

3. Sulfur oxides must also be considered when assessing effects of particulates because of synergistic effects.

More will be said about these effects in the discussion of sulfur oxides.

Effects on Materials The fallout of particulates produces an obvious unsightliness on buildings and streets (Fig. 4.6). Apart from this, particulates, especially when accompanied by sulfur oxides and moisture, can literally attack structural materials. For example, steel and zinc samples corroded three to six times faster in New York City than in State College, Pennsylvania, where the particulate concentrations were about 180 and $60\mu g/m^3$, respectively.

Effects on Vegetation   There is little evidence of the direct effect on plant life of the mixture of particulates typically found in the air. There is, however, evidence that specific types of particulates do produce damage. One of many examples is the case in which a marked reduction in the growth of poplar trees one mile from a cement plant was observed after cement production was doubled.

Federal Standards for Particulate Concentrations

The determination of an unacceptable pollutant level was a decision left to a state or a city until the enactment of the 1970 Clean Air Act. The local standards varied from none to some at least comparable with the new federal standards, and enforcement was difficult because the legal machinery was usually not provided. However, it appears that there are sufficient legal provisions for the federal standards. Federal agents have the power to close down factories if the Environmental Protection Agency deems it necessary.

The Environmental Protection Agency promulgates primary and secondary categories of national ambient air quality standards. Primary standards define levels of air quality that are judged necessary, with an adequate margin of safety, to protect the public health. Secondary standards define levels of air quality which are judged necessary to protect public welfare, for example, by protecting property, materials, and economic values. Meeting the standards is left to the jurisdiction of individual states. If a state so chooses, it may impose standards more stringent than the federal standards. The federal standards for particulates are the following.

The maximum average 24-hour concentration that is not to be exceeded more than once per year is

primary   260 micrograms per cubic meter of air,
secondary   150 micrograms per cubic meter of air.

The annual geometric mean* concentration cannot exceed

primary   75 micrograms per cubic meter of air
secondary   60 micrograms per cubic meter of air.

---

* The most familiar mean (or average) value is formally called the arithmetic mean and is obtained by summing the quantities of interest and dividing by the number of quantities. For example, the average class grade for an examination is determined by adding all the grades and dividing by the number of students. The geometric mean is obtained by multiplying the quantities of interest and taking the Nth root of the product where N is the number of quantities. For example, if 25 students took an examination, the geometric mean would be the 25th root of the product of all 25 grades.

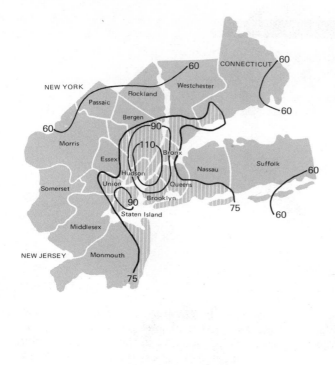

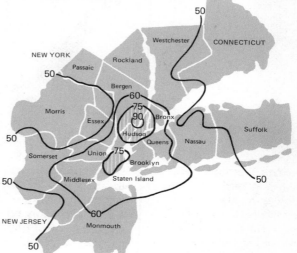

Figure 4.7 These two maps show how particulate levels decreased in the New York City metropolitan area between 1971 and 1974. A solid line outlines a region having the same particulate concentration measured in micrograms per cubic meter. The cross-hatching denotes areas where the annual mean total suspended particulate concentration exceeded the annual primary standard of 75 micrograms per cubic meter. Most American cities witnessed a similar decline in particulate concentrations in this time period.

(a) 1971   (b) 1974

While only about one third of the nation's 247 air quality-control regions were able to meet an implementation deadline of May 31, 1975 for compliance with the primary standards, considerable progress was and continues to be made. Figure 4.7 illustrates the reduction of particulate concentrations in the New York City metropolitan area between 1971 and 1974. Particulate reductions in New York City and many other urban areas have evolved from implementation of available, but expensive, particulate collection devices. The functioning of these devices provides an opportunity for elucidating some important physical principles.

Particulate Collection Devices

Particulate collection devices are categorized as gravitational, cyclone, wet collectors, electrostatic precipitators, fabric filtration, afterburner (direct flame), and afterburner (catalytic). The most common and the most appropriate for insight into the principles involved are the gravitational, cyclone, and electrostatic precipitator.*

---

* The details, including advantages and disadvantages, are given in "Control Techniques for Particulate Air Pollutants," NAPCA Publication AP-51, January 1969. Available from EPA, 401 M Street S.W., Washington, D.C. 20460.

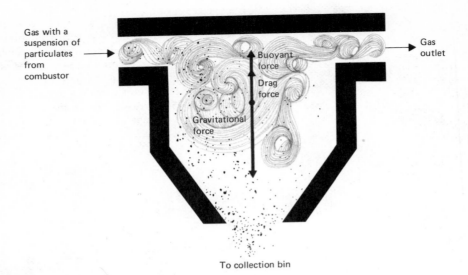

Gas with a suspension of particulates from combustor

Gas outlet

Buoyant force

Drag force

Gravitational force

To collection bin

Figure 4.8 Schematic illustration of a gravitational collector. Particulates are influenced by forces exactly like particulates in the atmosphere. The gravitational force pulls them downward; a drag force and a buoyant force due to the air medium opposes the downward motion.

Gravitational Collector The principle of the gravitational collector is shown in Fig. 4.8. The gases containing a suspension of particulates from the fuel combustion enter a large chamber. Like particulates in the atmosphere, they are influenced by both the gravitational force of the earth (pulling down) and retarding forces (drag and buoyant) due to the gaseous medium. Some of the particles will settle enough so that they will fall through the bottom of the container and be collected. As expected from our discussion of particulates settling in the atmosphere, the ones of larger diameter will settle fastest and are the ones most likely to be collected. Gravitational collectors function efficiently only for particulate diameters greater than about 50 micrometers. They are extremely useful because of their simplicity and ease of maintenance.

Cyclone Separator Any rapidly rotating air mass is loosely referred to as a cyclone. Such a condition is created with the combustion gases in a cyclone separator. As a result the particulates are forced to the outer edge of the cyclone where they can be collected. The principle is as follows. Any mass in circular motion is accelerating (Sec. 3.3). Therefore, some net force must act on it to produce the acceleration (Newton's second law). For example, if a mass attached to a string is swung in a vertical circle, the force of the string on the mass, combined with the gravitational force on the mass, provides a net force which keeps the mass in its circular path. Following Newton's third law, the mass exerts an opposite force on the string which produces a tension in the string. If the tension exceeds the breaking strength of the string, the mass flies outward from its circular orbit. If a gas with a suspension of particles is rotated by a cyclonic wind action, then the particles will subscribe to the circular motion only if there is some force available to accelerate it. This must be a force exerted on the particles by the gas. If the force is insufficient to hold the

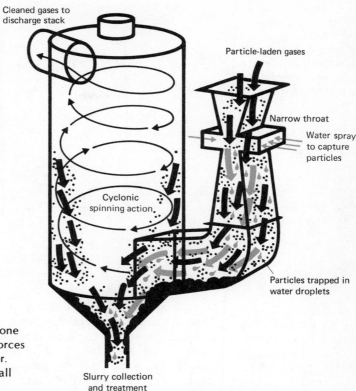

Cleaned gases to discharge stack

Particle-laden gases

Narrow throat

Water spray to capture particles

Cyclonic spinning action

Particles trapped in water droplets

Slurry collection and treatment

Figure 4.9 Schematic illustration of the cyclone particle collector. The cyclonic air motion forces particulates toward the walls of the collector. When the particulates strike the wall, they fall into a collection bin.

particles in a circular orbit, the particles move outward. In a cyclonic separator, the gas containing the particulates is forced into cyclonic motion down through a tapered cylinder (Fig. 4.9). The particulates move toward the walls of the cylinder. Some strike the walls and fall into a collection bin. This type of device is somewhat better than the gravitational system for collecting particulates of a smaller diameter—the lower limit for the diameter is about 5 μm for efficient collection.

Electrostatic Precipitator The electrostatic precipitator is the most efficient device for removing particulates with diameters less than 5 μm. It is based on the principle that unlike electric charges attract each other. In principle, the precipitator is a hollow container, say a cylinder, with wires located within it (Fig. 4.10). Provisions are made to make the wires strongly electrically negative and the cylinders strongly electrically positive. The electric force produced on electrons within the wires is large enough to literally pull electrons from the wires. The freed electrons then migrate toward the positively charged cylinder wall.

Figure 4.10 Schematic illustration of an electrostatic precipitator. The particulates having acquired a net negative charge are attracted toward the wall of the precipitator. When the particulates strike the wall, they become electrically neutral and fall into a collection bin.

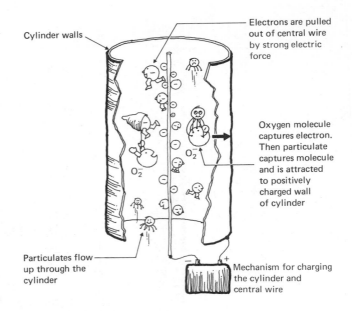

Cylinder walls

Electrons are pulled out of central wire by strong electric force

Oxygen molecule captures electron. Then particulate captures molecule and is attracted to positively charged wall of cylinder

O_2^-

O_2^-

Particulates flow up through the cylinder

Mechanism for charging the cylinder and central wire

Some of the gas molecules, such as oxygen (O_2), will capture an electron and acquire a net negative charge (O_2^-). These molecules will be accelerated toward the wall of the positively charged cylinder. During transit, the O_2^- ion may attach itself to a particle and the particle-O_2^- composite will migrate toward the wall. Once the composite strikes the wall, it becomes electrically neutral and the electric force vanishes. The particulates are then removed from the walls of the container.

The electrostatic precipitator is the most efficient of the systems discussed and is capable of removing 99 percent of the total *mass* concentration. However, the efficiency for particulates less than 0.1 μm in diameter is poor. Even though the electrostatic precipitator may remove 99 percent of the mass, only about 5 percent of the *number* of particulates are accounted for because there are many more of the smaller particulates in the gas effluent (see Fig. 4.4). The electrostatic precipitator has difficulty in extracting the small particulates because the chance of charge capture by a particulate decreases with the *effective size* of the particulate. The effective size is determined by the degree of *polarization*. Polarization of electric charges means a separation of positive and negative charges. The positive and negative charges in a piece of electrically neutral material can be separated by placing the material near an electrically charged object. This occurs when a comb is charged by combing and then placed near a bit of paper (Fig. 4.11). Particulates in the gas effluent are polarized because of the positively charged walls and negatively charged wires. As a result, the effective size for capture of a charged particle is increased. Because the small particulates are extremely difficult to polarize, they are very difficult to remove. It is difficult to assess the effect of not being able to remove these small particulates, but, as mentioned earlier, it may be the actual *number* of particulates rather than the *mass* which is ultimately of importance.

Figure 4.11 A bit of paper can be electrically polarized by bringing it near a comb which has been charged by combing hair.

Bit of paper has an equal number of positive and negative charges

If comb has a net negative charge,

negative electrons move away from the negatively charged comb forcing an accumulation on one side of the paper

Paper is polarized

Electrostatic precipitators inoperative

Electrostatic precipitators operative

That electrostatic precipitators are effective is illustrated by the pair of photographs in Fig. 4.12.

The Disposal Problem

The collection of fly ash has solved one problem but created another—disposal of the collected matter. Nearly 20 million tons of fly ash are collected annually and, for the most part, are pumped away from the precipitators into ponds where they settle to the bottom. Some commercial use is made of the by-product but it amounts to only a few percent of that collected. Some of the uses are as an aggregate for building bricks, improved traction material for car tires, fire-quenching material for coal mines, and fertility improver for soils.

General Properties

Sulfur dioxide (SO_2) and sulfur trioxide (SO_3) are the major sulfur oxides produced when coal is burned. However, the production of sulfur dioxide is some 40 to 80 times more probable than the production of sulfur trioxide. Sulfur dioxide is a nonflammable, nonexplosive, colorless gas. Most people can taste it in concentrations of about $2500 \,\mu g/m^3$ in air. It produces a pungent, irritating odor for concentrations greater than about $8000 \,\mu g/m^3$ in air.

It is easy to estimate how much SO_2 is produced from burning a ton of coal with a given sulfur content. However, because of the large variation in sulfur content, it is difficult to estimate how much is being produced in the entire country. An often quoted amount is 20 million tons for 1970, which is expected to reach 94.5 million tons in 2000.* In 1974, sulfur dioxide contributed about 16 percent of the weight of all atmospheric pollutants (Table 4.1).

* "Sulfur Oxide Control: A Grim Future," *Science News,* vol. 98, no. 187 (August 29, 1970).

Figure 4.12 The effectiveness of electrostatic precipitators for removing particulates emanating from a smokestack is illustrated in these pictures taken without and with the electrostatic precipitators in operation. The white clouds that appear in both photographs are composed of water vapor. (Photographs courtesy of Eastman Kodak Company.)

4.4 SULFUR OXIDES

Environmental Effects

Health Effects A number of experiments have been conducted on the effects of sulfur dioxide on animals. Such research is meaningful in the environmental sense to the extent that the results can be extrapolated to humans. The studies show that the respiratory system is most susceptible to damage. This information is important and has some ultimate use when trying to set standards. However, it is the effect of sulfur dioxide in an atmosphere containing a variety of other pollutants that is of real significance because the sulfur dioxide reacts chemically and produces other species more toxic than itself. For example, sulfur dioxide can be converted to highly corrosive sulfuric acid (H_2SO_4). One might think of this as happening by combining SO_2 and oxygen to form SO_3:

$$2\,SO_2 + O_2 \longrightarrow 2\,SO_3.$$

This reaction would be followed by SO_3 combining with water (H_2O) to form H_2SO_4:

$$SO_3 + H_2O \longrightarrow H_2SO_4.$$

This reaction does occur but only slowly because the formation of SO_3 by this proposed mechanism is slow. A more probable method of forming SO_3 is through a catalyst. (A catalyst modifies the reaction rate but is not consumed in the process.) One possible catalyst is nitrogen dioxide (NO_2), which is also a combustion product:

$$NO_2 + SO_2 \longrightarrow NO + SO_3.$$

The NO_2 used to form SO_3 is regenerated by the reaction

$$2\,NO + O_2 \longrightarrow 2\,NO_2.$$

The role of the many catalysts in a polluted atmosphere in the formation of sulfuric acid is an extremely important environmental problem.*

The effects of SO_2, like particulates, on human health are obtained from epidemiological studies. Although these effects are difficult to assess quantitatively, they do have the distinct advantage of being derived from cases involving human subjects; in this sense, they are extremely valuable for deciding on allowable concentrations of various pollutants. Three fairly reliable conclusions from studies of this type are as follows:

1. At concentrations of about 500 $\mu g/m^3$ averaged over a 24-hour period, with low particulate levels, *increased mortality rates* may occur.
2. At concentrations of about 715 $\mu g/m^3$ averaged over a 24-hour period, accompanied by particulate matter, *a sharp rise in illness rates* for patients over age 54 with severe bronchitis may occur.

---

* "Acid Rain: A Serious Regional Environmental Problem," G. E. Likens and F. H. Borman, *Science,* vol. 184, no. 1176 (1974).

3. At concentrations of about 120 μg/m³ averaged over a year, accompanied by smoke concentrations of about 100 μg/m³, *increased frequency and severity of respiratory diseases* in school children may occur.*

It is important to note that (1) the elderly, the young, and persons suffering from respiratory ailments such as bronchitis and emphysema are particularly bothered; and (2) there is an established synergistic effect between sulfur dioxide and particulates.

Figure 4.13 Birch leaves. The leaf on the left shows injuries caused by sulfur dioxide. (Photograph courtesy of the U.S. Department of Agriculture.)

Effects on Vegetation Plants, in general, are easily damaged by sulfur dioxide (Fig. 4.13). There is, however, a wide variation in the amount of SO_2 needed to damage different types of plants. Acute injury occurs in alfalfa (a common food crop for livestock) at concentrations of about 3300 μg/m³ for one hour. About 50,000 μg/m³ for one hour is the concentration required to produce an equivalent damage in privet (a type of shrub commonly used for hedges). Damage has been reported for chronic exposures as low as 80 μg/m³. The formation of sulfuric acid, as described earlier, that attacks plants is also a problem as are synergistic effects of sulfur dioxide with ozone and nitrogen dioxide.

---

* More of the details leading to these and several other similar conclusions are given in "Air Quality Criteria for Sulfur Oxides," EPA Publication AP-50.

Effects on Materials Combinations of particulates and sulfur dioxide do significant damage to materials. This is especially true if conditions are favorable for the formation of sulfuric acid. Building materials and statues often are discolored and disfigured from attacks by sulfuric acid. In addition, metals corrode and materials such as cotton, nylon, rayon, and leather are damaged by sulfur dioxide.

Federal Standards for Sulfur Dioxide Concentrations

The federal standards for sulfur dioxide concentrations are the following:

Primary

The maximum average 24-hour concentration that is not to be exceeded more than once per year is 365 micrograms per cubic meter.

The maximum average annual concentration is 80 micrograms per cubic meter.

Secondary

The maximum average 3-hour concentration that is not to be exceeded more than once per year is 1300 micrograms per cubic meter.

Table 4.4 gives some representative annual average sulfur dioxide concentrations for American cities for 1970. As seen, some exceed the federal standards and several are close. As for the control of particulates, substantial improvement has been made since the 1970 Clean Air Act. The situation in Cleveland is typical (Fig. 4.14). Unlike the nationwide reduction of particulate concentrations, the reduction of sulfur dioxide concentrations has not evolved entirely from implementing proven technology. Rather it stems mostly from burning coal containing less than 1 percent sulfur and substituting oil and natural gas for coal. Because the low-sulfur coal is mostly in the Western United States, an economic penalty is being paid for the reduction of sulfur dioxide concentrations in Eastern United States. While there is promise for controlling sulfur dioxide emissions without resorting to low-sulfur coal, the technology is in a development and testing stage, and meeting the federal standards poses aggravating problems for individual states and their industries.

Table 4.4 Sulfur dioxide annual average concentrations (in micrograms per cubic meter of air) at selected cities, 1970. From EPA. Air Quality Control Office.

Philadelphia	84	Nashville	15
Baltimore	54	Grand Rapids	13
New York City	74	Tacoma	9
Cleveland	64	Dallas	6
Chicago	119	Denver	12
Pittsburgh	57	San Francisco	7
Miami	5	Los Angeles	34

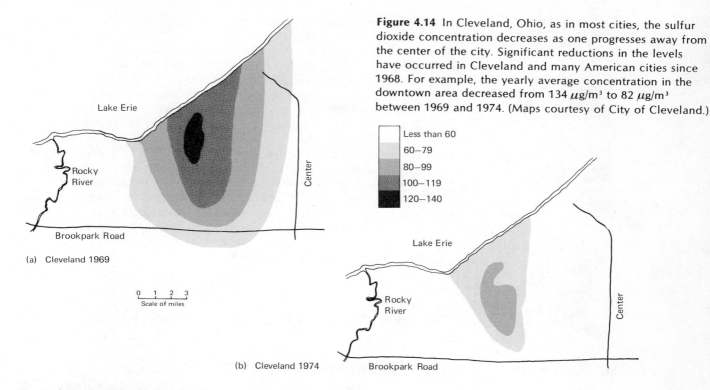

Figure 4.14 In Cleveland, Ohio, as in most cities, the sulfur dioxide concentration decreases as one progresses away from the center of the city. Significant reductions in the levels have occurred in Cleveland and many American cities since 1968. For example, the yearly average concentration in the downtown area decreased from 134 μg/m³ to 82 μg/m³ between 1969 and 1974. (Maps courtesy of City of Cleveland.)

Less than 60
60–79
80–99
100–119
120–140

Lake Erie

Rocky River

Center

Brookpark Road

(a) Cleveland 1969

0 1 2 3
Scale of miles

Lake Erie

Rocky River

Center

(b) Cleveland 1974

Brookpark Road

Control Methods for Sulfur Oxides

As a result of the ambient air quality standards mandated by the Clean Air Act, each state has established emission limitations for sulfur oxides. Enforcement of the emission limitations has produced a classic industrial–governmental confrontation. While it is essential to protect the public from the harmful effects of sulfur oxides, there is governmental pressure for substantially increasing the use of coal, our most abundant energy resource. Industry claims that the technology for controlling the sulfur oxide emissions is not at hand and implementation of the emission limitations is impractical in the short run. It is probably fair to say that the emission limitations will be enforced ultimately, but the exact date will depend on national needs and development of control technology.

The most effective method of minimizing sulfur oxide emissions has been to burn coal having less than 1 percent sulfur content. As pointed out before, most of this coal is to be found in Western United States, far from the Eastern markets. While Western coal is low in sulfur, it is high in ash and, generally, low in energy content. Thus the ash disposal problem is enhanced. Although burning low-sulfur Western coal is a viable interim solution, a technological solution that allows the use of high-sulfur Eastern coal must come eventually. Two promising schemes that have been developed but not proven in the long run are termed (1) flue gas desulfurization, and (2) coal beneficiation.

The flue gas desulfurization mechanism, commonly called scrubbing, allows the stack gases to pass over a wet slurry of lime (CaO) or limestone ($CaCO_3$). Chemical reactions producing calcium sulfite remove around 90 percent of the sulfur oxides. A sludge requiring disposal or other usage is produced. Scrubbers are being installed in a number of power plants. Their long-term worth is yet to be proven, but as more and more experience is gained, there are indications that scrubbers will prove worthwhile.

Coal beneficiation is a process that removes part of the sulfur before the coal is burned. The part that is removed is in the form of a mineral pyrite (FeS_2). The pyrite, which has a density 3½ times that of coal itself, is removed by crushing the coal and using mechanical separation methods. Burning the product with the pyrite removed reduces sulfur oxide emissions by 40 to 50 percent. In some cases this is sufficient to meet emission limitations. Coal beneficiation combined with flue gas desulfurization places a smaller burden on the desulfurization process and lessens the lime or limestone requirement as well as alleviating the sludge disposal problem. Most importantly it is the most economical way of meeting the emission limitations.

EMISSIONS FROM MOBILE FUEL COMBUSTION SOURCES

4.5 FUNDAMENTAL ORIGIN OF EMISSIONS

Transportation vehicles, primarily automobiles, account for a little over half of the weight of all air pollution. This does not necessarily mean, though, that motor vehicles are responsible for over half of all air pollution effects because some pollutants are more detrimental than others. Though the pollutants, for the most part, differ from those produced by burning coal, they still result from burning hydrocarbons. The hydrocarbons are refined products of crude oil, mainly gasoline and diesel fuel, and the burning takes place in the engines of automobiles, trucks, and buses. A glance under the hood of an American car will quickly convince you that the engine with all its auxiliaries for running an air conditioner, electric generator, windshield washer and wipers, hydraulic pumps, etc., is a complicated system. Although intricate, its essential function is to convert the intrinsic chemical energy of gasoline into kinetic energy of motion. It does this by igniting a mixture of gasoline vapor and air in the cylinders of the engine (Fig. 4.15). Since the burning takes place within a closed region, namely between the tops of the cylinder and piston, this type of engine is often called an internal combustion engine. The expanding gas from the explosion pushes down on the pistons within the cylinders. The linear motion of the pistons is converted to rotational motion by rods connected to a rotating shaft. This rotational motion is ultimately transferred to the wheels of the automobile by an appropriate linkage. The cycle of one of the pistons in an engine is shown in Fig. 4.16. All of the carbon monoxide and nitrogen oxides and 60 percent of the hydrocarbons generated are combustion products. Twenty percent of the hydrocarbons released are a result of evaporation from the fuel tank and carburetor. The remaining 20 percent are from vapor "blowing by" the pistons and the cylinder walls of the engine. Let us look now at the formation of these pollutants at the burning level.

A class of hydrocarbons known as alkanes has the chemical form C_nH_{2n+2}, where n is a number from 1 to 10. For example, if $n = 5$, the molecule is C_5H_{12}, called pentane. Gasoline with no special-purpose additives is a mixture of alkanes with from 5 to 10 carbon atoms. A special molecular form of octane (C_8H_{18}) called isooctane is particularly useful as an automotive fuel. Isooctane has become the standard for comparison of other automotive fuels. The reason is as follows. With some gasolines, ignition occurs more as an explosion rather than a smooth burning. This not only reduces the efficiency, but also produces an audible "knock" in the engine. A fuel of pure isooctane produces little knock while one of pure heptane (C_7H_{16}) knocks badly. The octane rating of a gasoline is a measure of the knock produced when used as an automotive fuel. Isooctane and heptane have been assigned octane ratings of 100 to 0, respectively. A mixture of 90 percent isooctane and 10 percent heptane would have an octane rating of 90. When any other fuel, regardless of its chemical composition, is burned in a special engine and produces the same knock as 90 percent isooctane and 10 percent heptane, it would also have an octane rating of 90. Since the inception of this procedure, fuels have been developed with antiknock properties superior to isooctane and these will have octane ratings greater than 100.

The pollutants evolve when the fuel, whatever it may be, is burned in the engine. Like the burning of coal in an electric power plant, the pollutants would be minimal if (1) the combustion were complete, and (2) there were no side effects. To illustrate, consider the combustion of isooctane, which we might consider the standard fuel. Ideally, isooctane reacts with oxygen to produce carbon dioxide, water, and energy:

$$2\,C_8H_{18} + 25\,O_2 \longrightarrow 16\,CO_2 + 18\,H_2O + \text{energy}.$$
$$\phantom{2\,C_8H_{18} + 25\,O_2 \longrightarrow 16\,}\text{carbon}\quad\text{water}$$
$$\phantom{2\,C_8H_{18} + 25\,O_2 \longrightarrow 16\,}\text{dioxide}$$

Observe that the reaction involves a rearrangement of atoms with no change in the number of atoms of any type. Except for long-term climatic effects (Chapter 8), the CO_2 is harmless. Similar, but undesirable, reactions produce carbon monoxide (CO). For example,

$$2\,C_8H_{18} + 25\,O_2 \longrightarrow 14\,CO_2 + 2\,CO + O_2 + 18\,H_2O + \text{energy}.$$

Oxygen for combustion is obtained from air which normally contains 78 percent nitrogen and 21 percent oxygen. The nitrogen does not enter into the burning process, but does react with the oxygen at the high temperatures produced in the combustion chamber of the engine. The reactions which produce most of the nitrogen oxide pollutants are

$$N_2 + 2\,O_2 \longrightarrow 2\,NO_2 \text{ (nitrogen dioxide)},$$

$$N_2 + O_2 \longrightarrow 2\,NO \text{ (nitric oxide)}.$$

Other oxides are produced and normally these, as well as NO and NO_2, are lumped together for environmental purposes and called nitrogen oxides, symbolized NO_x.

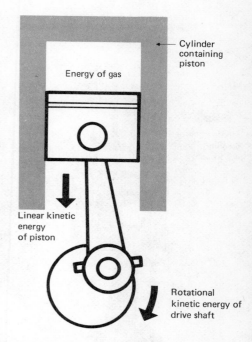

Energy of gas

Cylinder containing piston

Linear kinetic energy of piston

Rotational kinetic energy of drive shaft

Figure 4.15 The energy conversion process in an internal combustion engine.

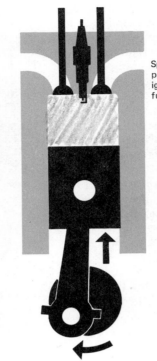

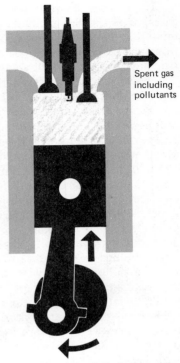

Valves that open and close to let gases into and out of the combustion chamber

Sparkplug for igniting fuel mixture

Intake valve

Exhaust valve

Mixture of gasoline vapors and air

Spark plug ignites fuel

Spent gas including pollutants

Intake stroke. A vapor–air mixture is pulled in from the carburetor through the open intake valve.

Compression stroke. The vapor–air mixture is compressed by the upward moving piston. Both valves are closed.

Power stroke. At the top of the compression stroke, the sparkplug ignites the vapor–air mixture, forcing the piston down to produce linear kinetic energy.

Exhaust stroke. At the bottom of the power stroke, the exhaust valve opens and the spent gas is forced out through the exhaust system.

Figure 4.16 The cycle of the internal combustion engine.

The efficiency of the internal combustion engine is directly related to the octane rating of the fuel. Throughout the years, the trend has been to increase the octane rating. One way of making a better fuel is to produce molecular structures with better burning characteristics—an accomplished and expensive process. The most common method of achieving a higher octane fuel is to add a compound called tetraethyl lead, $(C_2H_5)_4Pb$, in amounts of about 2 milliliters* per gallon. In addition to increasing the octane rating, the tetraethyl lead provides some lubrication and reduces valve burning. However, it produces deposits such as metallic lead and lead oxides which remain in the engine and impair the functioning of the valves and spark plugs. To combat this, ethylene dibromide $(C_2H_4Br_2)$ is added to convert the lead to lead bromide (PbBr) that escapes as a gas out the exhaust. Present day gasolines also contain

---

* A milliliter is about the volume occupied by three pennies.

antioxidants, metal deactivators, antirust and anti-icing compounds, detergents, and lubricants. Modern gasolines are far from the ideal isooctane suggested earlier.

Since the advent of 1975 models, American automobiles have been equipped with a pollution control device called a catalytic converter (Sec. 4.9). Catalytic converters can be rendered ineffective by the lead compounds resulting from tetraethyl lead in gasoline. To ensure that only unleaded gasoline is used in vehicles having catalytic converters, the opening in the filling port of the gasoline tank will not accept the nozzle from pumps issuing leaded gasoline.

Few metropolitan areas escape the eye-irritating effects of photochemical smog* produced from chemical reactions consummated from automobile effluents in the presence of sunlight. Often, the chemicals evoking the eye irritation are vastly more complex than the relatively simple molecules produced in the combustion of gasoline. While we are not prepared to understand the complex chemistry, we can comprehend the basic mechanisms leading to the formation of the irritants.

The most stable form of oxygen is the two-atom combination (O_2). Atomic oxygen (O) and a three-atom form called ozone (O_3) are extremely reactive in the presence of other atoms and molecules. Because of the high reactivity, normal air contains very little atomic oxygen and ozone. It happens that nitrogen dioxide (NO_2), one of the effluents from automobiles, strongly absorbs solar energy and separates into nitric oxide and atomic oxygen:

$$NO_2 + \text{solar energy} \longrightarrow NO + O.$$

More than 99 percent of the atomic oxygen created by the breakdown of NO_2 combines with molecular oxygen to form ozone:

$$\underset{\text{atomic oxygen}}{O} + \underset{\substack{\text{ordinary} \\ \text{oxygen molecule}}}{O_2} \longrightarrow \underset{\text{ozone}}{O_3}.$$

However, some atomic oxygen reacts with hydrocarbons from automobile exhausts in a series of complex chemical reactions. Nitrogen dioxide and ozone are regenerated and some very irritating chemicals such as formaldehydes, peroxyacyl nitrates (PAN), and acroleins are produced. The original high NO concentration from automobile emissions is diminished by reactions such as

$$NO + O \longrightarrow NO_2.$$

$$NO + O_3 \longrightarrow NO_2 + O_2.$$

Figure 4.17 shows how this photochemical smog process takes place in a large city. These processes are responsible for the characteristic haze over many cities (Fig. 4.18) and it is the chemicals that produce the eye-irritating effects.

---

* The word smog is a combination of parts of the words **smoke** and **fog** and was used to designate the type of polluted conditions that often occurred in London. Smog now describes a chemical condition which is much more complex than this relatively simple mixture.

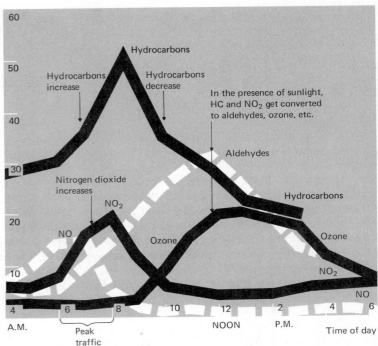

Figure 4.17 Hourly variations of the constituents of an intense photochemical smog. When commuter traffic materializes between 6 and 8 A.M., hydrocarbon and nitrogen oxide concentrations build up. The nitric oxide (NO) levels exceed the nitrogen dioxide (NO_2) levels because the combustion process strongly favors the formation of NO. In the presence of the morning sunlight, NO_2 begins to break down into NO and O and the photochemical smog process begins. The hydrocarbon levels decrease because they get converted to the eye-irritating chemicals such as aldehydes. The peak smog concentration occurs in early afternoon when the sun's rays are most intense. (From *Photochemistry of Air Pollution* by Philip A. Leighton, Academic Press, Inc., 1961.)

Figure 4.18 A visual illustration of clear conditions and an intense smog formation in Los Angeles. (United Press International photograph. Reproduced by permission.)

Table 4.5 Some distinguishing characteristics of photochemical (or Los Angeles) and London smogs.

Characteristic	Photochemical (or Los Angeles)	London
Major fuels involved	petroleum	coal and petroleum products
Principal constituents	O_3, NO, NO_2, CO, organic products	particulates, CO, sulfur compounds
Types of reactions	photochemical and thermal	thermal
Time of maximum occurrence	midday	early morning
Principal effects	eye irritation	bronchial irritation, coughing
Visibility	about 0.5 to 1 mile	less than 100 yards

Table 4.5 summarizes the characteristics of photochemical smog and gives the corresponding London smog data for comparison.

Cities are naturally the areas where smog tends to be problematical for it is here that industry and large concentrations of motor vehicles exist. Today, every sizeable city in the United States has smog problems. Yet the West Coast of the United States is particularly troubled. The reason is not that it is more industrialized than many other areas, but that it is subject to a characteristic weather condition which produces an atmospheric lid on the air near the ground for about 100 days of the year (see Chapter 8). The situation in Los Angeles is complicated by mountains and hills to the north, east, and south which prevent flushing of the trapped smog. Denver, Colorado, with ever increasing industrial and urban expansion, has similar problems.

Carbon monoxide is a colorless, odorless, and tasteless gas. Yet exposure to an atmosphere of air in which only one part out of each 1,000 parts (by volume) is carbon monoxide can produce unconsciousness in one hour and death in four hours. The problem stems from the fact that the bloodstream, which transports oxygen throughout the body, has a much greater affinity for carbon monoxide than for the life-supporting oxygen. When one breathes air containing an excessive amount of carbon monoxide, the body literally suffocates because its chemistry requires oxygen and not carbon monoxide. The burning of any carbon-based material—gasoline, coal, charcoal, tobacco, etc.—produces carbon monoxide to some extent. Therefore, one should ensure adequate ventilation for the effluents. It is extremely dangerous to operate an automobile in a closed garage or to tolerate a faulty exhaust system on an automobile or to use a charcoal grill in closed quarters.

4.7 EFFECTS AND EXTENT OF AIR POLLUTANTS

Figure 4.19 Above: Ozone damage to a White Cascade petunia. Right: Leaf of an ozone-sensitive variety of tobacco shows white spots characteristic of air-pollutant damage called weather fleck. (Photographs courtesy of the U.S. Department of Agriculture.)

The direct effect of nitrogen oxides and hydrocarbons on plant and animal life is not of great concern. Rather it is the consequences of photochemical smog which is the major concern. The effects generally attributed to smog are

1. eye irritation
2. characteristic damage to vegetation
3. respiratory distress and even death
4. reduction in visibility
5. objectionable odors
6. excessive cracking of rubber products and damage to other materials.

Some specific effects of air pollutants of concern are briefly summarized in Table 4.6. Using the specific effects as criteria, the Environmental Protection Agency has arrived at the standards shown in Table 4.7. For these pollutants, the primary and secondary standards are identical. While the standards are being questioned by reliable authorities, the general consensus is that they are very stringent.

Although Los Angeles has come to be identified with automotive air pollution in this country, it is by no means the only city with problems. Most cities will have trouble meeting the standards set by the Environmental Protection Agency.

Table 4.6 Brief summary of air pollutant effects.

Effects associated with oxidant concentrations in photochemical smog

Effect	Exposure (micrograms/m³)	Duration	Comment
Vegetation damage	100	4 hours	leaf injury to sensitive species
Eye irritation	Exceeding 200	peak values	Such a peak value would be expected to be associated with a maximum hourly average concentration of 50 to 100 μg/m³.
Impaired performance of student athletes	60–590	1 hour	exposure for 1 hour prior to race

Effects associated with carbon monoxide

Effect	Exposure (milligrams/m³)	Duration	Comment
Bodily damage	30	8 hours or more	impairment of visual and mental acuity
	200	2–4 hours	tightness across the forehead, possible slight headache
	500	2–4 hours	severe headache, weakness, nausea, dimness of vision, possibility of collapse
	1,000	2–3 hours	rapid pulse rate, coma with intermittent convulsions
	2,000	1–2 hours	death

SOURCE: The data were taken from the Air Quality Criteria Handbooks by the U.S. Department of Health, Education, and Welfare and the Environmental Protection Agency.

Table 4.7 Federal standards for carbon monoxide, hydrocarbons, photochemical oxidants, and nitrogen oxides.

Pollutant	Maximum average concentration not to be exceeded more than once a year (micrograms per cubic meter of air)	Averaging period
Carbon monoxide	10,000	8 hours
	40,000	1 hour
Hydrocarbons	160	3 hours (6–9 A.M.)
Photochemical oxidants*	160	1 hour

The annual arithmetic mean concentration of nitrogen oxides cannot exceed 100 micrograms per cubic meter of air.

*Photochemical oxidants are primarily ozone but they also include other chemical compounds created by the smog conditions.

4.8 THE FEDERAL EMISSION STANDARDS

If air quality is to be improved in critical areas and the stringent federal standards are to be met, then motor vehicle emissions, in particular those from the automobile, must be controlled. Such a program was begun in California in 1961 and nationwide in 1968. Both programs have called for a gradual decrease in allowable automobile emissions over a period of a few years. One provision of the 1970 Clean Air Act is to reduce, ultimately, carbon monoxide and hydrocarbon emissions from light-duty vehicles by 90% of the measured 1970 values and nitrogen oxide emissions by 90% of the measured 1971 values. These ultimate values are recorded in the first column of Table 4.8. While the ultimate emission limits appear to be firm, the implementation dates, which are federally legislated, have been subject to change. Amendments in 1977 of the 1970 Clean Air Act established the emission limits shown in Table 4.8. The dates refer to the model year for light-duty vehicles. It is important to realize that even if these maximum emission requirements were met, the federal air quality criteria could be exceeded if there were sufficient automobiles. The likely solution then would be to limit the number of vehicles on the roads.

Table 4.8 Federal timetable for maximum allowable light-duty vehicle emissions as established by Amendments in 1977 to the 1970 Clean Air Act. The data are expressed in grams emitted per mile of travel.

	Ultimate	1977	1978	1979	1980	1981
Carbon monoxide	3.4	15.0	15.0	15.0	7.0	3.4
Hydrocarbons	0.41	1.5	1.5	1.5	0.41	
Nitrogen oxides	0.4	2.0	2.0	2.0	2.0	1.0

4.9 APPROACHES TO MEETING FEDERAL STANDARDS

There are several possible approaches to meeting the maximum emission requirements. These would include the following:

1. Modification of the present internal combustion engine.
2. Development of a new fuel for the present internal combustion engine.
3. Removal of the pollutants from internal combustion engine effluents.
4. Development of a new power plant to replace the internal combustion engine.

All of these are being pursued. However, the major thrusts are toward removal of the pollutants from the effluents and, to a lesser extent, modification of the internal combustion engine and some of its auxiliary components.

Removal of Pollutants from the Internal Combustion Engine Effluents

Figure 4.20 shows some of the devices used on automobiles to curb pollutant emissions. All those devices excepting the catalytic converters are engine modifications. The bulk of the pollutant control, however, is done by the catalytic converters. Let us now look a little more deeply into these control devices.

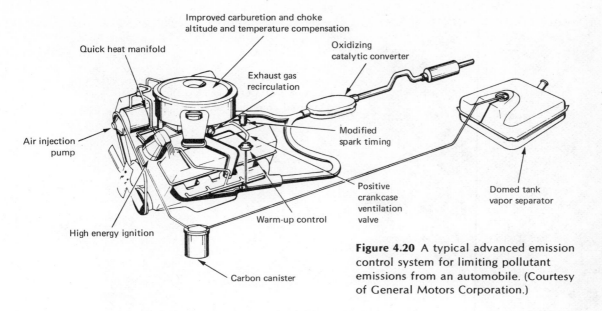

Quick heat manifold

Improved carburetion and choke altitude and temperature compensation

Oxidizing catalytic converter

Exhaust gas recirculation

Air injection pump

Modified spark timing

High energy ignition

Warm-up control

Positive crankcase ventilation valve

Domed tank vapor separator

Carbon canister

Figure 4.20 A typical advanced emission control system for limiting pollutant emissions from an automobile. (Courtesy of General Motors Corporation.)

Figure 4.21 provides a view inside a typical automobile engine. The crankshaft is used to convert the linear kinetic energy of the piston to rotational energy. Every piston (normally 4, 6, or 8 pistons) is connected to this shaft by a connecting rod. The crankcase serves as a reservoir of oil for lubricating the crankshaft, connecting rods, and other parts of the engine. Proper operation of the engine requires a tight, but moving, seal between the piston and the cylinder walls. However, a certain amount of vapor leaks out of the combustion chamber into the crankcase. This is called crankcase blowby. It was standard practice until 1961 to simply vent the crankcase to the atmosphere allowing these emissions to escape. However, the emissions can be recycled back into the air-intake system of the engine (Fig. 4.22). California required such controls in 1961, and in 1963 all American cars had them as standard equipment.

The characteristic smell associated with the filling of the fuel tank in an automobile is evidence of the evaporative ability of gasoline. The fact the smell is more noticeable on a hot day indicates that the degree of evaporation is related to temperature. As long as the fuel tank is vented to the atmosphere, gasoline vapor, and therefore hydrocarbons, will escape to the environment. Significant evaporative losses also occur at the carburetor, the function of which is to convert the liquid gasoline to a mixture of air and gasoline vapor. During operation of the engine, this mixture is injected into the combustion chambers and ignited. When the engine is shut down, the temperature of the carburetor will rise from the operating temperature of about 150°F to about 200°F because the cooling system is off. As a result, the liquid fuel bowl of the carburetor becomes hot and the fuel may even boil. If the carburetor is vented to the atmosphere, these vapors escape to the environment.

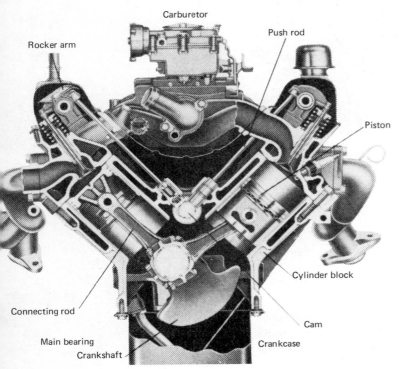

Carburetor

Rocker arm

Push rod

Figure 4.21 Cutaway view of a typical internal combustion engine.

Piston

Connecting rod

Cylinder block

Main bearing

Cam

Crankshaft

Crankcase

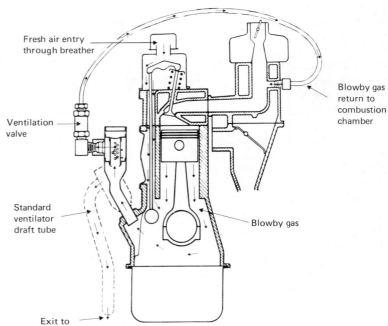

Fresh air entry through breather

Blowby gas return to combustion chamber

Ventilation valve

Figure 4.22 Typical recirculation pattern for crankcase blowby gases.

Standard ventilator draft tube

Blowby gas

Exit to atmosphere

There are two methods being used to control these evaporative losses. One utilizes the crankcase as a storage volume for the vapors. Then the vapor is pumped out of the storage into a condenser which returns liquid fuel to the gasoline tank. The second method utilizes a filter, namely activated carbon, that traps the vapor, and holds it until it can be fed back into the carburetion system.

Attack on reduction of the remaining emissions takes place with catalytic converters at the engine's exhaust. Recall (Sec. 4.4) that a catalyst is a substance that modifies the rate of a chemical reaction but is not consumed in the overall process. The modification may be either a decrease or an increase in the rate. The basic idea in a catalytic converter is to use an appropriate catalyst to convert carbon monoxide (CO) and hydrocarbons to harmless carbon dioxide (CO_2) and water vapor (H_2O) and the nitrogen oxides to molecular nitrogen (N_2) and oxygen (O_2). It is important to note that the operations on carbon monoxide and hydrocarbons are distinct from the operation on the nitrogen oxides. The former are converted to different molecular forms containing more oxygen. They are said to be oxidized. The latter are broken down from a molecular form containing oxygen to forms that do not contain oxygen. They are said to be reduced. Thus there are two distinct types of catalytic converters—oxidizing and reducing. A variety of oxidation catalysts are possible but the ones mentioned most often are platinum and palladium metals. The search for a suitable reducing catalyst continues.

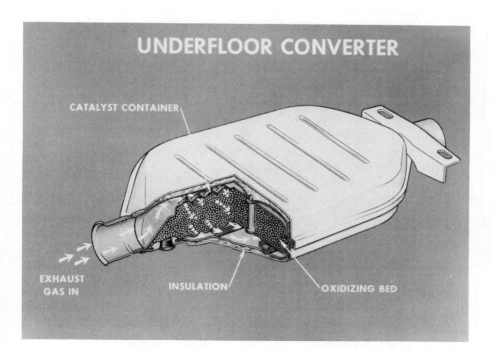

Figure 4.23 Cutaway view of a catalytic converter that can be placed in the exhaust line of an automobile to limit the emission of carbon monoxide and hydrocarbons. The exhaust gases flow through some appropriate catalytic material and are converted to some harmless gas (or gases). (Courtesy of General Motors Corporation.)

The chemistry involving the oxidizing catalyst is complex but the basic idea is as follows. Molecular oxygen (O_2) reacts with the catalyst and separates into atomic oxygen species bound to constituents of the catalyst:

molecular oxygen + catalyst $\longrightarrow$ atomic oxygen bound to constituents of the catalyst,

$$O_2 + \text{catalyst} \longrightarrow \underbrace{(O\text{---catalyst})} + (O\text{---catalyst}).$$

This symbol means that an oxygen atom is bound to a constituent of the catalyst.

The bound oxygen–catalyst system then reacts with carbon monoxide to produce carbon dioxide. The catalyst is freed in the process and the carbon dioxide escapes out the exhaust system:

$$CO + (O\text{---catalyst}) \longrightarrow CO_2 + \text{catalyst}.$$

Note that the catalyst used to instigate the process is regenerated in the oxidation of the carbon dioxide molecule. This process requires a comparatively low temperature (about 800°F). The platinum and palladium metals are deposited on a porous inert substrate. The exhaust gases circulate over the substrate and the catalytic reactions take place (Fig. 4.23). Catalytic converters first appeared on the 1975 model cars. Interim experience indicates that the oxidizing catalytic converter is very effective for reducing carbon monoxide and hydrocarbon emissions.

4.10 POSSIBLE ALTERNATIVES TO THE CONVENTIONAL INTERNAL COMBUSTION ENGINE

No one questions the convenience and reliability of the modern automobile with its internal combustion engine. Yet, if this engine is rendering the air environment unsuitable for living and contributing to the rapid depletion of petroleum resources, we can't help but wonder why it is so difficult to replace this engine with a nonpolluting and more energy efficient counterpart. Economics is a major factor. An engine with low emissions is useless if no one can afford to buy it. There are, however, some intrinsic characteristics that we demand of an automobile, and the internal combustion engine is one of the very few that can do the job. It is easy to assess these features. Let us start by considering what is asked of a contemporary automobile. For the most part it takes us to work, to shop, to visit friends and relatives, and to sundry other places. In fact, about 60 percent of the mileage is accumulated in trips of less than 3 miles. However, the contemporary 3000 pound car powered by an internal combustion engine has the energy and power to travel nonstop for 200–300 miles on an interstate highway at speeds in excess of 70 miles per hour. Any other engine that is to provide the versatility of the internal combustion engine must at least have comparable energy and power characteristics. The prospects, though, are extremely limited and like the internal combustion engine are not pollution-free. Steam engines and gas turbines are sometimes looked upon as possibilities. Steam engines have a number of disadvantages. They are bulky, massive, expensive, require a start-up time, and, because of

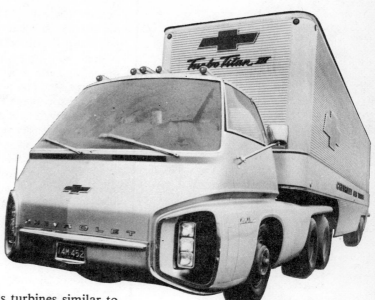

Figure 4.24 The characteristics of a gas turbine engine are best suited for large trucks and buses. The truck shown here is a prototype powered by a gas turbine engine. (Photograph courtesy of General Motors Corporation.)

the necessary boilers, are dangerous to some extent. Gas turbines similar to those used in jet aircraft are extremely reliable but they are efficient only at high speeds. To take advantage of the efficiency, a complex linkage between the turbine and wheels is required. The future of the turbine in an automobile is questionable but possibilities exist for using turbines in buses and trucks for long-distance trips with few stops. Beginning in 1978 the Departments of Energy and Transportation will instigate a seven-year program that will evaluate the performance of 50 gas turbine buses on city streets. The alternate power sources that appear to be most likely to lighten the burden on the conventional internal combustion engine are the stratified charge engine, the diesel engine, the Wankel rotary engine, and the electric motor. The stratified charge and Wankel engines are novel internal combustion engines using gasoline for fuel. The diesel is an internal combustion engine that operates on a lower grade fuel than gasoline. All will generate the same type of pollutants as the conventional internal combustion engine. However, the engines have some unique characteristics that make the pollutants easier to control. Electric vehicles, to date, do not have energy characteristics comparable to conventional cars. Thus they cannot travel long distances. Nevertheless, they could be extremely useful for much of the short-trip driving. Let us examine the features of these possibilities for replacing the conventional internal combustion engine.

The Stratified Charge Engine

The stratified charge engine is a relatively simple modification of a conventional internal combustion engine. The modification involves a second combustion chamber above the main combustion chamber (Fig. 4.25). The gas which is ignited in the cylinder of an engine to begin a cycle is called a charge.

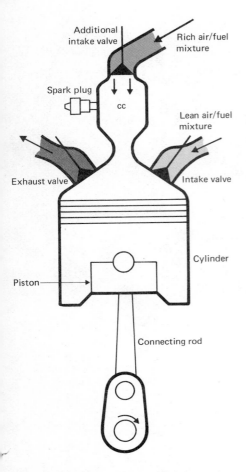

Figure 4.25 The stratified charge engine. This engine differs from the conventional internal combustion engine only in that an additional combustion chamber (cc) is attached to the top of each cylinder. An air/fuel mixture rich in fuel is first ignited in the small upper chamber. The fire spreads to the lean mixture in the main combustion chamber. Most exhaust pollutants are burned up inside the cylinder because of the prolonged combustion process.

Because there are different gasoline vapor–air mixtures in the two chambers of the engine, the charges are said to be stratified (i.e., layered). The amount of pollutants produced during ignition of the gasoline vapor–air mixture depends critically on the proportions of the mixture. Additionally, the reliability and operating smoothness also depend on the mixture. If the ratio of air to fuel is too large, the engine misfires, sputters, and is balky in starting. Under these conditions, though, the nitrogen oxide and hydrocarbon emissions are reduced significantly. The idea in the stratified charge engine is to inject into the upper chamber an easily ignitable mixture rich in fuel but lean in air. As the flame from this ignition spreads, a mixture rich in air but lean in fuel is injected into the lower chamber. This mixture is ignited easily by the first flame. With this principle, a smoothly running engine with low pollutant emissions results. This engine was first used in the Japanese Honda automobile which met the 1975 federal emission standards. An effort is being made to adapt this principle to larger American engines. It is unlikely, though, that these engines will be commercial before 1980.

The Diesel Engine

Likely, you have seen a large tractor-trailer truck with smoky, smelly exhausts cruising on a highway. Probably, this vehicle was powered by a type of internal combustion engine called a diesel engine. The diesel engine is constructed and functions much like a gasoline-powered internal combustion engine with reciprocating pistons. However, rather than using an electric spark to ignite an air–fuel mixture in the combustion chamber, the diesel engine takes advantage of the heating of a gas when compressed. This heating by compression is observable when you push the piston of a bicycle pump to force air into a tire by compression. The cycle of a diesel engine starts by compressing air in the combustion chamber. When the temperature of the air rises sufficiently, fuel is injected into the combustion chamber and is ignited by the high-temperature environment. The remainder of the cycle is identical to that of the gasoline-powered internal combustion engine. The diesel engine has several attractive features. It operates on a lower grade fuel; you may have noticed that diesel fuel is cheaper than gasoline at truck stops. The nonnecessity of spark plugs simplifies the ignition system. Finally, the diesel engines are generally more efficient than gasoline engines. While diesel engines are extremely popular in American trucks, buses, and trains, they have not been popular in American cars. It is probably the relatively long "warm up" time of the diesel engine that has contributed to its unpopularity. Mercedes Benz diesel-powered automobiles, introduced to the United States in 1952, are becoming increasingly popular. Because the government has mandated automobile fuel economy measures, the American automobile industry is reevaluating the diesel engine for family car use. Both American and foreign-produced models were introduced in 1978. The pollution and fuel economy characteristics are attractive.

The Wankel Engine

The Wankel engine (Fig. 4.26) does not constitute a revolutionary new combustion principle for it is also a four-cycle, gasoline fuel, internal combustion device operating much like the conventional internal combustion engine. Its type, but not quantity, of emissions is no different. It produces more hydrocarbons and carbon monoxide but fewer nitrogen oxides. Technologically it is easier to get rid of hydrocarbons and carbon monoxide than nitrogen oxides. The Wankel engine also has the extremely attractive feature of being able to get an equivalent amount of horsepower in a much smaller volume with much less weight. It is this feature which provides space for pollution controls, a space not available in most contemporary automobiles. Because of the engine's simplicity, it can conceivably be manufactured for half the cost of a conventional engine and the savings used for pollution abatement. Unlike the conventional internal combustion engine, the Wankel engine has no valves and it operates on low-octane fuel. All these features add up to low maintenance costs.

The Wankel engine was first used commercially in the Japanese Mazda automobile. Later, the French used it in the Citroën automobile. The Mazda was able to meet the 1975 federal emission standards. It took advantage of the low nitrogen oxide emissions of the Wankel engine and used a thermal reactor to convert the hydrocarbons and carbon monoxide to water vapor and carbon dioxide. Because carbon monoxide and hydrocarbons are the result of incomplete fuel combustion, the idea in a thermal reactor is to complete the combustion in another area of the engine. The exhaust manifold where the hot exhaust gases escape is chosen for the combustion area because combustion is facilitated by high temperatures. The thermal reactor is primarily a hot baffle and is designed to fit into or near the manifold. Oxygen required for the combustion is pumped into the reactor from the atmosphere.

Whether or not the Wankel rotary engine lives up to expectations remains to be seen. The Wankel-powered Mazda automobiles have been marred by failure of the seals between the engine rotor and combustion chamber wall. The fuel economy of the early models was only average. The American automobile industry has built prototype Wankel-powered cars. However, no production vehicles have been built at the time of this writing and no production plans have been announced. General Motors has abandoned plans for production of cars using the Wankel engine. Although the Wankel engine is still a contender for the low-polluting engine sought by the automobile industry, it may lose out to the conventional reciprocating internal combustion engine with catalytic converters if the converters are as effective as they seem to be.

The Electric Vehicle

All vehicles discussed thus far require a fossil fuel as the primary energy source. The electric vehicle utilizes electric energy which is stored in a battery. An electric motor converts the electric energy to mechanical energy. The

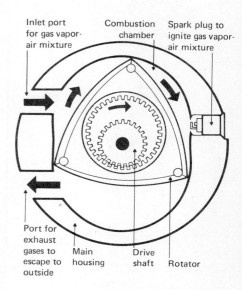

Figure 4.26 The essential components of the Wankel engine. The triangular-shaped rotator is forced into motion by gaseous explosions taking place between the faces of the rotator and main housing.

principle is not new. Special purpose electric vehicles have been around as long as the conventional automobile. However, the development of more energetic batteries, the quest for low-polluting vehicles, and the concern for depletion of our petroleum reserves have revived interest in electric vehicles.

Although an electric vehicle does not emit pollutants in its operation it does contribute to pollution indirectly. The electric energy for the batteries would likely come from a conventional electric power plant. If this were a coal-burning plant, then sulfur oxides and particulates would be produced. These pollutants differ in type from those from the internal combustion engine but they would be comparable in amount. For example, burning 10 gallons of gasoline in an automobile with no emission controls produces about 30 pounds of carbon monoxide. The energy equivalent of 10 gallons of gasoline is about 370 kW-hrs. The production of 370 kW-hrs of electric energy would produce about 24 pounds of sulfur oxides. Thus the amount of sulfur oxides produced is roughly the same as the amount of carbon monoxide produced from burning 10 gallons of gasoline in an internal combustion engine. This example does not mean that the pollution aspects of electric and internal combustion vehicles are the same. Rather it shows that electric vehicles would contribute to pollution and to oil depletion if the electric energy for the batteries is produced in an oil-burning electric power plant.

The power and energy requirements of a conventional automobile limit the utility of batteries in an electric vehicle. Conventional lead-acid batteries like those used for electrical services in present-day automobiles have energies

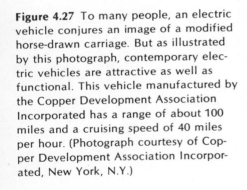

Figure 4.27 To many people, an electric vehicle conjures an image of a modified horse-drawn carriage. But as illustrated by this photograph, contemporary electric vehicles are attractive as well as functional. This vehicle manufactured by the Copper Development Association Incorporated has a range of about 100 miles and a cruising speed of 40 miles per hour. (Photograph courtesy of Copper Development Association Incorporated, New York, N.Y.)

of about 10 watt-hours per pound. Gasoline has 1000 watt-hours per pound or about 100 times as much. This means that, for the same weight, batteries have 100 times less energy. If a car can go 300 miles at 60 miles per hour on a tank of gasoline, it could go only three miles at the same speed using batteries of the same weight as the gasoline. Certainly, additional energy could be obtained by utilizing more batteries, but this also becomes more difficult. For example, a 15 gallon tank of gasoline weighs about 100 pounds. Thus the amount of batteries required to produce the same energy would be 10,000 pounds. If the speed requirements were eased, the energy requirements would also decrease. However, the electric vehicle with proven batteries is relegated to the class of special purpose vehicles for short trips (Fig. 4.27).

Experimental batteries with energies of 100 watt-hours per pound are being built. Like any new product they are expensive, but if their technical worth proves out, the price will surely decrease. These batteries could increase the range to 150 miles in urban traffic on a single charge which would handle all but major interstate trips. There is reason to believe that batteries with 300 watt-hours per pound are feasible but the timetable for development is around the year 2000.

TOPICAL REVIEW

1. What do the words "pollutant" and "pollution" mean to you?
2. What are nitrogen oxides, carbon monoxide, hydrocarbons, sulfur oxides, and particulates and how are they formed? What is the environmental concern for them?
3. Why is the micrometer unit of length useful for specifying particulate sizes?
4. Why is the settling time for particulates important?
5. Present a few arguments for the desirability of reducing pollution levels in urban areas.
6. What are the constituents of atoms and molecules?
7. What pollutants are attributable mainly to transportation? to industry?
8. What is a bound system?
9. Give a reason why a limit on the mass of particulates per cubic meter of air may not be completely appropriate.
10. Name three devices used to collect particulates in a coal-burning power plant.
11. What is a hydrocarbon? Give an example.
12. What is the electric force? What is its role in atoms and molecules?
13. To date, what has been the most effective way of controlling sulfur oxide emissions?
14. What is the purpose of tetraethyl lead in gasoline?
15. How is photochemical smog formed? Why is smog particularly bad in Los Angeles?

16. What is the approach being taken for controlling air pollution from automobiles?
17. What is a catalyst? catalytic converter?
18. Distinguish between oxidizing and reducing catalytic converters.
19. What alternatives are there for the conventional internal combustion engine?
20. What is a Wankel engine and why is it of interest?
21. Describe the stratified charge concept.
22. What are the attractive features of the diesel engine?

REFERENCES

Nuclear Model of the Atom

Any standard physics or physical science text will devote a number of pages to models of the atom. Three such texts that would be useful for expansion of the material presented in Chapter 4 are

1. *Physics for Society*, W. B. Phillips, Addison-Wesley, Reading, Mass. (1971). Chapters 4 and 8.
2. *Physical Science,* Arthur Wiggins, Houghton Mifflin, Boston (1974). Chapter 2.
3. *The Atom and the Universe,* James S. Perlman, Wadsworth, Belmont, Calif. (1970). Chapters 19–23.

Air Pollution

1. A comprehensive treatment of all aspects of air pollution is given in *Fundamentals of Air Pollution* by Samuel J. Williamson, Addison-Wesley, Reading, Mass. (1973), and "Air Pollution: Threat and Response" by David A. Lynn, Addison-Wesley, Reading, Mass. (1976).
2. Considerable attention is given to the problems and control of air pollution in *Environmental Pollution*, 2d ed., Laurent Hodges, Holt, Rinehart, and Winston, N.Y. (1977).
3. The mechanism and problems of implementing clean-air standards are given in "Enforcing the Clean Air Act of 1970" by Noel de Nevers, *Scientific American* **228**, No. 6, June 1973.

The problems and control of particulate matter, sulfur oxides, carbon monoxide, nitrogen oxides, and hydrocarbons are given in detail in the following publications available from the Environmental Protection Agency, 401 M Street S.W., Washington, D.C.

1. "Air Quality Criteria for Particulate Matter," NAPCA Publication No. AP-49, January 1969.
2. "Control Techniques for Particulate Air Pollutants," NAPCA Publication No. AP-51, January 1969.

3. "Air Quality Criteria for Sulfur Oxides," NAPCA Publication No. AP-50.

4. "Air Quality Criteria for Carbon Monoxide," NAPCA Publication No. AP-62.

5. "Air Quality Criteria for Nitrogen Oxides," NAPCA Publication No. AP-84.

6. "Air Quality Criteria for Hydrocarbons," NAPCA Publication No. AP-64.

7. "Air Quality Criteria for Photochemical Oxidants," NAPCA Publication No. AP-63.

8. "Control Techniques for Carbon Monoxide, Nitrogen Oxide, and Hydrocarbon Emissions from Mobile Sources," NAPCA Publication No. AP-66

The following articles are useful for more information on the topics covered in Chapter 4.

Air Pollution

1. "Air Pollution and Public Health," Walsh McDermott, *Scientific American* **205**, 49 (October 1961).

Control of Sulfur Oxides

1. Coal Research (IV): Direct Combustion Lags Its Potential," *Science* **194**, 172 (8 October 1976).

2. Sulfur Oxide Control Technology, U.S. Department of Commerce (1975).

The Wankel Engine

1. "The Wankel Engine," David E. Cole, *Scientific American* **227**, 14 (August 1972).

The Electric Vehicle

1. "Advanced Storage Batteries: Progress but Not Electrifying," *Science* **192**, 541 (7 May 1976).

2. "Electric Cars: The Battery Problem," Victor Wouk, *Bulletin of the Atomic Scientists* **27**, No. 4, 19 (April 1971).

4.1 Motivation

1. We have characterized a pollutant as being something that affects something that we value. Why is sulfur dioxide considered to be a pollutant?

2. Table 4.1 shows that refuse disposal creates significant amounts of all five air pollutants mentioned except sulfur oxides. What is the reason for this?

4.2 Fundamental Origin of Sulfur Oxides, Carbon Monoxide, and Nitrogen Oxides

3. A model airplane attached to a tethering cord flies in a circular path. In this sense it is bound to the controller like an electron is bound to the nucleus of an atom. What exerts the force on the plane that holds it in its circular path?

4. Two persons tugging at opposite ends of a rope form a bound system analogous to two atoms bound together to form an oxygen molecule. How could you use three persons and three ropes to simulate a bound system similar to a CO_2 molecule?

5. A racing car that travels at constant speed around a circular track is bound to the track something like a circulating electron is bound to a nucleus. What is the force that binds the car to the track?

6. Rubbing a glass rod with a piece of silk produces a net positive charge on the glass. Describe the mechanism by which the glass acquires the positive charge.

7. When there is little moisture in the air, it is common for a person to acquire a net electric charge (sometimes called static electricity) when getting out of a chair. How would you explain the acquisition of this charge?

4.3 Particulate Matter

8. Particles falling in the atmosphere are retarded by a viscous force. The larger the size of the particle, the larger the viscous force. How is this consistent with what you know about a parachutist floating gracefully to the ground?

9. Sometime when you are so inclined, take a handful of sand which has a wide variety of grain sizes and drop it into a clear bottle of water. Observe which particle sizes reach the bottom of the bottle first. Then draw an analogy between this effect and the settling of particulates in the atmosphere.

10. What are some of the geographical factors considered when choosing a site for an industrial city? How do some of these factors contribute to pollution problems?

4.4 Sulfur Oxides

11. Synergism exists around us in a number of ways. A parent knows that the net mischievousness of two children is often far greater than the sum of the effects that each might produce alone. The combination of makeup and a lady's face produces a synergistic effect. What are some other examples of synergism in our everyday lives?

12. It is very hard to visualize the smallness of a millionth of a gram (μg). As a help, imagine that you had one million sticks each having a length of one-half inch, and that these sticks were placed end to end in a straight line. If all these sticks are of one color except one which is placed at one end, how far would you have to walk from the opposite end to reach this odd-colored stick?

4.5 Fundamental Origin of Emissions

13. Which one of the five common air pollutants shown in Table 4.1 is *not* produced from a component of the burned fuel?

14. Since unleaded gasoline does not contain tetraethyl lead, why is it that unleaded gasoline generally costs as much, or more, as leaded gasoline?

4.6 Photochemical Smog

15. Why do severe photochemical smog conditions occur in mid-afternoon even though automobile emissions are greatest in the early morning hours?

16. Why does the breakup of NO_2 into NO and O require energy?

17. What is the environmental advantage to using pure oxygen rather than air in a car's carburetor?

4.7 Effects and Extent of Air Pollutants

18. Why do producers of charcoal for outdoor grills often warn consumers not to burn the charcoal in a closed area like a basement or a garage?

19. Why is it important to continually circulate fresh air into a busy automobile tunnel?

20. Attendants at toll gates on busy city highways often complain about exhaust fumes from vehicles passing through the gates. What component of the fumes is most likely the origin of their complaints?

4.8 The Federal Emission Standards

21. To what extent do nationwide economic conditions influence the public's desire for pollution controls on automobiles?

4.10 Possible Alternatives to the Conventional Internal Combustion Engine

22. It is common to use electric-powered service vehicles inside merchandise warehouses and manufacturing facilities. What advantages do electric vehicles have over gasoline-powered vehicles for this function?

23. Why isn't a battery-powered motorcycle practical?

24. The fuel gauge on a gasoline-powered car registers the amount of gasoline remaining in the tank. What would the fuel gauge on an electric car register?

25. There was a time when electric vehicles outnumbered gasoline-powered automobiles and the future for electric cars looked brighter. What do you think led to the demise of the electric car?

26. Small electric-powered carts are very popular for carrying golfers around a golf course. What characteristic makes these vehicles attractive for this function?

NUMERICAL PROBLEMS

4.1 Motivation

1. Table 4.1 shows that 20 million tons of particulates were emitted in 1974. To get a feel for the magnitude of such a quantity, consider the following. Suppose that a highway 50 ft wide extended for 3000 mi across the United States. If these 20 million tons were spread uniformly over this highway, how deep would the layer be? Assume that one cubic foot of particulates weighs 100 lb.

4.2 Fundamental Origin of Sulfur Oxides, Carbon Monoxide, and Nitrogen Oxides

2. In a particular demonstration, rubbing a balloon on a person's hair produces a net negative charge of 0.0000016 coulombs on the balloon. How many electrons were transferred in the process?

3. An oxygen atom has about half the mass of a sulfur atom. Knowing the composition of the sulfur dioxide molecule (SO_2), explain why nearly 100 pounds of sulfur dioxide are produced from the oxidation of 50 pounds of sulfur.

4.3 Particulate Matter

4. A micrometer, symbolized μm, is a millionth of a meter. Express the 0.001 centimeter diameter of a particulate in μm.

5. A sample of air is contained in a cubic container, each side having a length of 20 cm. Measurements show that the air contains 0.1 grams of particulate matter. Express the particulate concentration in micrograms per cubic meter of air.

6. A certain classroom measures 10 m × 15 m × 4 m. The suspended particulates in this room were determined to be 55 μg/m^3. What is the total mass (in grams) of suspended particulates in the room?

7. A high-quality coal from Eastern United States produces about 13,000 Btu of heat per pound. A low-sulfur Western coal may produce 10,000 Btu per pound. If a power company can meet its daily demands with 5000 tons of Eastern coal, how much Western coal would be required?

8. For the sake of argument, suppose that a sample of air contains 10,000 particles having a diameter of 0.1 μm and 100 particles having a diameter of 1 μm. All particles are made of the same material.

   a) Show that the total mass of 1 μm diameter particles is 10 times greater than the total mass of the 0.1 μm diameter particles.

   b) If for health reasons, you wanted to remove 90% of the 0.1 μm diameter particles, would removing 90% of the total mass do the desired job?

9. The white chalk commonly used to write on classroom blackboards is about 1 cm in diameter. A 1 cm length of the chalk has a mass of about 1¼ grams. If the chalk were pulverized to a fine powder and uniformly distributed throughout a classroom having dimensions 10 m × 15 m × 4 m, what would be the concentration of suspended chalk dust? How does this concentration compare with the primary standard for suspended particulates?

10. The concentration of suspended particulates in the atmosphere can be measured by forcing air through a semiporous filter paper that separates the particulates from the air. Determine the suspended particulate concentration from the measurements presented below:

    air flow rate = 0.71 m³/min,
    air sampling time = 24 hours,
    weight of filter paper before sampling = 4.5722 grams,
    weight of filter paper after sampling = 4.7424 grams.

11. In a certain area, the average particulate concentrations for three consecutive days were 343, 216, and 343 micrograms/m³.

    a) Was the federal primary standard violated during this period?

    b) Show that the arithmetic mean concentration is 301 μg/m³ and the geometric mean concentration is 294 μg/m³.

12. A certain particulate has a mass of 10 μg and a density of 1.25 g/cm³. What would a length of one of its sides be if it were a cube? Using Fig. 4.5 as a guide, what sort of particulate might this be? Remember, the graph uses micrometers for particulate size.

13. The differences in settling times for various sizes of particulates is striking. Show that this is true by calculating the time required for particles of diameters of 0.1, 1, 10, 100, and 1000 micrometers to settle a distance of 1 km. Data for the settling velocities are given in Table 4.3.

4.4 Sulfur Oxides

14. Sulfur trioxide (SO_3) can be formed by combining oxygen molecules (O_2) with sulfur atoms. Write a chemical reaction for this process.

15. There is considerable environmental concern over the formation of sulfuric acid (H_2SO_4) starting from sulfur dioxide (SO_2) released from the

burning of coal. Fill in the steps in the equations below for a possible mechanism for producing sulfuric acid:

$$\underline{\hspace{2cm}} + SO_2 \longrightarrow NO + SO_3,$$
$$SO_3 + \underline{\hspace{2cm}} \longrightarrow H_2SO_4.$$

16. The total electric energy production in 1974 was 1.87 trillion kW-hr. Assume that this was produced by a coal-burning power plant with an overall efficiency of 40 percent. Estimate how much coal was burned. Assuming that the coal contained two percent sulfur, estimate the amount of sulfur dioxide produced. Compare with that quoted in Table 4.1.

4.5 Fundamental Origin of Emissions

17. Two gasolines have octane ratings of 50 and 100. What is the octane rating of a gasoline having equal portions of these two gasolines?

18. About five pounds of nitrogen oxides are produced for every ten gallons of gasoline burned in an internal combustion engine. An average car will use 750 gallons of gasoline each year and there are about 100 million cars in the United States. Estimate the amount of nitrogen oxides produced in a year.

19. The complete oxidation of heptane (C_7H_{16}) produces only carbon dioxide (CO_2) and water (H_2O). Knowing this, complete the following reaction:

$$C_7H_{16} + 11\, O_2 \longrightarrow \underline{\hspace{2cm}} + \underline{\hspace{2cm}}.$$

4.9 Approaches to Meeting Federal Standards

20. A possible way of rendering nitric oxide (NO) harmless is to convert it to molecular nitrogen (N_2) and water vapor (H_2O) through thermal reactions with ammonia (NH_3). Fill in the blanks in the chemical reaction below for this process.

$$\underline{\hspace{2cm}} NO + 4\underline{\hspace{1cm}} \longrightarrow 5\underline{\hspace{2cm}} + \underline{\hspace{1cm}} H_2O.$$

21. The two reactions below can be used to convert SO_2 to SO_3. The end product of the first reaction is used as part of the input for the second reaction.

$$2\, NO + O_2 \longrightarrow 2\, NO_2$$
$$2\, NO_2 + 2\, SO_2 \longrightarrow 2\, NO + 2\, SO_3.$$

How many molecules of NO and NO_2 are consumed in the execution of these two reactions? What, then, is the role of NO and NO_2 in the reactions?

4.10 Possible Alternatives to the Conventional Internal Combustion Engine

22. A certain car has a 15 gallon gasoline tank. With a full tank, the car weighs 2500 pounds. What percentage of the total weight is due to the fuel? Assume a gallon of gasoline weighs 8 pounds. If the car were electric

and the batteries had an energy capacity (watt-hours per pound) 10 times less than gasoline, what percentage of the 2500 pounds of weight would be due to batteries?

23. The transformation of energy from fossil fuels to mechanical energy in an automobile via electric generation and battery-powered motors involves several steps with characteristic efficiencies.

1. Power plant efficiency: 40 percent
2. Ten percent energy loss through transformers and transmission to the city
3. Ten percent energy loss through transformers and distribution within the city
4. Battery-charging efficiency: 80 percent
5. Motor efficiency: 90 percent

Show that the overall efficiency is 23 percent. How does this compare with the efficiency of a typical automobile?

ELECTRIC ENERGY AND POWER

5

The metallic cables that transmit electric energy from a generating station to a distribution center often extend over hundreds of miles of countryside. They bring the most versatile energy of all to our homes and industries. Photograph by David Falconer. (Courtesy of the EPA.)

5.1 MOTIVATION

The word current suggests the motion of something such as water. When pertaining to an electric current, it is charges that are in motion. A motion of charges, and therefore an electric current, is created by the discharge of static electricity when you touch a metallic door knob or when a thundercloud discharges to produce a lightning flash. Uncontrolled electric currents such as these have been curiosities for centuries. It took the ingenuity of Thomas Edison to control electric currents in metallic conductors on a scale useful to mankind. Under Edison's guidance, the first commercial electric power generating facility was installed in New York City in 1881. In the interim, the United States and most of the world have become totally dependent on electricity. Any extensive man-made or naturally induced loss of electric power (as in New York City in 1965 and 1977) renders communities helpless and vulnerable. It is easy to understand why we are so dependent on, and yet so vulnerable, to a loss of electricity. Electricity is enormously convenient and flexible. Small children quickly learn to turn on lights and appliances and to insert the power plug of radios, electric train sets, electric toothbrushes, etc., into receptacles. But most children and adults do not realize that an electric power company does not have provisions to store electric energy. They produce electric power on demand. If the demand exceeds their generating capacity or if electric generators are rendered inoperative, trouble arises.

It is probably not surprising that the use of electric energy is the fastest growing segment of our energy economy. While overall energy use grew by 5.1 percent in 1976, electric energy use increased by 5.8 percent. Twenty-nine percent of the gross energy in 1976 was used to generate electricity. About two thirds of this energy converted to make electricity was lost in the conversion and transmission processes. This means that it takes about three units of input energy to make one unit of electric energy. For every unit of energy saved at the consumer end, three units will be saved at the generating end. Input energy for electric power plants was furnished in 1976 by coal (46.3%), oil (15.7%), natural gas (14.5%), nuclear (9.4%), and flowing water (13.9%). Although electric energy is in great demand, generating problems exist. These problems range from pollution to meeting societal demands and developing alternative methods of conversion. As consumers, it is essential that we understand these problems and do our part in conserving the use of electric energy. This chapter is devoted to helping achieve this goal.

5.2 ELECTRICITY

When water flows over a dam and impinges on a paddle wheel, work is done by the moving water on the paddle wheel. The initial motion of the water is provided by the gravitational force pulling the water from the top to the bottom of the dam. In an analogous fashion, electric charges moving through a metallic conductor and through a connected light bulb do work within the bulb and produce light energy and heat energy. The motion of the charges is provided by an electric force. We can characterize the water flow rate by gallons per minute, for example. Similarly, we can speak of a rate of flow of charge in a wire. Because the wire is composed of atoms having both positive (nuclei) and negative (electrons) charges, conceivably the electric current could involve both positive and negative charges. However, the nuclei are structurally bound to the conductor and cannot move. Only the outermost electrons are free to contribute to an electric current. In a copper wire, for example, only about one of the 29 electrons possessed by each copper atom contributes to an electric current. In principle, the electric current is measured by determining the amount of charge flowing through a cross section of the conductor in a given time interval:

$$\text{electric current} = \frac{\text{amount of charge flowing through a cross section}}{\text{time required for the flow}},$$

$$I = \frac{Q}{t}. \tag{5.1}$$

Charge (Q) measured in coulombs and the time interval (t) measured in seconds yields coulombs per second as the units of current (I). A coulomb per second is called an ampere (A). If 100 coulombs of charge flow by some position in a conductor in 10 seconds, we would say the current is 10 amperes. A force must be exerted on the electrons to cause them to flow through a conductor just as a force must be applied on water to keep it moving through a pipe. The force does work on the electrons just as work is done on an automobile when it is forced some distance. It is the role of an electric power company to supply this electrical ''pressure.'' This electrical ''pressure'' is called the potential difference and can be defined in terms of the amount of work done on a charge as it moves between two points:

$$\text{potential difference} = \frac{\text{work required to move an amount of charge}}{\text{amount of charge moved}},$$

$$V = \frac{W}{Q}.$$

Potential difference has units of joules per coulomb. A joule per coulomb is called a volt (V). If 100 joules of work are done on one coulomb of charge when it moves between two points, the potential difference is 100 volts. The potential difference between the two ''holes'' in the electric outlets in a house is effectively 115 volts. Batteries for automobiles provide either a 6- or a 12-volt potential difference.

It is important that you understand this notion of a potential difference. An everyday analogy may help. If you were to tell a friend that a building is 500 feet high, he would probably assume that you mean the difference in ''height'' between the top and bottom is 500 feet. Clearly, though, if the building were on top of a mountain its height would be more than 500 feet relative to the base of the mountain. Potential differences like heights, are also relative. When you say that the potential difference across an electrical outlet in a house is 115 volts it means that one of the contacts is 115 volts relative to the other contact. This other contact is at the same potential as any other part of the room, including yourself. You can, in fact, touch this contact and not get an electrical shock because there is no difference in potential between you and that contact. This contact is called the ground because it is connected electrically to a metal stake driven into the ground (earth). The other contact is often referred to as "hot." Informally, the word voltage is used often instead of potential difference. If we say the voltage at an electrical outlet is 115 volts, strictly speaking this means the potential difference between the terminals of the outlet is 115 volts.

When an appliance like a lamp is plugged into an outlet and the switch turned "on," an electrical circuit is said to be completed. It is complete in the sense that a complete (or closed) path is provided for the flow of electrons. It is entirely analogous to connecting a hose to a pump that pumps water out of the bottom of a container and returns it to its top (Fig. 5.1). The electric power company provides the electrical pump. The wire plays the role of the water hose. The switch functions as a shutoff valve.

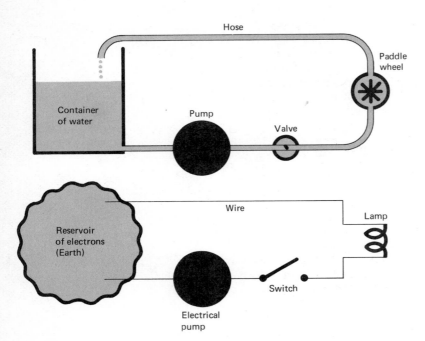

Figure 5.1 Schematic comparison of a fluid circuit and an electrical circuit. The water pump forces water to circulate through the water circuit. The electrical pump forces the free electrons to circulate in the electrical circuit. Electric charge is always conserved in the process just as water is conserved in the mechanical process.

5.3 ELECTRICAL RESISTANCE

In the water circuit (Fig. 5.1), the hose presents a certain resistance to the flow of water through it. If obstructions were placed in the hose, then surely it would be more difficult to force water through the hose. We would say that the resistance of the hose increased. Similarly, a wire offers resistance to the flow of electrons. Just as some hoses are better conductors of water than others, some wires are better conductors of electrons than others. This resistance depends not only on "obstructions" in the path of the electrons but also on the size of the conductor. If two similarly constructed hoses of different lengths are connected to the same water tap, then the flow is greater out the open end of the shorter hose. Replacing these two hoses with ones having the same length but different diameters shows that the flow is greater out the larger diameter hose. Electrical resistance, like hose resistance, increases as the conductor length increases and decreases as the conductor diameter increases. Electrical resistance of a conductor or electrical device is defined quantitatively as

$$\text{resistance} = \frac{\text{potential difference across the element measured in volts}}{\text{current through the element measured in amperes}},$$

$$R = \frac{V}{I}. \tag{5.2}$$

Resistance has units of volts divided by amperes. This unit is called an ohm. The relation $R = V/I$ is called Ohm's law—if R does not depend on V or I. In a household where the potential difference is fixed at 115 volts, the current through a device will depend on its resistance. As the resistance decreases, the current increases. The energy that is used to overcome the electrical resistance appears as heat. For example, the heat required for a toaster is produced by electrons overcoming the electrical resistance provided by the wire elements of the toaster.

> A light bulb glows when connected to the "hot" terminal of an electrical outlet and a cold-water faucet. This is because the pipe connecting the faucet passes through the earth thereby completing the electrical circuit. Electrical shock occurs when a person touches the "hot" terminal while standing on the earth or something connected to the earth. Because of the possibility of electrical shock, you should not attempt the demonstration shown in this photograph.

5.4 ELECTRIC POWER AND ENERGY

Like water flowing over a dam, electrons flowing in a conductor acquire kinetic energy as a result of the work done on them. This is another example of the work–energy principle. The electrons convert this energy in a variety of ways. In a lamp the energy is converted to heat and light. In an electric motor, the energy is converted to rotational mechanical energy. The rate at which moving charges convert energy is called electric power:

$$\text{power} = \frac{\text{energy converted measured in joules}}{\text{time required for conversion measured in seconds}},$$

$$P = \frac{W}{t}. \tag{5.3}$$

Because electric energy, like mechanical energy, is expressed in joules and time is in seconds, electric power is expressed in joules per second, or watts. Many

Table 5.1 Power requirements (in watts) of some common household appliances.

Cooking range	12,000
Heat pump	12,000
Clothes dryer	5,000
Water heater	2,500
Air conditioner (window)	1,600
Broiler	1,400
Frying pan	1,200
Toaster	1,100
Electric space heater	1,000
Television (color)	330
Food mixer	130
Radio	70
Razor	14
Toothbrush	7
Clock	2

electrical devices use power in the thousands-of-watts range and electric power plants generate power in the millions-of-watts range. You will often see the terminology, kilowatts and megawatts, symbolized kW and MW, meaning thousand and million watts, respectively. From the definitions of voltage $V = W/Q$ and current $I = Q/t$ it follows that electric power can also be written as $P = VI$. For two electrical appliances operating on the same voltage, the one using the greater current will consume energy at the greatest rate. For those devices that generate heat, the power equation can be written in terms of the electrical resistance. In terms of resistance and current,

$$P = I^2 R. \tag{5.4}$$

In terms of resistance and voltage

$$P = V^2/R. \tag{5.5}$$

Any conductor will have some resistance, and there will be some energy loss in the form of heat when it carries a current. If the conductor is very long so that the resistance is large, these losses can be significant. This is the situation with conductors used to transmit electric power from a generating facility to a customer. We will say more about this in Section 5.8.

The companies that provide the potential difference for forcing electrons through our appliances are often referred to as power companies. You might suspect, then, that they are paid for power. But they are not. They are paid for energy. If this is not obvious, a little thought should convince you. Power is the rate of using energy. A 100-watt light bulb consumes energy at the rate of 100 joules per second regardless of how long it is turned on. Clearly, though, the cost of operation depends on how long the bulb is on. Using Eq. 5.3, it follows that energy, power, and time are related by

$$\text{energy} = \text{power} \cdot \text{time},$$
$$W = P \cdot t. \tag{5.6}$$

So any proper unit of power multiplied by a unit of time yields a unit of energy. Watts (or kilowatts) and hours are proper units of power and time. Thus the watt-hour (or kilowatt-hour, the unit for paying the utilities company) is a unit of *energy*. It is a peculiar unit because the units of time (hours) do not cancel the time units of seconds in power.

$$W = P \text{ (watts)} \cdot t \text{ (hours)}$$

$$= P \left(\frac{\text{joules}}{\text{second}} \right) \cdot t \text{ (hour)}.$$

Units of hours do not cancel the units of seconds.

Table 5.1 lists the power requirements for several household appliances. Note that those that involve the generation or removal of heat require the most power. These devices tend to have low resistance.

The energy consumed by a 12,000 watt (12 kW) range operating for one hour would be W = 12 kW • 1 hr = 12 kW-hrs. If electric energy costs three-and-a-half cents per kW-hr, it would cost forty-two cents to operate it each hour.

We have discussed electric current and the idea that an electrical "pressure" is required to produce the current but have said nothing about how the pressure is produced by the power company nor how it is transmitted to homes and factories. To do this we must understand some of the fundamentals of the magnetic force and magnetism.

We can replace the balloons in the electric force model (Fig. 4.2) by two straight lengths of wire each carrying a current, and we would find that the wires experience mutual forces (see Fig. 5.2). This is due to the magnetic force and is attributed to the property of a *charge in motion*. The magnetic effects seen in a toy magnet are due to motions of charges in the atoms comprising the magnet. Studies of magnets show that the magnetic effects are concentrated near the ends. Historically, these concentrated magnetic regions have been called poles. For centuries, man has known that the earth possesses magnetic poles located near the north and south geographic poles. If a pole of one magnet is placed near the poles of another magnet it will be attracted toward one pole and repelled away from the other pole. We distinguish the fundamental difference between the poles by calling one N and the other S. An N pole is that pole of a suspended magnet such as a compass that points toward the geographic north pole. Just as an uncharged bit of paper can be attracted to a charged comb (Fig. 4.11) an unmagnetized bit of iron can be attracted to the pole of a magnet. An unmagnetized piece of iron contains a large number of tiny atomic magnets having a random orientation. As a result, the iron shows no *net* magnetic effect.

5.5 MAGNETIC FORCE AND MAGNETISM

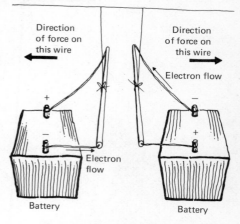

Figure 5.2 If two straight lengths of wire are suspended and a current is produced in each, they will experience mutual forces as a result of a magnetic force created by the moving charges in the wires.

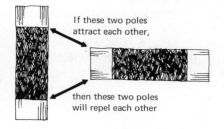

When a magnet is brought near the unmagnetized iron, the poles of the atomic magnets experience a magnetic force and align. In the following illustration the repulsive force between S poles causes the atomic magnets to align with their S poles away from the S pole of the magnet. The S pole of the magnet is attracted to the N pole in the iron.

A large number of iron bits sprinkled in the vicinity of a magnet will align in a characteristic pattern of lines as shown below.

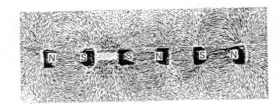

These lines are called magnetic field lines. Although these lines are imaginary, we find it useful to visualize the space between the poles of a magnet as being filled with magnetic field lines. For example, if a wire is moved between the poles of a magnet it will intercept these imaginary lines and we say that the wire "cuts" the magnetic field lines. The needle of a compass will align tangent to a magnetic field line. The direction to which the N pole of the compass points determines the direction of the magnetic field line at the position of the compass.

A gravitational field is said to exist in a region of space in which a *mass* experiences a gravitational force. For example, a gravitational field exists in a room. If we release a pencil held above the floor, it is accelerated downward by the gravitational force. Similarly, a magnetic field is said to exist in a region of space if a magnet (or a moving charge) experiences a magnetic force. For example, a magnetic field exists around a magnet because another magnet experiences a force if brought into its vicinity.

If a charge moves in a magnetic field, it experiences a force due to a magnetic interaction between a *moving* charge and a magnetic field. This principle is employed to force electrons through a conductor like a metallic wire to create an electric current. If a wire is forced through a magnetic field (see Fig. 5.3), so as to "cut" the magnetic field lines, all charges experience a force by virtue of being moving charges in a magnetic field; but only the free electrons move. It is the motion of these electrons that gives rise to an electric cur-

Figure 5.3 A current is induced in a wire when it is forced through a magnetic field so as to "cut" the magnetic field lines. If the wire were moved directly from the N to the S pole, no magnetic field lines are "cut" and, therefore, no current is induced.

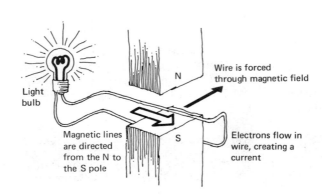

Light bulb

Magnetic lines are directed from the N to the S pole

N

Wire is forced through magnetic field

S

Electrons flow in wire, creating a current

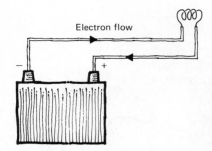

Electron flow

Figure 5.4 When a device such as a light bulb is connected to a battery, electrons flow out of the negative (−) terminal, through the device, and into the positive (+) terminal. This type of current is called direct current and is denoted d.c.

rent in the wire. It takes mechanical energy to force the wire through the magnetic field. In a practical electric generator used by a commercial electric power company, this mechanical energy is supplied by a steam turbine or moving water. While we have illustrated the production of the induced current by holding the magnet fixed and moving the wire, a current would also be induced if the wire were held fixed and the magnet moved. All that is required is that there be relative motion between the wire and the magnetic field.

If the wire were forced through the magnetic field in a direction opposite to that shown in Fig. 5.3, the direction of the electron current would change. If the wire were oscillated back and forth through the magnetic field, a current whose direction also oscillates would be produced. Such a current is called an alternating current, symbolized a.c. This current differs from that produced by a battery (Fig. 5.4). The current produced by a battery is called direct current, symbolized d.c., because the charge flows only in one direction.

5.6 ELECTRIC GENERATORS

Although the voltage generated by a commercial power plant is not produced by oscillating a single wire back and forth in a magnetic field, the voltage does have this a.c. character. A practical generator uses a coil with many turns of wire that is rotated in a magnetic field. To understand the principle let us look at a single loop of this coil as shown schematically in Fig. 5.5. The metallic slip rings are connected electrically to and rotate with the coil. The brushes rub against and make electrical contact with the slip rings. The coil, slip rings, brushes, and connected electrical device form an electrical circuit in which charges can flow. While this scheme involves circular motion of a loop of wire, it is essentially two parallel wires moving in opposite directions because only the sides labelled A and B are able to cut the magnetic field lines extending from the N pole to the S pole of the magnet. Because sides A and B always move in opposite directions, the induced current will always be in opposite directions in the two wires. The maximum effect occurs when the wires move perpendicular to the magnetic field lines. There is no induced current when the wire moves parallel to the magnetic field lines. When the portion of the loop designated A in Fig. 5.5a is at the top of its rotational path, it is moving horizontally to the right. Because the magnetic field lines are directed to the right from the N to the S pole, the wire is instantaneously moving parallel to the magnetic field lines. Hence at this position there is no current induced in

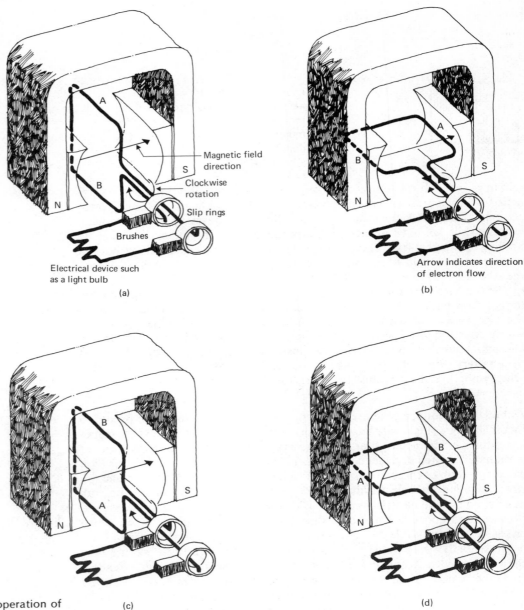

Figure 5.5 Principle of operation of an a.c. generator. A current is induced in the rotating coil when it moves through the magnetic field created by the magnet. The slip rings rotate with the coil and make electrical contact with the brushes that are connected to an electrical device such as a light bulb.

the loop. As the loop rotates, the wires begin "cutting" the magnetic field lines and a current is induced in the loop. When the portion of the loop designated A has moved down to the position shown in Fig. 5.5b, it instantaneously is moving perpendicular to the magnetic field lines. In this position the induced electron current is a maximum and has the direction shown by the arrows on the loop. The electron current decreases to zero when the coil reaches the position shown in Fig. 5.5c. As the coil rotates further, the electron current again increases but it also changes direction. It reaches a maximum when the coil achieves the position shown in Fig. 5.5d. Finally the electron current drops to zero when it returns to the initial starting position. A plot of electron current versus position of the coil, or equivalently time since the position depends on time, is shown in Fig. 5.6.

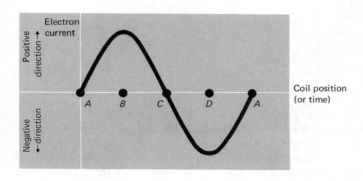

Figure 5.6 Plot of induced current in the rotating coil of an a.c. generator. Positive and negative currents are used to distinguish the two directions of charge flow. The letters (A,B,C,D) on the horizontal axis refer to the coil positions designated by parts a,b,c,d of Fig. 5.5.

A generator that provides electricity for a headlamp on a bicycle functions very much like the one illustrated here. Mechanical energy is provided by the bicycle wheel rubbing against and rotating a shaft connected to the coil of the generator. The magnetic field is provided by a magnet. A generator like this produces about 5 watts of electric power. While the generator in a commercial electric power plant is enormously larger in physical size and often produces a billion watts of electric power, the basic physical principle is the same. The commercial generator uses a steam turbine for the input mechanical energy and a massive stationary coil with a circulating electric current to provide the magnetic field. For both the bicycle and power plant generators, there is no mechanism to store electric energy. If you turn off the bicycle headlamp, the generator stops producing electric energy. Likewise, if the demand stops for electric energy from a power plant, the generator stops producing electricity. Although a commercial electric power plant is most economical when it is operating at maximum output, it cannot always operate this way because the public demands for electricity vary throughout the day. A power company often uses a large generator to provide the average demand and smaller units to assist when the demand peaks. In order to keep the power plant operating most efficiently, a company will often offer reduced rates during the off-peak periods.

5.7 TRANSFORMERS

A compass used to determine the direction of geographic north is often used to determine the presence and direction of a magnetic field. Figure 5.7 shows how a compass points when it is placed in the magnetic field produced by a current in a wire. The magnetic field lines are circles that are concentric with the wire. If the terminals of the battery are reversed so that the direction of the current reverses, the direction of the magnetic field lines reverses. When the battery is

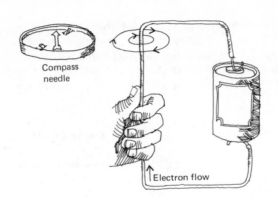

Figure 5.7 An ordinary compass used to determine the presence and direction of a magnetic field around a current-carrying wire. If the direction of the current is changed, the compass needle will point in the opposite direction.

Compass needle

Electron flow

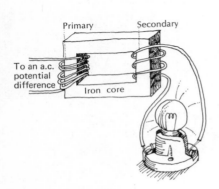

Primary   Secondary

To an a.c. potential difference

Iron core

Figure 5.8 Rudiments of a transformer used to change the size of an a.c. voltage. The magnetic field lines produced by an a.c. current in the primary coil are guided by the iron core into the secondary coil.

replaced by an alternating potential difference, the magnetic field around the wire also alternates. If a loop of wire is placed in this changing magnetic field, an alternating current will be induced in it because of the relative motion of the magnetic field and wire. This principle is used in transformers that change the size of an a.c. voltage. You often see these transformers on the utility poles that support the wires bringing electricity to a home. An actual transformer uses coils of wire (Fig. 5.8) rather than single wires as in the example. The winding to which the a.c. potential difference is connected is called the primary winding. The other winding is called the secondary winding. The iron core guides the magnetic field lines from the primary winding to the secondary winding. Experiment reveals that the ratio of the primary and secondary voltages is equal to the ratio of the number of turns of wire in the primary and secondary windings of the transformer. In symbols,

$$\frac{V_s}{V_p} = \frac{N_s}{N_p}. \tag{5.7}$$

Thus the size of the voltage on the secondary is determined by the turns ratio of the transformer and the voltage applied to the primary. For example, if there are five times as many secondary turns as primary turns, the voltage on the secondary will be five times the voltage applied to the primary. It would be to our benefit if at the same time the current in the secondary were also five times the current in the primary. This would mean that the power in the secondary would be 25 times greater. However, this would violate the principle of conservation of energy; you cannot get more power out of the system than you put in. Ideally you can get the same power out that you put in. The efficiency for energy transfer is about 99% in a well-designed transformer. Thus if the voltage on the secondary is five times the primary voltage, the secondary current is one fifth the primary current, assuming 100% efficiency for the transformer.

5.8 TRANSMISSION OF ELECTRIC POWER

Electric power is often transmitted over wires which are hundreds of miles long. Wires of this length have considerable electrical resistance. If the current in these wires is large, significant energy losses in the form of heat can occur. This follows from the relation $P = I^2R$. Using transformers (Fig. 5.9) to change the voltage and current it is possible to transmit the desired power using a small current thereby reducing heat losses. Voltages at the generating facility typically are stepped-up to 120,000 volts for transmission. In some cases, the voltage is as high as 400,000 volts. From the relation $P = VI$ it follows that the current goes down since the power is constant and the voltage has increased.

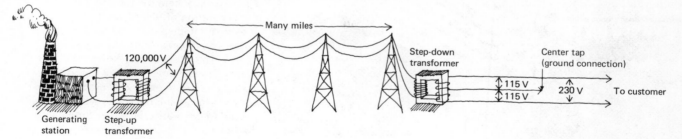

Figure 5.9 Elements of an electric power transmission system.

At a home where the power is used, the voltage is stepped-down by a series of transformers to 230 volts. By tapping the secondary of the transformer at the midpoint of the windings, 115 volts can be obtained between the center tap and either of the outer leads. The center tap is the ground connection (see Section 5.2). The 115-volt connections are used for small units such as light bulbs, mixers, vacuum sweepers, and toasters. The 230-volt connection provides energy for large elements such as electric stoves, water heaters, and air conditioners. In addition to the 115 and 230 voltages used in a home, a factory will often use still higher voltages for some operations.

The steam turbine is the major source of mechanical energy for commercial electric generators in the United States. Water vaporized by the heat generated from burning fossil or nuclear fuels is forced onto blades attached to the shaft of the turbine (Fig. 5.10). It is somewhat like blowing air onto a pinwheel. With this basic model in mind, we can expand the system to that shown in Fig. 5.11. Several of the major components of this process can be identified in the commercial facility shown in Fig. 5.12. These diagrams serve as a flow chart for energy and are useful for obtaining quantitative information. So let us "dump" 1000 pounds of coal (1000 pounds of coal would occupy a cubic space about three feet on a side) into the hopper and see how much electric energy comes out and what by-products evolve along the line. For this system, sulfur oxides, particulate matter, and heat are of most interest. The release of carbon dioxide is discussed later.

Finding coal for the hopper or, more appropriately, data for the model is no problem.* There is, however, a wide variation in the energy available per pound (heat of combustion), the sulfur content, and the ash content. A "typi-

5.9 ENERGY AND POLLUTION MODEL OF A FOSSIL-FUEL ELECTRIC POWER PLANT

---

* *Handbook of Chemistry and Physics,* Chemical Rubber Co., 18901 Cranwood Parkway, Cleveland, Ohio 44128.

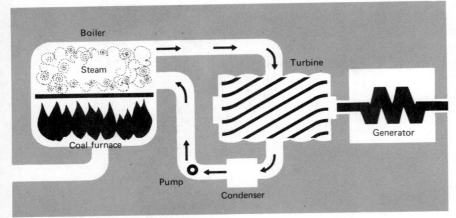

Figure 5.10 Schematic representation of the generation of electricity. High pressure steam impinges on the blades of a turbine producing rotational mechanical energy. The turbine rotates the generator shaft and mechanical energy is converted to electric energy. The condenser converts the steam back into water and the pump circulates the water through the boiler.

Figure 5.11 Major energy components of an electric power plant

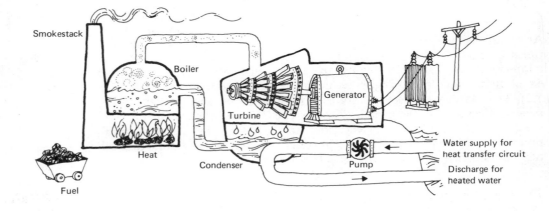

Figure 5.12 The Cincinnati Gas and Electric Company Beckjord Station located on the Ohio River. Several of the components shown schematically in Fig. 5.11 are identified in the photograph.

cal'' coal might be one from the vast West Virginia coalfields. For each pound burned, this coal produces about 13,000 Btu of heat, 0.1 pound of sulfur oxides and 0.1 pound of fly ash. Let us now follow the 1000 pounds of coal through the power plant. We will assume that the boiler, turbine, and electric generator have energy conversion efficiencies of 88, 47, and 99 percent, respectively.*

Boiler

Energy available at the boiler = energy available per pound of coal
$\times$ number of pounds

$$= 13,000 \ \frac{Btu}{\cancel{lb}} \cdot 1000 \, \cancel{lb}$$

$$= 13,000,000 \ Btu.$$

Energy out of the boiler and into the turbine = energy in the boiler
$\times$ energy conversion efficiency of boiler

$$= 13,000,000 \ Btu \times 0.88$$

$$= 11,440,000 \ Btu.$$

The energy that doesn't enter the turbine escapes, or is released, mostly through the smokestack.

Energy out the smokestack = energy into boiler − energy into turbine

$$= 13,000,000 - 11,440,000$$

$$= 1,560,000 \ Btu.$$

The sulfur oxides are produced when the coal is burned. The amount of sulfur oxides produced when 1000 pounds are burned is

sulfur oxides (lbs) = 1000 $\cancel{\text{lbs of coal}} \times 0.1$ lb $\frac{\text{sulfur oxides}}{\cancel{\text{lb of coal}}}$

$$= 100 \ lbs.$$

These oxides do not necessarily escape out the smokestack if there is some mechanism installed to extract them. Few power plants are equipped to remove even part of the sulfur oxides produced. We will assume that all 100 pounds go out the smokestack and into the atmosphere.

The fly ash is simply an unburnable product in the coal.

Fly ash = pounds of coal $\times$ pounds of fly ash for each pound of coal

$$= 1000 \times 0.1$$

$$= 100 \ pounds.$$

---

* The efficiencies used in these calculations were taken from "The Conversion of Energy" by Claude M. Summers, *Scientific American*, vol. 224, no. 3 (September 1971): p. 148. The definition of efficiency is given in Section 3.8.

Most power plants do have facilities for removing about 99 percent of the fly ash before it goes out the smokestack. Hence one percent goes out the smokestack.

Amount leaving smokestack = 100 (.01) = one pound.

Turbine

Energy out the turbine and into generator = energy into turbine
$$\times \text{ energy conversion efficiency of turbine}$$
$$= 11,440,000 \ (0.47)$$
$$= 5,377,000 \text{ Btu.}$$

The low value for this efficiency is probably the most difficult thing to accept in this discussion. However, there is a practical limit to the efficiency of a steam turbine which is imposed by the physical laws of thermodynamics. This is discussed in detail in Chapter 6.

Condenser

Heat energy rejected to condenser = energy into turbine – energy into generator
$$= 11,440,000 - 5,377,000$$
$$= 6,063,000 \text{ Btu.}$$

Generator

Energy out of generator = energy into generator $\times$ energy conversion efficiency of generator
$$= 5,377,000 \ (0.99)$$
$$= 5,323,000 \text{ Btu}$$
$$= 5,323,000 \ \cancel{\text{Btu}} \left(1055 \ \frac{\cancel{\text{joules}}}{\cancel{\text{Btu}}} \right) \cdot \frac{1}{3,600,000} \ \frac{\text{kW-hr}}{\cancel{\text{joules}}}$$
$$= 1560 \text{ kW-hr.}$$

Thus, this 1000 pounds of coal would produce enough electric energy to run a 1560-watt air conditioner for 1000 hours.

With these numbers, the model of Fig. 5.11 can be quantified as in Fig. 5.13.

This procedure accomplishes several things:

1. Numerical results are obtained for a given set of conditions. This gives a feel for the emissions from a typical power plant.

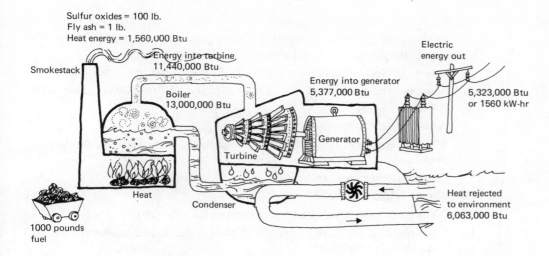

Sulfur oxides = 100 lb.
Fly ash = 1 lb.
Heat energy = 1,560,000 Btu

Smokestack

Energy into turbine,
11,440,000 Btu

Boiler
13,000,000 Btu

Energy into generator
5,377,000 Btu

Electric
energy out

5,323,000 Btu
or 1560 kW-hr

Generator

Turbine

Heat

Condenser

Heat rejected
to environment
6,063,000 Btu

1000 pounds
fuel

Figure 5.13 Typical outputs for conversion of 1000 pounds
of coal into electric energy.

2. The origin of the pollutants in the energy conversion process has been established.

3. Most importantly, a model applicable to any system of this type has been deduced. Only the numbers are different for another system of the same type.

The magnitude of the pollution cannot be fully appreciated because the element of time has not been considered. A 1000-megawatt unit operating at capacity would use the 1000 pounds of coal in about five seconds. This means that about 10,000 tons of coal are used per day and this would release 1000 tons of sulfur oxides to the atmosphere!

Comment on Thermal Pollution

The energy model for generation of electric energy (Fig. 5.13) shows that over 50 percent of the energy injected into the turbine is rejected as heat to the environment. Such a loss of energy is of major concern and must be studied. It is important, though, to try to understand first why so much heat has to be rejected. This requires some basic knowledge of the thermodynamics of the thermal to mechanical energy-conversion process. It is the quest for this understanding which motivates the next two chapters, "Thermodynamic Principles and the Steam Turbine" and "Disposal of Heat Energy from Steam Turbines."

TOPICAL REVIEW

1. What are some reasons for studying electric energy and power?
2. Describe the function of an electric generator in energy terms. How efficient is the generator in performing its stated function?
3. How is an electric current similar to the flow of water in a pipe?
4. What are the units of electric current, potential difference, power, and energy?
5. What is an electrical circuit?
6. Why are electrical devices that produce heat more expensive to operate than household devices that produce mechanical motion?
7. Distinguish between electric power and electric energy.
8. What is a kilowatt-hour?
9. Distinguish between alternating current and direct current.
10. What physical principle is involved in the production of an electric current by an electric generator?
11. What is a transformer? Why is a transformer useful in electric power systems?
12. Why is electric power transmitted at very high voltage?
13. Name the pollutants of significance produced by electric power plants using coal for input energy.
14. What is the most *inefficient* energy conversion device in an electric power plant?

REFERENCES

Electric Energy and Power

Extensive discussions of electricity, electric energy, and electric power are given in most physics and physical science texts. Two such texts that would be useful for expansion of the material presented in Chapter 5 are

1. *Physics: A Descriptive Analysis,* by Albert J. Read, Addison-Wesley, Reading, Mass. (1970).
2. *Conceptual Physics,* by Paul G. Hewitt, Little, Brown, Boston, Mass. (1977).

Two federal documents are particularly useful as sources of information on the environmental impact of producing electric power. These are

1. "Selected Materials on Environmental Effects of Producing Electric Power," Joint Committee on Atomic Energy, Congress of the United States (91st Congress, First Session), August 1969. Available from Superintendent of Documents, U.S. Government Printing Office, Washington, D.C. 20402.
2. "A Teachers' Guide for the Environmental Impact of Electric Power Generation: Nuclear and Fossil," prepared under contract AT(40-2)-4167, United States Atomic Energy Commission by Pennsylvania Department of Education, 1973. Available from Superintendent of Documents, U.S. Government Printing Office, Washington, D.C. 20402.

5.2 Electricity

1. For water to flow through a pipe, there must be a water pressure difference between the ends of the pipe. What is the analogous situation for the flow of charge through a wire?

2. The diagram in Fig. 5.1 for the circulation of water also applies to the circulation of blood in the human body. What constitutes the pump, hose, valves, and container in the human system? How can fluid resistance occur in the human circulatory system?

3. An ammeter can register the number of amperes (current) in the wires bringing electricity into a house. A measure of the current is a measure of how much electric energy is being used. What kind of meter would be appropriate for measuring the amount of natural gas used in a gas-heated house?

4. A friend claims that you have a 50% chance of not getting shocked if you touch one of the terminals of a household electrical outlet. Does your friend know what he is talking about?

5. When a battery is connected to a small light bulb (Fig. 5.4), the direction of the flow of charge through the bulb depends on which terminal of the bulb is connected to the positive terminal of the battery. If you distinguish the two directions by calling one positive and the other negative, what would a graph of current and time look like if you rapidly changed the battery connections every second? How is this current similar to the a.c. current in a bulb connected to a household outlet?

6. An electrical shock results when a person gets connected between the ground and "hot" leads of the electric power source in a household. The better the electrical connection, the more severe the shock. Knowing this, why is it especially dangerous to touch or use an appliance such as a hair drier while in electrical contact with the water pipes in a bathroom? (A wire service report March 12, 1978 reported that the French singer Claude Francois (Clo-Clo) was electrocuted in the bathroom of his Paris apartment.)

5.3 Electrical Resistance

7. Define the fluid resistance of a pipe similar to the way that electrical resistance of a wire was defined.

8. It is logical that the larger the diameter of a given wire, the easier it is to force electrons through it. Thus one expects the resistance to decrease as the diameter increases. If the diameter doubles, why will the resistance decrease by a factor of four?

9. If a device obeys Ohm's law, how would the current in the device change if the potential difference across it were cut in half?

10. Why can a light switch in the open position be thought of as having very high (essentially, infinite) electrical resistance?

11. Suppose the voltage at the position where electric power comes into a house is 115 volts. Why will the voltage always be less than 115 volts at an outlet in a house when something is plugged into the outlet?

12. Electrical resistance is a measure of the opposition presented to the flow of charge through a conductor. For a given potential difference, the current decreases as the resistance increases. We could just as well have defined a conductance (C) that measures the ability of a conductor to conduct charge. How would you expect the current to change when the conductance is increased? What would Ohm's law be in terms of conductance?

5.4 Electric Power and Energy

13. A person sliding down a rope from the top of a building loses gravitational potential energy that is converted into heat—his hands and the rope get warm. Charges flowing through a resistor lose electric potential energy. Energetically, how is the electrical situation similar to the mechanical situation?

14. How does the wattage rating of a light bulb enter into its cost of operation?

15. Electric power companies are paid on the basis of the number of kilowatt-hours used. Would you find that an appliance is labeled in units of kilowatt-hours? What unit would be used to rate an appliance?

16. A power company will often reduce the generating voltage when demands for electric power exceed the capacity for production. Why does reducing the voltage tend to conserve power?

17. Electric stoves are usually designed to operate on 230 volts. If 115 volts were used on a stove designed to operate on 230 volts, does it mean that the power used would drop by one half?

18. What is the fallacy in the following statement taken from a newspaper description of a new type of electronic device? "The device should save about 26,000 watts of electric power per hour."

5.5 Magnetic Force and Magnetism

19. If the N pole of a compass points toward the north geographic pole, what kind of a magnetic pole is located near the north geographic pole? What merit would there be for calling one pole of a magnet positive and the other pole negative?

20. Two pieces of iron bar are identical in appearance. One is magnetized, the other is not. By touching one bar to the other, how could you determine which piece is magnetized?

21. Automobiles are sometimes equipped with magnetic compasses for determining direction. Why might these compasses give erroneous readings when the car passes under electric transmission lines?

5.6 Electric Generators

22. Many bicycles are equipped with a light that is powered by a small electric generator. The shaft of the generator is turned by allowing it to rub against one of the tires. Describe the basic operation of the generator. What energy transformations are involved in the operation of the generator? Why would the bicycle be slightly easier to pump if the electric switch for the light were off?

23. A small electric motor produces rotational energy when connected to a flashlight battery. If the shaft of the motor is connected to the shaft of an identical motor, then the second motor functions as a generator producing a potential difference across its leads. If a small bulb is connected across the leads, it will light up. Why can't you get as much electric energy into the bulb as was extracted from the battery to run the motor?

24. Figure 5.5 shows how a current is induced in a coil of wire that is being rotated in a magnetic field. Suppose that the coil were held fixed and the magnets were rotated. Why would a current still be induced in the coil? Would there be any simplifying feature of the setup if this latter method were used to induce the current in the coil?

5.7 Transformers

25. What is being "transformed" by an electric transformer?

26. The input and output powers are identical in a perfect transformer. This is nearly the case for a practical transformer. What power losses are involved which would make the output power smaller than the input power in a practical transformer?

27. A friend claims that if you connect a flashlight battery to the primary of a step-up transformer that a voltage appears across the secondary as long as the battery is connected. Does your friend know what he is talking about?

28. As an outdoor, visual exercise, see if you can identify near a house or building (1) a transformer, (2) a watt-hour meter, and (3) the ground wire near an electric utility pole.

5.8 Transmission of Electric Power

29. Why are the wires used to transmit electric power separated by several feet when considerable space could be saved by placing them close together?

30. Why isn't a bird electrocuted by sitting on a conductor of an electric power transmission line?

5.9 Energy and Pollution Model of a Fossil-fuel Electric Power Plant

31. Home heating using electric energy is often advertised as "clean." Comment on the rationale of this statement.

32. An electric heater used to heat a house converts electric energy to heat with nearly 100 percent efficiency. The conversion efficiency for the burning of fossil fuel to heat a house is somewhat less than 100 percent. However, it is usually cheaper to heat a home by burning a fossil fuel than by using electric heat. Why?

33. What are some of the economic and environmental considerations for siting a fossil-fuel electric power plant?

34. Air pollution in cities by electric power plants could be minimized by moving the plants many miles away. What other environmental and economic problems would this produce?

NUMERICAL PROBLEMS   5.2 Electricity

1. If a current of one ampere exists in a wire, how many electrons flow through any cross-section of the wire each second?

2. The lifetime of a battery depends on the amount of charge that is drawn from it: the more the current, the shorter the lifetime. The quality of a battery is often specified by the number of ampere-hours (that is, amperes multiplied by hours) that it can provide.

   a) What physical quantity would an ampere-hour measure?

   b) If a battery is rated at 100 ampere-hours, how long would it last if it delivers 0.25 amperes?

5.3 Electrical Resistance

3. If the current in a light bulb is 1 ampere when connected to a 115-volt household electrical outlet, what is the resistance of the bulb? How much electric power does the bulb require?

4. When the potential difference across some device is 10 volts, the current through the device is 1 ampere. When the potential difference is increased to 20 volts, the current increases to 2.5 amperes. Does this device obey Ohm's law? Explain.

5. A circuit breaker is a device placed in an electrical circuit to open the circuit like a switch when the current exceeds some given amount. In a house, circuit breakers for a line that feeds power to small appliances are usually designed to open when the current exceeds 15 amperes. If the line voltage is 115 volts, what is the least resistance that you could connect to a line and not have the circuit breaker open? How many 100-watt light bulbs could you connect to a line and not have the circuit breaker operate?

6. An electric toaster requires 1100 watts of electric power. How much electric current is required when the toaster is connected to a 115-volt household electrical outlet? What is the resistance of the toaster?

7. What is the electric current in a 100-watt light bulb connected to a 115-volt outlet in a home?

8. A typical electric range uses 12,000 watts of power. How many amperes of current are produced when the range is connected to a 230 volt source?

9. A certain household clothes dryer uses 5000 watts of power. If electric energy costs four cents per kW-hr, how much does it cost to run the dryer for 4 hours?

$$\frac{5 \text{ KW Hrs}}{4}$$

10. a) A typical household electric clothes dryer consumes electric energy at a rate of 6000 W. Estimate the time required to dry a load of clothes and figure the total cost if the rate is four cents/kW-hr.

    b) A night-light in a child's bedroom typically uses a 5-W bulb. Estimate the cost per year if the bulb is left on continuously. Assume the rate is four cents per kW-hr.

11. A certain 100-watt light bulb costs 75¢ and lasts for 120 hours. Another longer-lived 100-watt light bulb costs $1.20. How much longer would the more expensive bulb have to last to make it a "good buy"?

12. Selling electric energy is big business. To illustrate, assume that the electricity from a 1000-megawatt power plant is sold for 3¢/kW-hr. How much revenue is generated in a day's time if the plant operates at capacity the entire day?

13. Suppose that each of two electric clothes dryers require 5000 watts of electric power. However, one costing $200 requires a half hour to dry a load of clothes and the other costing $150 requires one hour to dry an identical load. If electric energy costs 2.5 cents per kW-hr, how many loads would you have to dry with the more expensive model to make up the $50 difference in price between the two machines?

14. Light bulbs for home movie projectors typically have a lifetime of operation of 25 hours and a power rating of 500 watts. How much electric energy is used in the lifetime of one of these bulbs?

15. A certain 50-watt light bulb costs 50¢ and, on the average, lasts for 100 hours before burning out. Another longer-lived 50-watt light bulb costs $1.50 and has a lifetime of 250 hours. Is the longer-lived bulb a good buy? Explain.

16. A certain room is illuminated with 30 fluorescent lamps each having a light conversion efficiency of 10 percent and requiring 50 watts of electric power. It is discovered that a new 50-watt lamp that is 5 percent more efficient at producing light can be purchased. How many of the new lamps would be required to produce the same amount of light as the 30 older-type lamps?

17. In the operation of a two-battery flashlight, the electric power developed by the batteries is about 5 watts.

    a) Express the power in kilowatts.

b) If the flashlight burns continuously, the batteries will "go dead" in about 1 hour. How many kilowatt-hours of electric energy will have been used in the lifetime of the batteries?

c) If the batteries cost 50¢, how much per kW-hr did the energy cost? How does this cost compare with the cost per kW-hr from an electric power company?

5.6 Electric Generators

18. Bicycling at a moderate speed requires an energy expenditure of about 25 food calories per hour on the part of the rider. If an electric generator on the bicycle produces 5 watts of electric power when in operation, by how much will the bicyclist have to increase the power output (measured in food calories per hour) to maintain the same speed? (From Table 3.2, 1 food calorie = 4,186 joules and 1 joule per second = 1 watt).

5.7 Transformers

19. Household door bells and chimes use a transformer to reduce the 115-volt house voltage to 6 volts. Does the primary winding contain more or less turns than the secondary in these transformers? If the secondary winding contains 1000 turns, how many turns are there on the primary?

20. The input power to a certain "perfect" transformer is 1000 watts. How much current is in the secondary if the secondary voltage is 10 volts? From the information given, can you determine whether this is a step-up or a step-down transformer?

21. Using a step-up transformer, a student applies a potential difference of 10 volts to the primary and creates a current of 2 amperes in the primary. On the secondary, which is connected to a lightbulb, he reports that the potential difference is 20 volts and the current is 2 amperes. What advice do you have for the student?

5.8 Transmission of Electric Power

22. A certain electric utility produces 120,000 kilowatts of power at 12,000 volts.

a) How much current exists in the output of this utility?

b) If the voltage is stepped-up to 240,000 volts for transmission, what is the current in the transmission line?

23. There is a trend toward increasing the voltage for transmitting electric power. The reason is that heat losses decrease because the current in the transmission lines decreases. Referring to problem 22, if 10 percent of the power is lost when the 120,000 kilowatts of power produced are transmitted at 240,000 volts, how much power will be lost if it is transmitted at 480,000 volts?

5.9 Energy and Pollution Model of a Fossil-fuel Electric Power Plant

24. Assume that a 1000-megawatt power plant has an overall efficiency of 33⅓ percent and that on the average it operates at 50 percent of capacity in a day's operation. Using coal having an energy content of 13,000 Btu per pound, show that a daily supply amounts to about 4700 tons. (Remember 1 Btu = 1055 joules and 1 watt = 1 joule/second).

25. How many pounds of sulfur are contained in one ton of coal having five percent sulfur content?

26. Calculate the overall efficiency for the electric power plant model used in the text.

27. If a railroad coal car can transport 100 tons and an electric generating facility requires 10,000 tons of coal per day, how many cars are required to supply the coal for one day's operation? What insight does this give you into the importance of coal transportation facilities and long-term labor agreements?

28. A 1000-megawatt electric power output means that electric energy is being produced at a rate of 1,000,000,000 joules per second. If the plant has a 33⅓ percent conversion efficiency, then energy is derived from coal at a rate of 3,000,000,000 joules per second. If 1000 pounds of coal having a heat content of 13,000 Btu per pound are burned in this plant, show that it takes about 4 seconds.

29. a) How much heat can be obtained from 5000 pounds of a typical coal (see Table 3.2)?

    b) If this heat is used to feed a steam turbine and the conversion efficiency is 85 percent, how much energy gets into the turbine?

    c) What happens to the energy that doesn't get into the turbine?

30. In 1900 it took about 20,000 Btu of heat to produce one kW-hr of electricity. Today it takes less than 9000 Btu. What are the corresponding efficiencies? (Be careful of units.)

31. In the electric power plant model discussed in Section 5.9 it was assumed that the coal came from a West Virginia coalfield. Do the same computation assuming the coal came from Montana and had the following characteristics:

    heat content = 10,000 Btu per pound

    sulfur oxide production = 0.03 pounds per pound of coal

    fly ash production = 0.12 pounds per pound of coal

THERMODYNAMIC PRINCIPLES AND THE STEAM TURBINE

6

Nature has provided a natural thermodynamic system in the form of geysers. In Northern California, the Pacific Gas and Electric Company produces electricity from energy derived from geysers. Photograph courtesy of the Pacific Gas and Electric Company.

6.1 MOTIVATION

For a very large fraction of the driving public, an automobile engine is a mysterious device that derives energy from gasoline and converts some of it to energy of motion. Many drivers know how many miles the car travels for each gallon of gasoline consumed, but few realize that only about 25 percent of the energy is converted to energy of motion. Most drivers also know that oil has to be added periodically to an automobile engine and that the oil serves as a lubricant to reduce friction between moving parts. Friction might be suspected as the cause for the low efficiency of the engine; we might suppose that if we were sufficiently ingenious or if we had a better lubricant, we could make an automobile engine with an efficiency close to 100 percent. The same conclusion might be reached when we read that the efficiency of a steam turbine used to convert heat to mechanical energy for an electric generator is about 50 percent. Friction is indeed a consideration, but a more fundamental cause is found in the thermodynamics of the energy conversion process. The steam turbine, like all heat engines, takes in heat energy from a source (a container of steam, for example), converts some of the heat energy into mechanical energy, and releases heat energy to a sink of some sort (a lake, for example). Typically, as much heat energy is released as is converted to mechanical energy. Because there are so many large steam turbines in operation, especially in the electric power industry, concerns are being registered for the biological health of our lakes, rivers, and streams that carry the burden of these heat loads.

Our goals in this and the next two chapters are several and they are correlated. First, we want to learn enough thermodynamic principles to understand why the efficiency of a commercial steam turbine is intrinsically low. Second, we want to use these principles to understand the methods being used to dispose of heat in more environmentally acceptable ways. Third, we want to understand from the thermodynamic viewpoint the stable atmospheric conditions that prevent the dispersal of pollutants released from the combustion of fossil fuels. Fourth, we want to investigate the possibility of increasing the average temperature of the earth as a result of the "greenhouse" effect produced by the accumulation of carbon dioxide in the atmosphere. Finally, we want to examine the very broad environmental implications of the laws of thermodynamics.

6.2 BASIC THERMODYNAMICS

Pressure, Volume, Temperature

Both the steam turbine and the steam engine are devices that convert the energy of an expanding gas to useful mechanical energy (Fig. 6.1). While the turbine has largely replaced the engine as a practical device, we shall discuss the steam engine because it is easier to visualize, and it is an ideal way to introduce some basic principles of thermodynamics. Happily, our conclusions will also apply to the steam turbine.

We start with the observation that an expanding gas pushing against a piston does mechanical work on the piston. The expanding gas does work in the same way you would do work by pushing on the piston with your hand (Fig. 6.2). If we knew the force exerted on the piston by the gas, then (as in

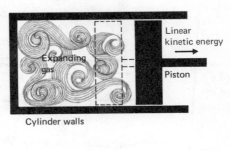

Figure 6.1 (a) Schematic illustration of the operation of a steam turbine. Gas or vapor exerts a force on the blades of a turbine producing rotational kinetic energy. (b) Schematic illustration of the operation of a steam engine. Gas or vapor exerts a force on a piston contained in a cylinder producing linear kinetic energy.

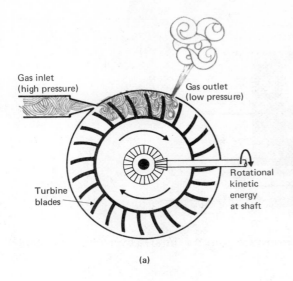

Gas inlet (high pressure)

Gas outlet (low pressure)

Turbine blades

Rotational kinetic energy at shaft

(a)

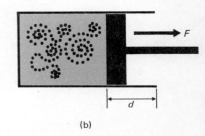

Linear kinetic energy

Piston

Expanding gas

Cylinder walls

(b)

Figure 6.2 (a) Work done by a person exerting a force F on a box. (b) Work done by an expanding gas pushing on a piston.

(a)

(b)

Section 3.5) we could relate the work to the force and the distance the piston moves ($W = F \cdot d$). To obtain an expression for the force, we use a model of the gas that employs the ideas of pressure, volume, and temperature (Fig. 6.3). For simplicity, we symbolize these as P, V, and T. Although a formal definition of these terms may not be familiar, their usage is common. Pressure implies a force. When we are pressured to study for an exam, it suggests that we are forced to do so. Automobile drivers are conscious of the air pressure in the tires. When a balloon bursts, it is because the air pressure within the balloon exceeds the breaking strength of the balloon. Anything that supports the weight of some object has a pressure exerted on it. The surface of the earth supports the weight of the air atmosphere above it; therefore the atmosphere exerts a pressure on the earth's surface. It might seem from these examples that the concepts of force and pressure are the same. However, there is an

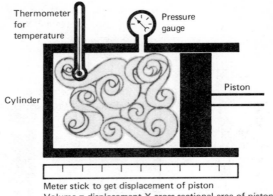

Figure 6.3 Illustration of an operational way of measuring the thermodynamic variables of a gas in a cylinder. The thermometer is similar to a glass thermometer often used to measure body temperature. The pressure gauge functions like a gauge used to record the air pressure in an automobile tire.

Thermometer for temperature

Pressure gauge

Cylinder

Piston

Meter stick to get displacement of piston
Volume = displacement X cross-sectional area of piston

important basic difference. Pressure accounts for the area over which the force acts:

$$\text{pressure} = \frac{\text{force on a surface}}{\text{area over which the force acts}},$$

$$P = \frac{F}{A}. \tag{6.1}$$

Pressure has units of force divided by units of area. Typical units are newtons per square meter (N/m^2) and pounds per square inch (lb/in^2). Atmospheric pressure at sea level is about 100,000 N/m^2 (15 lb/in^2). This means that each square meter on the surface of the earth at sea level experiences a force of 100,000 newtons. Note that pressure can be very large for a moderate force if the contact area is very small. A very large pressure can be exerted with a thumbtack with only a small force because the contact area of the point of the tack is very small.

We have already discussed volume in connection with the concept of density. Here the volume is that which is occupied by the gas. As the piston moves, the cylindrical volume occupied by the gas changes.

There is probably no scientific word that is more familiar than temperature. Nearly all houses have a thermostat which regulates the household temperature. Atmospheric temperature is a regular feature of TV newscasts and newspapers. Temperature is, however, a much subtler concept than pressure and volume. Unlike pressure and volume, which are ordinary mechanical quantities, temperature is a measure of the sensation of hot and cold. A thermometer is a measuring device used to assign a number related to this sensation. Thermometry is a science in itself and there are many ways to measure temperature. The basic idea is to find some physical quantity that changes when temperature changes. Because the height of a column of liquid such as mercury in a glass tube is dependent on temperature, a number designating the temperature can be assigned to the height of the mercury column when it is placed in some environment. This method requires that two reference temperatures be established. Customarily, these are the ice and steam points of water.

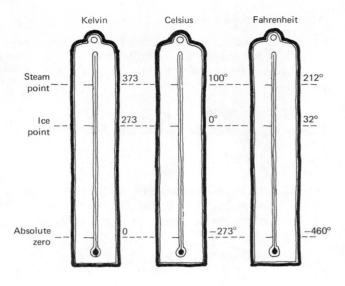

	Kelvin	Celsius	Fahrenheit
Steam point	373	100°	212°
Ice point	273	0°	32°
Absolute zero	0	−273°	−460°

Figure 6.4 Reference temperatures for the three common temperature scales.

The ice point is the temperature of an equilibrium mixture of ice and water at atmospheric pressure at sea level. The steam point is the temperature at which water boils at atmospheric pressure at sea level. The ice and steam points are assigned numbers 32 and 212 respectively, on the familiar Fahrenheit scale. They are assigned 0 and 100 and 273 and 373 on the less familiar, but more rational, Celsius and Kelvin scales, respectively (Fig. 6.4).

If a meter stick and a yardstick were used to measure the depth of water in a tub, different numbers would be recorded because the scales are different. The values are related, though, because they are measuring the same physical quantity, namely the depth of the water. Knowing that 2.54 centimeters equals one inch, the depths can be related as follows:

depth in centimeters = depth in inches × 2.54,
cm = in. × 2.54.

Similarly, if Celsius, Kelvin, and Fahrenheit mercury-type thermometers were placed in the same environment, the liquid would rise to the same level. But, because the scales differ, each thermometer would register a different number for the temperature. Yet the numbers must be related to each other because they are derived from the same physical situation. The relations are

$$K = C + 273, \tag{6.2}$$

$$F = \frac{9}{5} C + 32. \tag{6.3}$$

You might like to try to deduce these relations. You can check their validity by calculating the temperature at the ice point on the Kelvin and Fahrenheit scales knowing that this temperature is 0 on the Celsius scale.

Work Done by the Force of an Expanding Gas

Having deduced a formal relation between force and pressure, we can now relate the work done by an expanding gas to the pressure of the gas. By definition, the work is the product of the force (F) and the distance (d) the object is moved:

$$W = F \times d.$$

Because force is related to pressure by $F = PA$, the expression for work can be written

$$W = PAd, \tag{6.4}$$

where A is the cross-sectional area of the cylinder and A multiplied by d is the *change* in volume occupied by the gas. For example, if the cross-sectional area is 0.1 m² and the piston moves 0.1 m as a result of a 100,000 N/m² pressure, then the work done is

$$W = (100,000)\ (0.1)\ (0.1)$$

$$= 1000 \text{ joules.}$$

Heat

While work is a concept having broad everyday usage, the principles of physics allow us to define it precisely and quantitatively. Heat falls into the same category. When someone says, "The house is cold, we need heat," the request is clear. Similarly, we know that exercise produces both warmth and an increase in body temperature and that energy is derived from food. So we make some intuitive connection between heat, energy, and temperature. Placing your hand in water at 90°C produces a greater sensation of being hot than if your hand were placed in water at 40°C. Quite appropriately we would say that the 90°C water contains more energy and that energy was transferred to your hand by this water. We distinguish carefully the energy content of the water and the energy transferred. The energy *content* is called *internal* energy. It is energy that is within the system—water in this example. It is energy that arises from the kinetic energy of the atoms and molecules and whatever forms of potential energy that might be involved. As the temperature increases, the internal energy of a system increases. The energy in *transit* is called *heat*. Heat energy is transferred between two objects (hand and water, for example) only if a temperature difference exists. The energy flows from the object having the higher temperature. If your hand is at 30°C and you place it in ice water, heat flows from your hand to the water—your hand cools. The internal energy of your hand decreases; the internal energy of the water increases. If your hand is at 30°C and you place it in water at 90°C, heat flows from the water to your hand—your hand warms. The internal energy of your hand increases; the internal energy of the water decreases.

The calorie (cal) and British thermal unit (Btu) are units used to quantify this concept. If the temperature of one gram of water is raised by 1°C, the water is said to have absorbed one calorie of heat. The kilocalorie (kcal) or ordinary food calorie is equal to 1000 calories defined in this way. A Btu is the heat required to raise the temperature of one pound of water by 1 °F. Precisely, the one degree temperature change is from 14.5 °C to 15.5 °C for the calorie and from 63 °F to 64 °F for the Btu. For our purposes the differences are so slight that we need not be concerned. For example, if in a household hot-water system we want to know how much heat is required to raise the temperature of 20 gallons (165 pounds) of water from 80 °F to 140 °F, it suffices to say

$$\text{heat} = \text{weight} \times \text{temperature change} \times 1 \text{ Btu per pound} \bullet \text{ °F},$$

$$= (165)\,(60)\,(1)$$

$$= 9,900 \text{ Btu.}$$

Table 6.1 Specific heat of some common substances. The data were taken from *Handbook of Chemistry and Physics*, Chemical Rubber Publishing Co., 18901 Cranwood Parkway, Cleveland, Ohio 44128, and *Applied Solar Energy*, Aden B. Meinel and Marjorie P. Meinel, Addison-Wesley Publishing Co., Reading, Mass. (1976).

Substance	Specific heat $\left(\dfrac{cal}{g \bullet °C}\right)$
Water	1.00
Aluminum	0.22
Iron	0.11
Copper	0.093
Concrete	0.22
Brick	0.20
Dry earth	0.19
Wet earth	0.50

When we discuss in Chapter 12 the storage of thermal energy in solar heated houses, we will want to know how much heat is required to raise the temperature of a mass of rocks a certain number of degrees. Thus it is also meaningful to know how much heat is required to raise the temperature of one gram (or one pound) of any substance by one degree Celsius (or by one degree Fahrenheit). This is called the specific heat. Just as for water, the specific heat depends on temperature, but for most calculations it is assumed to be constant. The specific heats of some common substances are shown in Table 6.1. If in a solar energy application we want to know the heat required to raise the temperature of a 1000 kg block of concrete from 25 °C to 30 °C we would look up the specific heat of concrete (Table 6.1) and write

$$\text{heat} = \text{mass} \times \text{temperature change} \times \text{specific heat},$$

$$= (1,000,000 \text{ grams})\,(5°C)\,(0.22 \text{ cal/g} \bullet °C)$$

$$= 1,100,000 \text{ cal}$$

$$= 1,100 \text{ kcal.}$$

First Law of Thermodynamics

Rudolph Clausius was instrumental in the formal development of thermodynamics. In 1865, he stated that "the total energy of the universe is constant." This statement has come to be known as the first law of thermodynamics. Literally, it means that one type of energy may be transformed into another, but there is no change in the net amount of energy. The statement does not imply that the energy is accessible after the transformation. This is a crucial factor in the consideration of useful sources of energy.

The first law of thermodynamics for systems involving heat relates internal energy, work, and heat in a quantitative way. To deduce this formal relation, consider the following experiments. A cylinder with a fixed piston is placed in contact with a heat source (Fig. 6.5). The heated gas in the cylinder presses against the piston; however, the piston is held in place, and no mechan-

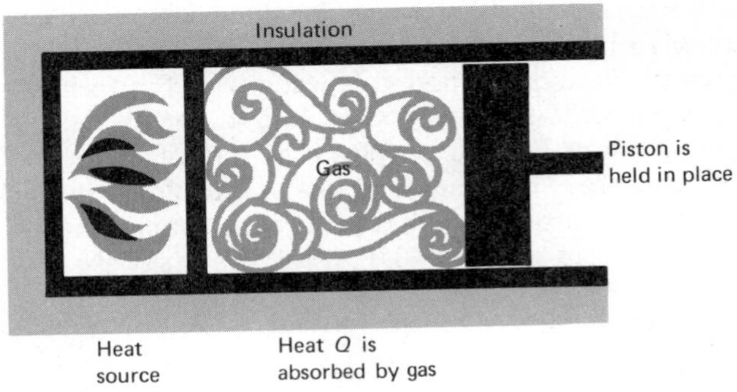

Figure 6.5 The gas in a cylinder with a fixed piston absorbs heat Q. No work is done by the gas, but the internal energy and temperature of the gas increase.

ical work is done. Where did the heat energy go? The heat (Q) was absorbed by the gas, increasing its temperature and its internal energy (U):

heat absorbed = *change* in internal energy,

$$Q = \Delta U.$$

(ΔU is spoken "delta U" and means "change in U.") This says nothing about how the energy is distributed in the gas or in what form it exists. It is a manifestation of the energy-conserving aspects of the first law of thermodynamics.

Now suppose that the heat source is removed and the cylinder is protected from the surroundings so that heat can neither enter nor leave the cylinder. If the piston is released, it will move and mechanical work W is done. Where does the energy for this work come from? It can come only from the internal energy of the gas, because the cylinder is thermally protected from its surroundings. Therefore, the temperature and the internal energy must decrease:

change in internal energy = work done on the piston,
$$\Delta U = -W.$$

(The minus sign in front of W is used to denote a decrease in internal energy.)

Suppose now that we allow heat to enter the cylinder and we allow the piston to move thereby doing work. Part of the heat energy taken in will be responsible for the work that is done and the remainder of the heat energy will increase the internal energy of the gas:

change in internal energy = heat taken in minus the work done

$$\Delta U = Q - W. \tag{6.5}$$

This is the formal mathematical statement of the first law of thermodynamics for systems involving heat, internal energy, and work. It says simply that the heat energy must go somewhere and in this case it goes toward doing work and changing the internal energy of the gas.

Very likely, this exercise seems complicated to you. If you view this as a budget involving energy rather than money, then it may be easier to understand. When a sum of money is spent, it can go for a variety of things. When you total the individual purchases, the end sum must equal the amount that you started with. Likewise, when you "spend" a certain amount of heat (Q) in a steam engine, it must go somewhere. There are two possible dispositions of the energy. It can cause the piston to move, thereby doing work, and/or it can increase the internal energy (ΔU) of the gas. In the end when you add the change in internal energy and the work done, the sum must equal the amount of heat spent in the first place. You might like to work through this same exercise for the situation where heat is taken out of the cylinder.

General Principles

These discussions of the thermodynamic principles have been fairly theoretical even though we have attached meaning to them by drawing on everyday experience. It is now appropriate to apply these ideas to the practical problem of the efficiency of a steam turbine. This application is extremely important because the efficiency is much lower than we might expect. The low efficiency is a contributing factor to the rapid depletion of our energy resources and to the concerns about thermal pollution. Before proceeding with the subtle details let us get a clear picture of the energy aspects of a steam turbine by drawing an analogy with the somewhat more familiar energy aspects of a hydroelectric power plant. In a hydroelectric power plant, the gravitational potential energy of water in a dammed river is converted to kinetic energy by allowing the water to fall through a vertical distance as it flows over the dam. The greater the difference in water level (i.e., the height of the dam), the greater the potential energy and subsequent kinetic energy of the water. (If there is no difference in levels, there is obviously no conversion of potential energy to kinetic energy.) The moving water strikes the blades of a turbine (just as the gas under high pressure is pictured as doing in Fig. 6.1a) and some, but not all, of the kinetic energy of the water is converted to rotational energy. This same language applies to a steam turbine by replacing gravitational energy with internal energy and a difference in water level with a difference in temperature. In a steam turbine, water is raised to a high level of internal energy by applying heat to the extent that the water vaporizes and becomes steam (Fig. 6.6). Like water flowing over a dam, the steam flows through the turbine to a *heat sink* having a lower temperature and, therefore, lower internal energy. (A heat sink is something that absorbs heat. A river which absorbs the heat rejected by a steam turbine is a heat sink.) The greater the difference in temperature of the steam and the sink, the greater the difference in internal energy and subsequent kinetic energy of the steam. If there is no difference in the temperature of the steam and sink, there is no conversion of internal energy to kinetic energy. When the steam strikes the blades of the turbine, some, but not all, of the kinetic energy of the steam is converted to rotational energy. To produce as much rotational energy out of the steam as possible, the temperature difference is made as large as possible. The steam temperature is limited for practical reasons like mechanical strength of pipes and containers and heat transfer. The sink temperature is determined essentially by the temperature of the

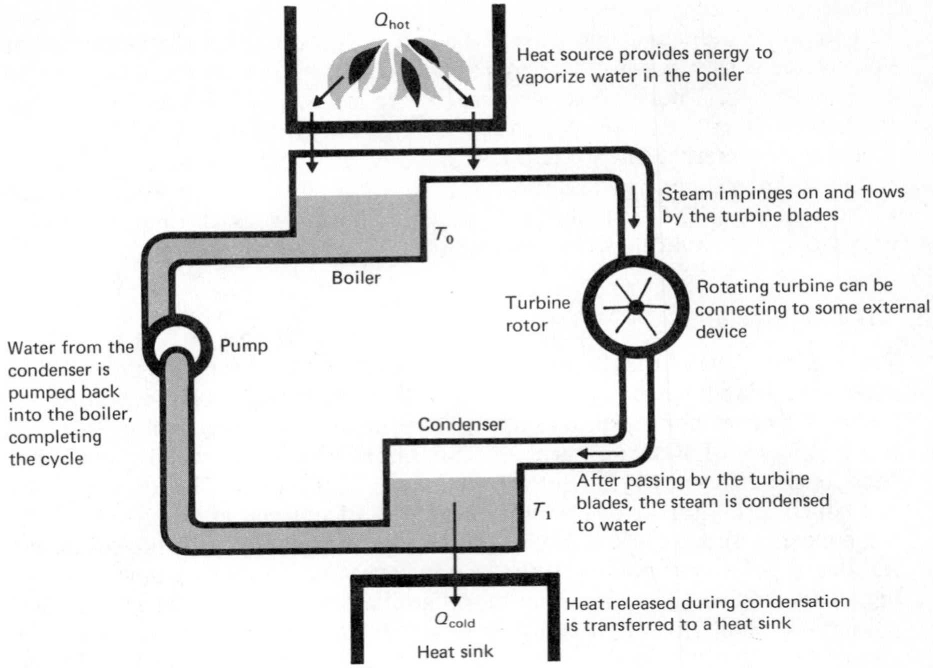

Figure 6.6 Schematic diagram of processes in a steam turbine. The boiler, turbine, condenser, and pump are separated to illustrate the cyclic flow of steam and water. The combination of all four components constitutes the steam turbine heat engine.

Labels in figure:

Q_{hot}

Heat source provides energy to vaporize water in the boiler

Steam impinges on and flows by the turbine blades

T_0

Boiler

Turbine rotor

Rotating turbine can be connecting to some external device

Water from the condenser is pumped back into the boiler, completing the cycle

Pump

Condenser

After passing by the turbine blades, the steam is condensed to water

T_1

Q_{cold}

Heat sink

Heat released during condensation is transferred to a heat sink

environment. At the temperature of the environment, the steam will condense back to water and heat will be released. The heat is taken away from the condenser (Fig. 6.6) and transferred to a river, lake, stream, or some other body of water. From the efficiency standpoint it is important to recognize that the efficiency for converting the internal energy of steam to mechanical energy will somehow depend on the temperature difference between the steam and the heat sink. As this temperature difference increases, the efficiency also increases. Recognizing this, let us proceed with the analysis of the physics of the process.

Like many important physical and environmental processes, the steam turbine is a cyclic process. In a cyclic process, a system starts from some initial state of affairs, proceeds through a series of other states, and ultimately returns to the conditions of the initial state. A clock is a cyclic system with time defining a state. Any given state (time) is repeated every 12 hours. A state of a steam turbine during any part of the cycle is specified by the pressure, volume, and temperature of the gas. The cycle is started by converting a certain quantity of water into steam. After impinging on and passing through the turbine blades, the steam is condensed back into water and returned to the starting point where the cycle is restarted. The pressure, volume, and temperature of the water is unchanged at the end of the cycle. Hence, regardless of how com-

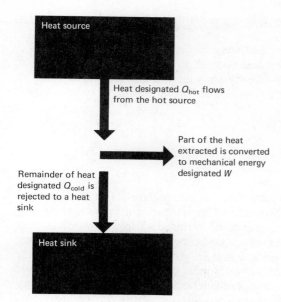

Heat source

Heat designated Q_{hot} flows from the hot source

Part of the heat extracted is converted to mechanical energy designated W

Remainder of heat designated Q_{cold} is rejected to a heat sink

Heat sink

Figure 6.7 In each cycle of its operation, a steam turbine draws heat (Q_{hot}) from a hot source, converts part of this heat to mechanical energy (W), and rejects the unused heat (Q_{cold}) to a sink at a temperature lower than the source.

plicated the energy transformations may be during the cycle there is no change in the internal energy of the water during the cycle. No change means $\Delta U = 0$. If $\Delta U = 0$, the first law of thermodynamics tells us that $W = \Delta Q$, which means that the amount of work done is the difference between the heat energy taken in from the heat source and the heat energy rejected at a lower temperature (Fig. 6.7).

$$W = \Delta Q = \text{heat taken in} - \text{heat released}$$

$$W = Q_{hot} - Q_{cold}. \tag{6.6}$$

The function of the steam turbine system is to convert as much of the heat taken in into mechanical energy. Hence the efficiency (Greek epsilon (ϵ)) for converting the heat taken in to mechanical energy is

$$\epsilon = \frac{\text{work done}}{\text{heat taken in}} \times 100\%$$

$$\epsilon = \frac{W}{Q_{hot}} \times 100\%.$$

Using the first law of thermodynamics (Eq. 6.6, $W = Q_{hot} - Q_{cold}$) to replace W, we have

$$\epsilon = \frac{Q_{hot} - Q_{cold}}{Q_{hot}} \times 100\%. \tag{6.7}$$

This expression is perfectly general and applies to all heat engines no matter how simple or how complicated.

There is a prescribed sequence of steps in the operation of the cycle of a heat engine. There are many physical systems in which a given sequence of events can be reversed. We have already encountered one such system in the form of chemical reactions. For example, the formation of the oxygen molecule O_2 from two oxygen atoms is

$$O + O \longrightarrow O_2.$$

The reverse of this (the separation of an oxygen molecule) would be

$$O + O \longleftarrow O_2.$$

The different direction of the arrow denotes the reversal. The energetics of these two processes are quite different. The latter requires that energy be added; the former requires that energy be released. The arrows on a heat engine can also be reversed. This means that heat is taken out of a *cold* source, some outside agent does work *on* the system, and heat is released to a *hot* reservoir. Household refrigerators, heat pumps, and air conditioners do just this. The electric motor on a household refrigerator provides the energy source for *doing* work to take heat out of the cold interior of the refrigerator. This idea is also employed in a heat pump that actually pumps heat from the cold exterior of a building into the warm interior for heating purposes. This turns out to be an efficient way of utilizing electric energy.

We measure the "effectiveness" of a refrigerator for doing its job as

$$\text{effectiveness} = \frac{\text{amount of heat removed from the cool region}}{\text{work required to remove the heat}},$$

$$E_R = \frac{Q_{cold}}{W}. \tag{6.8}$$

We measure the "effectiveness" of a heat pump for doing its job as*

$$\text{effectiveness} = \frac{\text{total amount of heat added to the warm region}}{\text{work required to add the heat}},$$

$$E_{HP} = \frac{Q_{hot}}{W}. \tag{6.9}$$

The Carnot Engine

One always likes to score 100% on an examination. But preparing for an examination invariably involves certain constraints and conditions. Perhaps there is insufficient time or you lost your textbook. With a given set of conditions, it would be of practical interest to know what is the maximum score you could attain. If this maximum attainable score were 60% and you scored 55%, you might consider your performance as good. In 1824, a French engineer

---

* The "effectiveness" as defined here is called "coefficient of performance" in many texts.

named Sadi Carnot considered the steam engine in this vein of thought. He posed, and answered, the question: "If you have a heat source and a heat sink at given temperatures and you want to operate a heat engine with this source and sink, what is the maximum efficiency that you can anticipate using a perfectly constructed engine?" This theoretical engine has come to be called a Carnot engine. It has no friction between the piston and cylinder walls. It can be reversed and operated as a refrigerator. During parts of the cycle, the walls of the cylinder are required to be perfectly insulated from the surroundings so that no heat can either come into or leave the working gas in the cylinder. A process like this that does not exchange heat with its surroundings is called an *adiabatic* process. The working gas used to drive the piston has an ideal or perfect behavior. For this theoretical engine, Carnot showed that the heat rejected divided by the heat taken in is equal to the Kelvin temperature of the cold heat sink divided by the Kelvin temperature of the hot heat source.

$$\frac{Q_{cold}}{Q_{hot}} = \frac{T_{cold}}{T_{hot}}. \tag{6.10}$$

Using the generalized efficiency equation for a heat engine (Eq. 6.7), the efficiency of the Carnot engine is

$$\epsilon = \left(\frac{T_{hot} - T_{cold}}{T_{hot}}\right) \times 100\%. \tag{6.11}$$

Note that the efficiency is zero if the temperatures of the steam and sink are equal and that the efficiency increases as the temperature difference increases. This is consistent with our analogy between a hydroelectric system and a steam engine (Sec. 6.3).

The efficiency of a Carnot engine depends only on the temperatures of the hot heat source and cold heat sink. In a steam turbine, energy is extracted from steam at a temperature typically 825 K. Heat is released to the environment at a temperature of about 300 K. Hence, a Carnot engine at these temperatures would have an efficiency of

$$\epsilon = \left(\frac{825 - 300}{825}\right) \times 100\% = 64\%.$$

Despite all the ideal conditions required for operation of a Carnot engine, its efficiency is not particularly impressive. But as unimpressive as this efficiency may be, we have reasons to believe that it is the greatest efficiency that can be obtained by any heat engine for two given operating temperatures.

The Second Law of Thermodynamics

From the standpoint of the first law of thermodynamics, a perfect heat engine would convert all the heat that it takes in into mechanical work. A perfect refrigerator would take heat from a source and release only that heat extracted to a source at a higher temperature. No work would be required to cause the heat transfer. Friction is always found in a mechanical system and so we cannot expect to build perfect engines and perfect refrigerators. Even if friction could be eliminated, our experience with thermodynamic processes leads us to the belief that perfect engines and perfect refrigerators still could not be built. In this concept lies the essence of the second law of thermodynamics. There are several ways of stating the law. The following two ways are equivalent.

1. For a system operating in a *cycle* between heat sources designated as T_{hot} and T_{cold}, it is impossible to extract a net amount of heat from the hot source and convert it entirely into work. Some heat must be rejected to the cold source. This precludes a perfect heat engine.

2. Heat will not spontaneously flow from a cold region to a hot region. It must be forced to flow and this results in work being done *on* the system. This precludes a perfect refrigerator.

These statements are based on experience and cannot be proven, but it is possible to prove that they are equivalent. They would be disproven if an exception were found; to date, no exception has been discovered.

Maximum Efficiency of a Heat Engine

If the Carnot engine cannot be constructed, it seems that the time spent considering it is purely academic. However, the Carnot engine serves an extremely useful function. We can show that within the framework of the second law of thermodynamics no heat engine can be more efficient than a Carnot engine operating between the same two temperatures. If a heat engine is required to operate between two given temperatures, we can immediately determine its maximum efficiency from the simple expression (Eq. 6.11) for the efficiency of a Carnot engine. We will illustrate this supposition concerning the maximum efficiency by taking a functional heat engine, that is, one that could actually be built, and letting it furnish the mechanical energy for a Carnot refrigerator. (Remember a Carnot refrigerator is nothing more than a Carnot engine running backwards.)

Let us use the diagram in Fig. 6.6 to represent a functional steam turbine. We will allow the rotating turbine to provide input energy for a Carnot refrigerator (Fig. 6.8). Both the steam turbine and the refrigerator operate between the same temperatures, which we will assume to be 600K and 300K. Because the refrigerator is coupled directly to the steam turbine, it is forced to operate on the same cycle as the steam turbine. Thus the steam turbine and Carnot refrigerator as a unit constitute a single cyclic system, and because the energy produced by the turbine is used to run the refrigerator there is no net work produced by this combined system. For the assumed temperatures we know from our discussion of the Carnot cycle that one unit of mechanical energy fed to the refrigerator will cause one unit of heat energy to be extracted from the cold region and two units of energy to be transferred to the warm region. We do not know how efficient the functional engine is but for this discussion all that is required is the assumption that it be more efficient than a Carnot engine

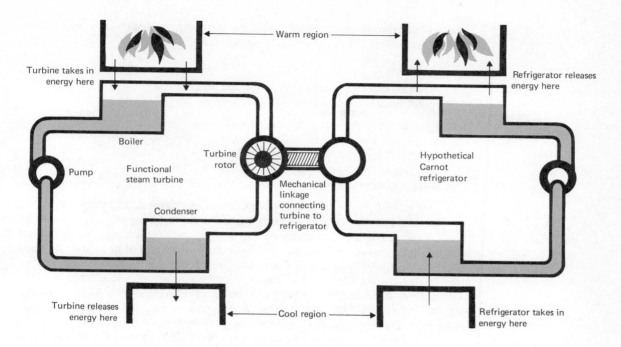

Figure 6.8 Schematic illustration of a functional steam turbine that provides mechanical energy to run a hypothetical Carnot engine. The turbine produces mechanical energy from heat energy that it takes in from the hot thermal energy reservoir. The refrigerator uses this mechanical energy to cause heat to flow out of the cold region. It is important to note that (1) the refrigerator is forced to operate on the same cycle as the steam turbine, (2) there is no net work involved in the engine–refrigerator combination, and (3) both the engine and refrigerator are functioning between the same two temperatures.

operating between 600K and 300K. For the sake of argument, we assume an efficiency of 60 percent. That is, if one unit of heat energy is taken into the heat engine it will produce 0.6 units of mechanical energy. A Carnot heat engine would produce only 0.5 units of mechanical energy for the same energy taken in. The goal is to show that our assumption violates the second law of thermodynamics and that this functional heat engine cannot do a better job than the hypothetical Carnot engine. We start by letting the heat engine take in one unit of heat energy, do 0.6 units of work, and then reject 0.4 units of heat energy at the lower temperature. The 0.6 units of work provide input energy to the refrigerator that, as we have shown earlier for the temperatures involved, will extract the same amount of heat energy from the cold region and transfer 1.2 units of heat energy to the warm region. The *net* work done by the engine–refrigerator combination is zero. But in accounting for the heat transfers we find that in one cycle of operation the engine–refrigerator combination has extracted from the cool region a *net* amount

$$0.6 \quad - \quad 0.4 \quad = 0.2 \text{ units,}$$

pulled out   put in   *net*
by refrigerator   by engine

and transferred to the warm region a *net* amount

$$1.2 \quad - \quad 1.0 \quad = 0.2 \text{ units.}$$

put in   pulled out   *net*
by refrigerator   by engine

Thus we have a cyclic system that has caused an amount of heat energy (0.2 units) to be transferred to a warmer region without doing any net work. The second version of the second law of thermodynamics is violated. We can avoid this dilemma by concluding that the fuctional engine cannot have an efficiency that exceeds the efficiency of a Carnot engine at the specified temperatures.

It seems that there is something very special about the Carnot engine. There is—in the sense that the calculations are relatively simple. However, any engine that can be reversed to make a refrigerator can theoretically have the same efficiency as a Carnot engine if operated between the same two temperatures. Therefore, we could have used any reversible engine in the exercise and we would have arrived at the same conclusion.

6.4 AVAILABLE ENERGY AND SECOND LAW EFFICIENCY

Some of the most famous persons in physics—Einstein, Kelvin, Clausius, Carnot—have contributed to the formulation of the science of thermodynamics. At some point, Einstein was inspired to say

> Thermodynamics is the only physical theory of universal content concerning which I am convinced that within the framework of the applicability of its concepts it will never be overthrown.

This universal validity of the laws of thermodynamics forces some fundamental, and often hard to accept, conclusions; for example, the conclusion that there is a maximum efficiency that can be obtained by any heat engine for a given set of operating temperatures. We now introduce a concept that we shall call the second law efficiency that capitalizes on the second law of thermodynamics and tells us in a numerical way just how close an actual, functional heat engine comes to the maximum possible efficiency. It is of fundamental significance and contains one of the most important messages in our entire study.

For a given set of operating conditions, no heat engine can be more efficient than a Carnot engine. Using this as a basis for comparison, we will compare the performance of a modern steam turbine with the theoretical performance of a Carnot engine. We will do this by dividing the efficiency of the turbine by the Carnot efficiency. Because this comparison involves the second law of thermodynamics, we will call the ratio second law efficiency. Thus

$$\text{second law efficiency} = \frac{\text{efficiency of the steam turbine}}{\text{Carnot efficiency}}$$

$$= \frac{\epsilon}{1 - \dfrac{T_{cold}}{T_{hot}}}. \tag{6.12}$$

Remember that the first law efficiency for a modern steam turbine is 47%. If it operates between temperatures of 825 K and 300 K, its second law efficiency is

$$\text{second law efficiency} = \frac{47\%}{1 - \dfrac{300}{825}} = 74\%.$$

Because the second law efficiency cannot exceed 100%, a modern steam turbine actually does a creditable job. As long as the task is one of converting thermal energy into mechanical energy, it is going to be difficult to find a system much more efficient than the conventional steam turbine.

Water flowing from the top of a dam eventually runs its course and comes to a stationary configuration. Then it is no longer able to drive a water turbine or to do some other type of mechanical work. Similarly, when an amount of heat (Q_{hot}) flows spontaneously out of a thermal energy reservoir it eventually runs its course and ends up in one grand thermal energy reservoir which is the environment. Although no energy is ever lost in this traversal from reservoir to environment, the ability to do further work has been lost because there is no other region at a lower temperature into which heat from the environment may flow. For the given amount of heat (Q_{hot}) the second law of thermodynamics ensures that there is a maximum amount of mechanical energy that can be garnered. It can be calculated from the maximum efficiency relation (Eq. 6.11).

$$\text{Efficiency} = \frac{\text{work performed}}{\text{heat extracted}} = \frac{W}{Q_{hot}} = 1 - \frac{T_{cold}}{T_{hot}}.$$

Therefore,

$$W = Q_{hot}\left(1 - \frac{T_{cold}}{T_{hot}}\right). \tag{6.13}$$

W is called the available energy. It is the energy that is theoretically available from the heat Q_{hot}. Even though the total heat energy (Q_{hot}) is never lost, the available energy is. Available energy is consumed; it is gone forever!

In a practical situation, a heat engine is selected to perform a certain task. One decides how much work must be done and then chooses or designs a heat engine that will do the job. For a given set of operating conditions, no heat engine will do a given task using less energy than a Carnot engine. This minimum energy, termed the available energy input, is Q_{hot} in Eq. 6.13. The available energy input to a Carnot engine will result in extraction of the available energy. Any real engine operating under the same conditions will require more energy to do the same task. We will call this energy Q_a, meaning the actual energy used to perform the same task. The second law efficiency in terms of the available energy is written

$$\text{second law efficiency} = \frac{\text{available energy input}}{\text{actual heat energy required to do the task}}$$

$$= \frac{Q_{hot}}{Q_a}. \tag{6.14}$$

The second law efficiency is a measure of how effective a heat engine is at utilizing the available energy. If, for example, thermodynamic considerations show that the available energy input is 100,000 joules for an assigned task and the engine actually used 200,000 joules for the task required, then the second law efficiency would be

$$\text{second law efficiency} = \frac{100,000}{200,000} = 0.5 = 50\%.$$

We would interpret this as meaning that, theoretically, the task could have been done with 50% of the energy actually used.

Although very subtle, the concepts of available energy and second law efficiency are extremely important in an energy-conscious society, and it is important that you understand the difference between the first law efficiency and the second law efficiency. Nature has incorporated nothing in the first law of thermodynamics that says you cannot convert all the thermal energy content of a reservoir such as an ocean into useful work. It says simply that there is no net gain or loss of energy in the process. Not only does the second law of thermodynamics forbid the conversion of all the energy content of a reservoir into useful work, but it establishes a limit on the amount of work that can be performed. From the first law standpoint, energy is never consumed—it is only converted into other forms. The second law recognizes that once work has been done by a system, the capacity to do further work has been diminished. Available energy is a measure of just how much that capacity has been reduced. The available energy is ultimately used in an energy conversion process. The second law efficiency measures how well a given system utilizes the available energy in performing an assigned task. It is a concept that applies not only to heat engines but to all energy conversion systems. We will say more about this in the concluding chapter on energy conservation when we utilize these ideas to evaluate a heat pump, a device that is becoming increasingly popular for providing household space heat.

Considerations of a heat engine tend to focus on the hot source. On the other hand, the cold sink is equally important. No matter how hot the source may be, heat will never flow spontaneously out of it unless there is a sink at a lower temperature to accept the heat. Increasing the temperature of the hot source or decreasing the temperature of the cold sink improves the efficiency. Because the temperature of the cold sink is usually that of the environment (about 300 K), over which we have little or no control, it is tempting to suggest a significant increase in the temperature of the hot source. There are, however, many nontrivial engineering aspects to this. For example, a typical metal like copper melts at about 1350 K.

Using the Carnot efficiency expression as a guide, we can show that temperatures significantly different from those used presently are required to produce modest changes in the efficiency. Figure 6.9 summarizes calculations made for various temperatures of the hot source and a cold sink temperature of 300 K. Significant improvement in efficiency results for *changes* between 300 and 900 K, but a change in temperature from 1200 to 1500 K produces only a 5 percent increase in efficiency. The practical aspects of making such a temperature increase are forbidding, and the reward in efficiency is small. What might be expected, then, in the way of increased efficiency of practical turbines? The projections in Fig. 6.10 are not encouraging and they indicate that as long as steam turbines are used vast amounts of heat will have to be disposed of or ways of utilizing it will have to be found.

6.5 PROSPECTS FOR IMPROVED EFFICIENCY OF STEAM TURBINES

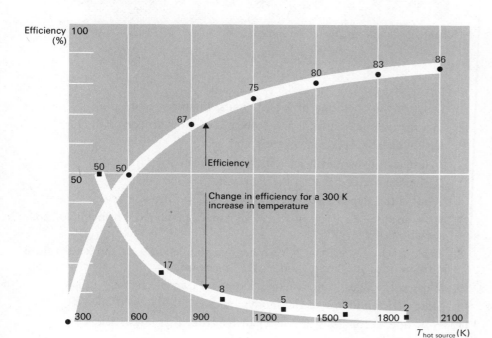

Figure 6.9 Efficiency and change in efficiency of a Carnot engine for various temperatures of the hot source assuming a cold reservoir temperature of 300 K. The number shown adjacent to each point is the value of the efficiency or change of efficiency represented by the point. Note that small changes in efficiency occur after the hot source has achieved a temperature of about 900 K.

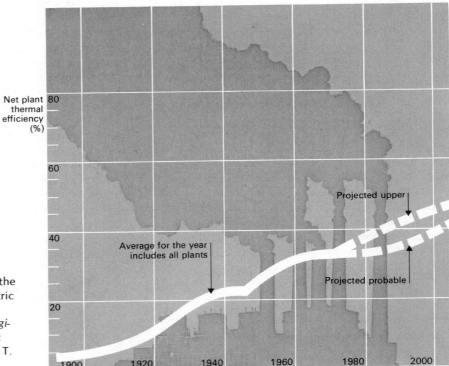

Net plant thermal efficiency (%)

80

60

40

Average for the year includes all plants

Projected upper

Projected probable

20

1900   1920   1940   1960   1980   2000

Year

Figure 6.10 Past and future trends of the overall thermal efficiency for an electric power plant. (From R. T. Jaske, J. F. Fletcher, and K. R. Wise, *Chemical Engineering Progress,* vol. 66, no. 11 (1970): p. 17. Reprinted with permission of R. T. Jaske.)

TOPICAL REVIEW

1. Describe the operation of a heat engine in energy terms.
2. In what way are pressure and force alike? How do they differ?
3. Distinguish between temperature and heat.
4. Distinguish between internal energy and heat.
5. Define calorie and British thermal unit.
6. What is the significance of the concept of specific heat?
7. How does the first law of thermodynamics involve the principle of conservation of energy?
8. Explain why the efficiency of a heat engine depends on the temperature difference of the "hot" and "cold" reservoirs.
9. What is the significance of the theoretical Carnot engine?
10. State the second law of thermodynamics.
11. How does second law efficiency differ from first law efficiency?

Basic Thermodynamics

Most physics and physical science texts will devote a number of pages to the study of thermodynamics. Two such texts that would be useful for expansion of the material presented in Chapter 6 are

1. *Physics: A Descriptive Analysis,* by Albert J. Read, Addison-Wesley, Reading, Mass. (1970).

2. *Conceptual Physics,* by Jae R. Ballif and William E. Dibble, Wiley, New York (1971).

Some interesting insight into thermodynamics and the role of the steam engine in the Industrial Revolution are presented in Chapters 10 and 11 of

3. *Physics and Its Fifth Dimension: Society,* by Dietrich Schroeer, Addison-Wesley, Reading, Mass. (1972).

The most detailed physics treatment of the second law efficiency is presented in

4. "Efficient Use of Energy: The APS Studies on the Technical Aspects of the More Efficient Use of Energy," American Institute of Physics, New York, 1975.

Less mathematical discussions are presented in

5. "Energy Efficiency: Our Most Underrated Energy Resource," Marc H. Ross and Robert H. Williams, *Bulletin of the Atomic Scientists* **32**, 30 (November 1976).

6. "Energy I," Barry Commoner, *The New Yorker,* February 2, 1976.

7. "Industry Can Save Energy Without Stunting its Growth," Tom Alexander, *Fortune,* May 1977.

6.2 Basic Thermodynamics

1. The electrical resistance of a wire changes with temperature. How could you conceive a thermometer based on this principle?

2. Why does a sharp knife cut better than a dull knife?

3. Why is a heavy chair or table having narrow legs often set on flat, circular plates when on a carpeted floor?

4. Why does a large tractor exert considerably less pressure on the ground than a hypodermic needle on the skin of a patient?

5. The pressure exerted by gas molecules on the walls of an automobile tire is greater than the pressure recorded by a "tire gauge" used to measure the tire pressure. On the basis of pressures (or forces) acting on the tire, explain why this is so.

6. Why is it that a body of water in sunlight warms more slowly than the land which contains it? Land is, of course, not completely composed of

iron, but you can give plausibility to your answer by calculating the temperature rise of 1000 g of iron and water if each absorbs 10,000 cal of heat. (The specific heats of iron and water are given in Table 6.1.)

7. When a gas is compressed by pushing on a piston in a cylinder, why will the internal energy of the gas increase?

8. What is the direction of heat flow when
   a) you place your hand in water at 40°C,
   b) you place your hand in ice water.

9. The rubber hose of a bicycle tire pump will get noticeably warm during the inflation of the tire. What is the origin of the energy producing the warming of the hose?

6.3 Efficiency of a Heat Engine

10. An object is placed in contact with another object at a lower temperature. Is there anything in the first law of thermodynamics that prevents a spontaneous transfer of energy from the cooler object to the hotter object? If this energy transformation did take place, what changes would result in the temperatures of the two objects?

11. Heat does not flow spontaneously from one object to another object at a higher temperature. But is it possible to contrive a set of conditions where heat can be transferred from one object to another at a higher temperature? Explain.

12. It is reported (*Technology Review*, July/August 1972, p. 51) that there is a perceptible buildup of heat in the ground surrounding a subway system. One of the factors is 600 new air-conditioned cars. Why should this tend to heat up the tunnels and the surrounding ground?

13. If a friend were to ask you to *prove* the validity of the second law of thermodynamics, what would be your response?

14. What would someone have to do to convince you that the second law of thermodynamics is not true in all instances?

15. Conceive a situation involving a refrigerator such that the second law of thermodynamics is violated but the first law of thermodynamics is not. Repeat for a heat engine rather than a refrigerator.

16. An animal is, in a sense, a machine. It takes energy in as food and converts it to other forms of energy at a certain efficiency. Suppose that each animal in a carnivorous food chain involving five animals has a conversion efficiency of 0.6. What is the overall conversion efficiency for the five-animal chain? What are the implications of eliminating one animal from the chain?

17. Discuss the idea of defining ecology as "the quantification of the first and second laws of thermodynamics for environmental processes."

18. Normally an advertisement for an air conditioner will characterize it as having a cooling capacity of so many Btu, for example 10,000 Btu. Although time has not been mentioned explicitly in the evaluation why is it very important?

19. The oceans constitute an enormous source of energy. Why, then, doesn't someone construct an engine that extracts heat from the surface of the ocean, converts some of this energy into useful work, and rejects the remaining energy back into the surface of the ocean?

20. At least in principle, the sequence of operation of many systems can be reversed. An electric motor that converts electric energy to mechanical energy is, in principle, an electric generator running "backwards." Using this idea explain why an air conditioner is simply a heat engine running "backwards." It might be helpful to draw a diagram for an air conditioner like that in Fig. 6.7 for a heat engine.

21. Since a motor is, in principle, a generator running backwards why couldn't the electric energy output of a generator be fed to a motor and the mechanical energy output of the motor fed to the generator so that a perpetual motion machine results? Use the first law of thermodynamics to explain your answer.

22. Suppose that an inventor with no knowledge of thermodynamics comes to you with a drawing of a heat engine that does not reject any heat. Since the two of you have no common basis for argument, what might you demand him to provide in the way of evidence supporting his thesis? (This is just what would happen if the inventor were to go to the Patent Bureau with his drawings.)

23. An irate person threatens to cool the house by opening the refrigerator door when the air conditioning system fails. How would you tactfully answer the person? What is the significance of placing the exhaust of a room air conditioner on the outside of a house?

6.4 Available Energy and Second Law Efficiency

24. There always seems to be one student in a class who outperforms all other students. Yet this student may not always score 100 on the examinations. If you were to compare the other students' scores with this student's score, how would this be similar to the concept of the second law efficiency for a heat engine?

25. To maintain the freezing compartment of a refrigerator at freezing temperature when the temperature outside of the refrigerator is above the freezing temperature requires that work be performed on the refrigerator (by an electric motor, for example). What determines the least amount of work that must be performed?

6.2 Basic Thermodynamics

1. Liquid nitrogen boils at 77 K. Determine the boiling temperature on the Celsius and Fahrenheit scales.

2. The temperature in a room is usually about 68°F. Show that this corresponds to 20°C and 293 K.

3. Suppose that you were not satisfied with any of the three temperatures scales mentioned in the text and you chose to assign numbers of 100 and 200 to the ice and steam points. If you characterized your scale by °M, show how you could convert °M to °C.

4. How does a temperature *change* of 1°C compare with a temperature *change* of 1°F? How does a temperature *change* of 1°C compare with a temperature *change* of 1 K?

5. Atmospheric pressure at sea level is about 15 lb/in². How big a square on the earth's surface would you need so that the total force exerted on this square would equal the weight of a 135-lb person?

6. A 10-ton tractor distributes its weight over two tracks each having an area of 20 ft². What is the pressure in lb/ft²? If the diameter of a hypodermic needle is 1/64 in., how much pressure is exerted on the skin by a one-lb force supplied by a nurse? (You will need to determine the cross-sectional area of the needle. The formula for this is $A = 0.785d^2$ where d is the diameter).

7. Assuming that the bottom of your shoe is a rectangle, estimate the pressure that you exert on a floor when you are standing flat-footed, on one foot, and on tiptoes.

8. a) A piston in a cylinder has a diameter of 10 cm. If this piston moves a distance of 4 cm, what is the change of volume (in m³) in the cylinder?
   b) If this displacement is caused by a 10,000-N force, how much work is done in the process?

9. Water, though ordinary in appearance, has many unusual physical properties. For example, very few substances have a higher specific heat. Water is especially useful for cooling the condenser of a steam turbine because of its high specific heat. To show why, determine the temperature rise of one kg of water and one kg of copper if each absorb 10,000 cal of heat. (Specific heats are given in Table 6.1.)

10. Hot water usage in an average home amounts to about 20 gallons per day for each occupant. Complete the table below to determine the daily hot water costs for a house using an electric hot water heater.

    Number of occupants = 5
    Intital water temperature = 25 °C
    Hot water temperature = 70 °C
    20 gallons of water = 75 kg
    Specific heat of water = 1 cal/g · °C
    1 kW-hr = 860,000 calories
    Cost = 4¢ per kW-hr

$$\frac{\underline{\hspace{1.5cm}} \text{ gallons}}{\text{day person}} \times \underline{\hspace{1.5cm}} \text{ persons} = \underline{\hspace{1.5cm}} \text{ gallons}$$

$$\underline{\hspace{1.5cm}} \text{ gallons} \times \underline{\hspace{1.5cm}} \frac{\text{grams}}{\text{gallon}} = \underline{\hspace{1.5cm}} \text{ grams}$$

$$\underline{\hspace{1.5cm}} \text{ grams} \times \underline{\hspace{1.5cm}} \frac{\text{calories}}{\text{gram} \cdot {}^{\circ}\text{C}} \times \underline{\hspace{1.5cm}} {}^{\circ}\text{C} = \underline{\hspace{1.5cm}} \text{ calories}$$

$$\underline{\hspace{1.5cm}} \text{ calories} \times \underline{\hspace{1.5cm}} \frac{\text{kW-hr}}{\text{calorie}} = \underline{\hspace{1.5cm}} \text{ kW-hr}$$

$$\underline{\hspace{1.5cm}} \text{ kW-hrs} \times \underline{\hspace{1.5cm}} \frac{\cent}{\text{kW-hr}} = \underline{\hspace{1.5cm}} \cent$$

Therefore, the daily cost is _____ ¢.

11. A household hot water heater generally has a reservoir of about 80 gallons. Assuming that water from the local supply is at a temperature of 65 °F, how many Btu of heat are required to raise the temperature of 80 gallons of water to 135 °F? (The weight of 20 gallons of water is 165 pounds.)

12. A 1000-kg concrete pillar in a house is used to store thermal energy that it acquires from sunlight entering through windows in the house. It is found that on a bright sunny day the temperature of the pillar rises to 30 °C.

    a) How many calories of thermal energy are available from this pillar when it cools to 24 °C? (See Table 6.1 for the specific heat of concrete.)

    b) If the nightly heat requirement to maintain the house at 24 °C is 7 million calories, what fraction of the heat load could be provided by the concrete?

6.3 Efficiency of a Heat Engine

13. A certain steam turbine takes in 1,000,000 Btu of heat and rejects 600,000 Btu of heat in each cycle. Determine the efficiency for converting heat to useful work.

14. An engine in each cycle takes in 50,000 Btu of heat and releases 20,000 Btu of heat.

    a) How much energy is converted to work?

    b) What is the efficiency of the engine?

    c) What physical principle did you use for part (a)?

15. In the expression for the maximum efficiency of a heat engine (Eq. 6.11)

$$\epsilon = \left(1 - \frac{T_{\text{cold}}}{T_{\text{hot}}}\right) \times 100\%,$$

it is necessary that the temperatures be expressed in kelvins. Taking $T_{cold} = 27\,°C$ and $T_{hot} = 527\,°C$, show that numerical results are different for temperatures in degrees Celsius and kelvins.

16. Assume that a steam turbine in a power plant operates at maximum efficiency. The temperature of the heat source is 900 K and heat is rejected at 300 K. The efficiency can be improved by either increasing the temperature of the heat source or decreasing the temperature of the heat sink. Given a choice of either a 25 K increase in heat source or a 25 K decrease in heat sink, what would your choice be? The heat sink for a steam turbine is usually a river or a lake. Do you think it would be significant to try to use water from the lower part of a lake as opposed to that from the upper part?

17. A heat engine having an efficiency of 40 percent is going to reject heat to a cool region having a temperature of 300 K. What is the minimum temperature for the hot reservoir of this system?

18. During the compression of a gas in a cylinder, 3100 joules of work are done on the gas and 500 joules of heat escape from the gas. What is the change in internal energy of the gas? Is the change an increase or a decrease?

19. The addition of 500 calories of heat to the gas in a cylinder containing a piston causes 250 calories of work to be done by the piston. What is the change in internal energy of the gas? Is the change an increase or a decrease?

20. In a laboratory exercise on heat engines a participant reports his measurements as

> heat taken in during each cycle = 4320 Btu,
> work produced in each cycle = 5,275,000 joules.

Using the laws of thermodynamics as a basis for discussion, comment on the results of the experiment.

21. A steam turbine operating at 50 percent efficiency and having a thermal power input of 2000 megawatts would dissipate 1000 megawatts of thermal power to the condenser. This means that each second the condenser must absorb 1,000,000,000 joules of heat energy.

a) If each second this energy were deposited to a tank of water and you allowed the temperature to rise 10 °C, how many kilograms of water would be required?

b) How many gallons of water would be required? (1 gallon of water has a mass of 3.8 kilograms.)

If a 10 °C temperature difference is maintained between the input and output water lines to the condenser, then water must flow through the condenser at the rate calculated in part (b).

22. Efficiency is a quantitative measure of what is expected of an engine:

$$\text{efficiency} = \frac{\text{work out}}{\text{energy in}}.$$

Using the laws of thermodynamics, explain why this quantity is always numerically less than one.

23. *If* in the operation of a refrigerator X Btu are taken from its cold interior, Y Btu are deposited in the surroundings, and Z Btu of work are required, determine for the data given below which of the laws of thermodynamics are violated.

a) $X = 1000$ $Y = 1000$ $Z = 0$

b) $X = 1000$ $Y = 2000$ $Z = 1000$

c) $X = 2000$ $Y = 1000$ $Z = 1000$

24. The total energy consumption in the United States in 1976 amounted to 7.42×10^{16} Btu whereas the electric energy consumption amounted to 2.04×10^{12} kW-hr.* (1 kW-hr $= 3413$ Btu)

a) What percent of the total energy is electric?

b) About 77 percent of this electric energy was generated by fossil fuel plants. What percent of the total energy was generated by fossil fuel plants?

c) If the average conversion efficiency for fossil fuel plants was 35 percent, what percent of the total energy was used at the input of the power plants?

d) If 45 percent of this energy is rejected as heat, what fraction of the total energy is rejected into the environment in the form of heat?

6.4 Available Energy and Second Law Efficiency

25. During each cycle of operation, a certain heat pump requires 34 Btu of input energy in order to deposit 112 Btu of heat into a house. What is its effectiveness?

26. In the design of a particular mechanical system, a heat engine producing a mechanical power output of 10,000 watts is required. The operating temperatures of the engine are established at 300 K and 600 K.

a) What is the theoretical least amount of thermal power that must be extracted at 600 K to do the desired task?

b) If it is found that in practice 40,000 watts of thermal power must be extracted to perform the desired function, determine the first and second law efficiencies.

27. In a certain situation thermal energy is available at 627 °C and a cold area having a temperature of 27 °C is available. If 1200 Btu of heat energy is extracted from the hot system, what is the theoretical maximum amount of work that can be performed by any heat engine operating between these given temperatures?

---

* The total energy value is from the Federal Energy Administration. The electric energy value is from the Federal Power Commission. Both values can be found in "Energy in Focus: Basic Data," published by the Federal Energy Administration (1977).

DISPOSAL OF REJECTED HEAT FROM STEAM TURBINES

7

Giant cooling towers such as this one at the Trojan Nuclear Power Plant on the Columbia River near Prescott, Oregon, are used to transfer heat from steam turbine condensers to the environment. Photograph by Gene Daniels. (Courtesy of the EPA.)

7.1 MOTIVATION

Usually, a pollutant is a biologically hazardous or otherwise undesirable particle or chemical compound in the environment. Thermal pollution, light pollution, and noise pollution are examples of pollution that add no mass to the environment. Thermal pollution is the addition of unacceptable quantities of heat to the environment. Knowing the many uses of heat in our everyday lives, one wonders why any heat is discarded. It is just that there are no economic incentives to capitalize on the rejected heat. Environmental and energy considerations may in time alter this state of affairs, and thermal pollution may change to thermal enrichment.

7.2 THERMAL POLLUTION

To optimize the efficiency of a steam turbine, the low pressure region (Fig. 6.1) is maintained at as low a temperature as possible. Generally, this temperature is that of the environment. This temperature is low enough that steam will condense to a liquid in the low-pressure region. The condensation process liberates heat that would otherwise cause a temperature rise in the low-pressure region. The region where the condensation takes place is called the condenser. Heat is extracted from the condenser by a continuous flow of water in thermal contact with the condenser. Water for this heat-transfer system comes from a river, lake, or some man-made reservoir (Fig. 7.1). The water may return to its source some 5°–10°C (10°–20°F) warmer. If the heat-dissipating abilities (the natural flow of a river, for example) of the source cannot cope with this heat load, the temperature of the source may rise. For example, the temperature of the entire Yellowstone River below Billings, Montana, rose 1°–1.5°F as a result of a 180-megawatt plant near Billings.

Increasing the temperature of a body of water above its normal value produces a variety of changes on the quality of the water. Those changes of concern to plant and animal life in the water are the following:

1. As the temperature of the water increases, the capacity to hold oxygen dissolved in the water decreases. At the same time, the need for oxygen by aquatic life increases.

Figure 7.1 Schematic diagram of the condenser cooling system of a steam turbine used in a commercial electric power plant. Steam condensing to water in the region of the condenser liberates heat that is carried away by water pumped from a river, for example.

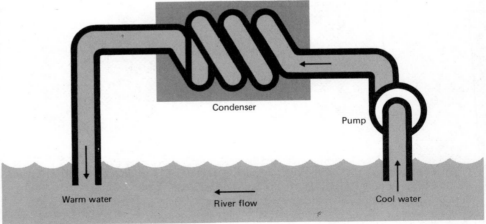

Condenser

Pump

Warm water

River flow

Cool water

2. Addition of heat to a body of water can cause stratification having different temperatures.

3. Chemical reactions are accelerated at elevated temperatures.

4. Concentrations of undesirable chemicals increase because of accelerated evaporation of water.

The anticipated effects of elevated temperatures on aquatic life include the following:

1. Decreased spawning success of fish.

2. Decreased survival probability of young fish.

3. Limitation of migration patterns.

4. Changes in processes that depend on biological rhythms.

5. Increased susceptibility to disease.

6. Alteration of the balance between plant and animal life.

Many laboratory studies show that some of the anticipated effects are very real. For example, 50 percent of a sample of brook trout will die in 133 hours in water at 25 °C (77 °F) if the fish have been acclimated to a temperature of 20 °C (68 °F).* Thermal effects in actual streams or lakes are less well known, but research is proceeding. Reports of studies in the estuaries of Maryland make it clear that the addition of heat to an aquatic environment could cause changes in its biota.† Another study‡ indicates that "there appears to be little doubt that effluents from the existing plants have done severe damage in Biscayne Bay (Florida)." Bodies of water that are already at temperatures near a critical level are especially vulnerable to thermal effects.

Present-day thermal pollution problems are minimal. However, precautionary measures are needed now to ward off future problems. For example, about 10 percent of the total energy presently converted in the United States appears as waste heat from electric power plants. By the year 2000 about 20 percent of the nation's average runoff of water would have to be used to cool power-plant condensers if present trends in power consumption continue and technology remains unchanged.§ Fears of thermal pollution are real and the utilities industry is being challenged. Construction of a power plant on Cayuga Lake in New York was halted partly because of potential thermal effects on the lake. However, the thermal effects from power plants already in existence should be carefully studied because the deleterious effects that have been anticipated may not in fact actually

---

* "Selected Materials on Environmental Effects of Producing Electric Power," Joint Committee on Atomic Energy, U.S. Government Printing Office, Washington, D.C. (1969), p. 305.

† "Report of the Thermal Research Advisory Committee to the Maryland Department of Water Resources," Department of Water Resources, Annapolis, Maryland (December 1969).

‡ *Science News,* vol. 98, no. 5 (August 1, 1970).

§ *Selected Material on Environmental Effects of Producing Electric Power,* Joint Committee on Atomic Energy, U.S. Government Printing Office, Washington, D.C. (1969), p. 30 (Data from Federal Power Commission).

occur.* There is some evidence that heated water actually resides in a fairly thin layer on top of a main body of water and does not affect aquatic life in the main body.

Efforts toward making thermal enrichment a reality are in progress. A major problem with using heat from present-day plants is that it is low grade, meaning low temperature (about 90°F), and consequently its applications are limited. The Scots find that in a fishery warmed by waters from their Hunterston nuclear power plant, sole mature in two years rather than the normal four. Similar uses of warm water are made in Japan. The feasibility of growing shrimp in a northern climate using waste heat is being studied at the Vermont Yankee Nuclear Power Plant. At the Oak Ridge National Laboratory, fish, clams, and crayfish are grown in heated ponds using algae feed. The algae is produced using animal waste and sewage, and reject heat to enhance the growth. A half-acre (about three fourths the size of a football field) greenhouse utilizing waste heat to stimulate produce production is being built at the Brown's Ferry Nuclear Power Station in Decatur, Alabama. Studies indicate that large-scale (300–500 acres) greenhouses are feasible. The temperature of the warm water can be increased by reducing the efficiency of the turbine. Water at higher temperatures is more useful, especially for heating purposes. The lower efficiency may be tolerable if greater use is made of the rejected heat.

7.3 ATOMIC DESCRIPTION OF INTERNAL ENERGY

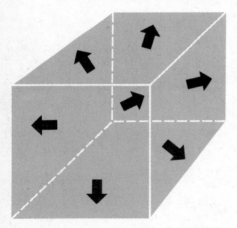

Figure 7.2 Schematic representation of molecules in a container exerting a force on the walls.

The rejected heat from a power plant ultimately ends up in the atmosphere. The problem is to transfer it from the heated water in the condenser to the atmosphere in an acceptable way. Heat can be transferred by three basic mechanisms—conduction, convection, and radiation. Heat is also removed from a liquid in the process of vaporization. All these processes are currently employed to some extent in industrial heat-transfer processes. Study of them at the atomic level provides us with an ideal opportunity to understand more fully the concept of heat, internal energy, and the second law of thermodynamics.

Atomic models are intrinsically difficult to formulate because the nature of the forces between atoms and molecules must be either known or assumed. Even if known, the physics and mathematics are usually difficult. We clearly cannot do a rigorous analysis, but some of the results of a successful model can be made plausible and acceptable.

Let us begin by considering a gas in a closed container such as an inflated football. The ball contains many trillions of molecules that bounce off its walls like balls rattling around in a vibrating box (Fig. 7.2). Pressure is exerted on the walls of the football by virtue of collisions of the molecules with the walls. It is easy to feel this pressure by pressing your fingers against the ball. There are several ways to change the pressure. If some of the molecules are let out of the football, the pressure decreases. If the football is squeezed, the pressure increases. Squeezing reduces the

---

* "The Calefaction of a River," by Daniel Merriman, *Scientific American,* vol. 222 (May 1970): p. 42.

volume of the ball and increases the frequency of the molecular collisions with the walls. Heating the football also causes the pressure to increase. This effect is commonly observed with automobile tires on hot days or when tires warm after being run on a highway for an extended time. Although the pressure changes resulting from letting gas out and changing the volume can be understood fairly easily, the connection between heat and pressure is rather subtle. A pressure increase, regardless of how it is made, means that the molecules are hitting the walls with greater frequency. This increase in frequency could come from reducing the volume as suggested earlier or by increasing the speed of the molecules. If the volume does not change, then the increased collision frequency would have to come from an increase in the speed of molecules. Greater molecular speed means greater kinetic energy. So when heat is added and the pressure increases, an increase in the kinetic energy of the molecules results. Knowing that other forms of energy exist we might suspect that some of the heat goes into forms other than kinetic energy. This is indeed the case but it is not necessary to include these other potential energy forms in these considerations since the behavior of many gases can be explained by assuming that all the molecular energy is kinetic.

Some thought should convince you that one direction of molecular motion is just as likely as any other—the motion of molecules is completely random. There is also ample evidence that all molecules do not have the same speed. This may not be obvious, but it is not hard to accept. The molecules of a gas are somewhat like bees buzzing around in a hive. Surely we would not expect every bee to be moving with the same speed. So we cannot talk meaningfully about the kinetic energy of a given molecule. But we can relate to the average kinetic energy of a molecule since this concept does not distinguish a particular molecule. There is considerable experimental and theoretical evidence to support the hypothesis that the average random kinetic energy of a molecule in a gas is directly proportional to the kelvin temperature of the gas. If the kelvin temperature doubles, the average kinetic energy of a molecule doubles. Summarized as an equation,*

$$E = \frac{3kT}{2}. \tag{7.1}$$

This random kinetic energy is called thermal energy. This concept now provides an atomic interpretation of the behavior of a gas. For example, the observed temperature increase when heat is added to a gas is interpreted as being due to an increase in the average kinetic energy (or average speed) of the molecules of the gas. Similarly, the energy that is derived from a gas to run a steam turbine comes from the random energy of the gas. If energy is extracted from the gas, the temperature decreases.

We have used a gas as the basis of our discussions because it is of particular interest at this time. However, liquids and solids, which are also

---

* The letter k in the formula for average random kinetic energy is called Boltzmann's constant. It is extremely small and has a numerical value of 0.00000000000000000000000138 (1.38×10^{-23}) joules/ (K • molecule).

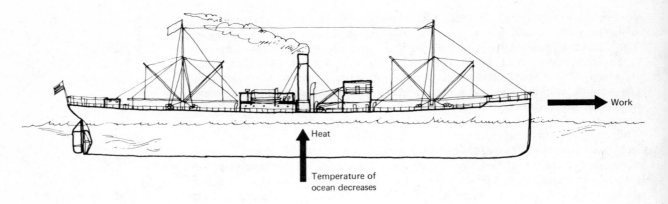

Work

Heat

Temperature of
ocean decreases

Figure 7.3 A ship that extracts heat energy from the ocean and converts it to work to move the ship. Energy taken from the ocean would lower the temperature of the ocean.

composed of atoms, or molecules as the case may be, will possess thermal energy. The higher the temperature of an object, the greater its thermal energy. Thus we might think that a high temperature is mandatory for a large quantity of energy. Yet if there is a large amount of some substance it can contain considerable energy even though its temperature is not particularly high. The oceans and the atmosphere are good examples. The thermal energy content of a gallon of water at the temperature of the environment (300 K) is roughly 1000 Btu. Therefore the thermal energy in 10 gallons of water at this temperature is roughly equal to the energy produced by burning a pound of coal. The oceans with billions and billions of gallons of water contain an enormous amount of thermal energy. Then why can't a heat engine be built to extract some of this energy and convert it to useful work, with the only environmental effect being a cooling of the source because energy has been extracted? (See Fig. 7.3.) The answer is that although energy is available, it is random. To do useful work, the random motion must be converted to organized motion. The conversion process requires a heat sink at a lower temperature so that energy can be transferred from a region of higher average kinetic energy to a region of lower average kinetic energy.

7.4 CONDUCTION AND CONVECTION

We now have the background to understand the heat transfer processes of conduction and convection at the atomic level. Let us start with conduction.

When some object is placed in contact with another object at a lower temperature, the hotter one cools down and the cooler one warms up until an equilibrium temperature is achieved. We say that *heat* has flowed (or been conducted) from the hotter to the cooler object. At the molecular level we would say that the average kinetic energy of the molecules in the hotter object has decreased and the average kinetic energy of the molecules in the cooler object has increased. At equilibrium, the average kinetic energy is the same (Fig. 7.4). The transfer of energy has taken place by molecular collisions in much the same way that energy is transferred by colliding billiard balls. Heat is appropriately defined, then, as energy transferred by molecular collisions. This is why we say that heat is energy in transit.

There is an enormous difference in the heat conducting properties of materials. With your bare hand you can hold a wooden-handled spoon in boiling water without eliciting any hot sensation. But you quickly remove your hand or the spoon if the spoon is made of metal. This is because heat is conducted from the water much more readily by metal than by wood. Metals tend to be very good heat conductors. The free electrons which make metals good electrical conductors are instrumental in heat conduction. Liquids and gases are generally poor heat and poor electrical conductors. Air conducts heat very poorly.

Heat conduction also depends on several factors not relating to the fundamental nature of the conductor. In wintertime, we don several layers of clothing to prevent bodily heat loss. Therefore, heat conduction depends on the thickness of the conductor—the thicker the conductor the smaller the heat loss. If you want to warm your cold hand by touching another object, you touch your entire hand rather than the tip of a single finger. The conduction of heat to your hand depends on the contact area between the conductor and source—the greater the area the greater the heat transfer. Finally, experience tells you that body heat loss in wintertime depends on the temperature of your environment. As the temperature difference between your body and its environment increases, the heat loss increases. We can relate all these factors to the rate of heat conduction with a single equation:

$$\text{rate of heat loss} = \frac{\left(\begin{array}{c}\text{heat conductive}\\\text{property of}\\\text{the material}\end{array}\right) \times \left(\begin{array}{c}\text{cross-sectional}\\\text{area of the}\\\text{conductor}\end{array}\right) \times \left(\begin{array}{c}\text{temperature}\\\text{difference}\end{array}\right)}{\text{thickness of the conductor}},$$

$$\frac{H}{t} = \frac{kA\,(T_{\text{warm}} - T_{\text{cool}})}{d}. \tag{7.2}$$

The heat conductive property symbolized by k is called the thermal conductivity. Thermal conductivity can be expressed in a variety of units. If heat flow is measured in calories per second, lengths in centimeters, and temperature in °C, then thermal conductivity is measured in calories per second per centimeter per degree Celsius. In short,

$$\left(\frac{\text{cal}}{\text{sec}}\right) \cdot \left(\frac{1}{\text{cm °C}}\right).$$

Table 7.1 lists some representative thermal conductivities. The wide variation in values represents the enormous difference in the heat conduction properties of materials. Note that air is the poorest heat conductor of all the materials listed. That is why in winter one wears loose-fitting coats and jackets that trap air spaces thereby diminishing conductive heat losses from the body. You might also wonder why the walls of a home are often filled with rock wool, which is not cheap, rather than with air which is a poorer heat conductor. The reason is that the rock wool forms a physical barrier which restricts convective heat losses.

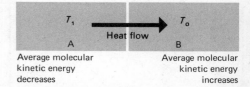

(a) Heat flows from region A to region B where the average molecular kinetic energy is lower. The energy transfer produces a decrease in the average molecular kinetic energy of the warmer region (A) and an increase in the average molecular kinetic energy of the cooler region (B).

Average molecular kinetic energy and temperature are the same

(b) An equilibrium temperature is established when the average molecular kinetic energy is the same in both regions.

Figure 7.4 Atomic interpretation of heat transfer. Initially, region A has a higher temperature than region B.

Table 7.1 Some selected thermal conductivities. A more complete list is found in *Handbook of Chemistry and Physics*, Chemical Rubber Publishing Co., 18901 Cranwood Parkway, Cleveland, Ohio 44128.

Material	Thermal conductivity $\left(\frac{\text{cal}}{\text{sec}}\right) \cdot \left(\frac{1}{\text{cm °C}}\right)$
Silver	0.97
Copper	0.92
Concrete	0.0022
Building brick	0.0011
Glass (window)	0.0025
Rock wool	0.00009
Wood (typical)	0.00009
Air	0.000057
White pine	0.00027
Oak	0.00035

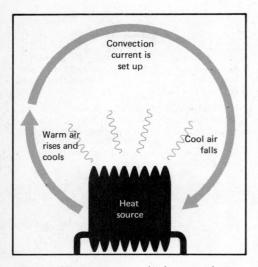

Figure 7.5 Manner in which a circulating air current is produced in a room.

Thermal conductivity values are of enormous practical value. A designer of a house needs to know how much heat will be conducted through the walls. Given the dimensions of the wall, the type of material, and the difference in temperature, the calculation is straightforward. Likely, you are not a house designer and such calculations are of little interest. As an energy conscious citizen, you should be interested in how different materials compare in heat conducting properties. For example, there is a significant difference between the amount of heat transferred through a single-pane glass window and through a four-inch thick rock wool insulation in the walls. For the same contact area and temperature difference, the heat loss through the glass is over 200 times that through the rock wool insulation. Heat loss through windows can be reduced by using two panes with an enclosed layer of air and by using storm windows. But even the best-designed window systems conduct heat about five times more rapidly than an insulated wall. We will say more about these heat losses in the final chapter on energy conservation.

It is also possible to transfer energy by literally moving material. If a mass of air with a given average molecular energy moves into a region with lower average molecular energy, then energy will be transferred. How might this be accomplished in practice? Suppose that a container of air, a room for example, is heated from the bottom (Fig. 7.5). The air in the vicinity of the source warms, the average kinetic energy increases, and the density of the air decreases. The less dense air rises in the room and is replaced by cooler (higher density) air coming down from the top. As a result, a current, or mass flow, of air is set up. This is the process of convection; it is the main mechanism by which heat is circulated in houses. Convection currents also give rise to wind. Because convection also occurs in liquids, it is partially responsible for ocean currents.

There is another heat transfer process called radiation (Chap. 8). It is the transfer of energy by electromagnetic waves. This mechanism is an important one but it is not primarily involved in any man-made system for disposing of heat from a steam turbine.

7.5 EVAPORATION

To conclude this discussion of heat-transfer processes, let us consider evaporation. Evaporation is literally the escape of molecules from a liquid. As an analogy, imagine bees in a box that has one end covered by a thin membrane that some energetic bees are able to penetrate (Fig. 7.6). Their motion is random, and there is a distribution of energies and speeds. All bees with sufficient energy can escape, but those energetic ones near the surface are more likely to escape. The same is true of molecules in a liquid. Surface tension, the result of the attraction of the molecules in the liquid for each other, tends to act like a membrane in keeping the molecules inside the container. Surface tension is sufficient to allow small water bugs to support themselves on the surface of water (Fig. 7.7). Only the more energetic molecules, especially those near the surface, are able to escape.

The origin of the molecular force that causes surface tension is understandable. Molecules experience attractive molecular forces from their neighbors. A molecule in the interior of a liquid (Fig. 7.8) is closely surrounded on all sides by similar molecules. Therefore, it experiences forces from all directions. However, a molecule at the surface has few or no neighbors beyond the surface. As a result, it is subject only to forces which tend to pull it back into the liquid.

If these forces did not exist, then every molecule would readily escape. Since only the most energetic molecules escape, the average kinetic energy of the liquid decreases as the temperature decreases. Evaporation is therefore a cooling process. Also, since evaporation is a surface phenomenon, it follows that the larger the surface area, the greater the cooling effect.

The reverse of evaporation is also possible. A molecule in the vapor above a liquid can be captured by the liquid. This process is called condensation; in any real situation, both evaporation and condensation are taking place simultaneously. If evaporation predominates, there will be a loss in the volume of the liquid and a cooling of it. The rate of evaporation depends strongly on the amount of vapor above the liquid. The smaller the amount of vapor, the greater the evaporation rate. The amount of vapor that can exist in the air before the vapor begins to condense is limited and this amount, as we would expect from everyday experience, depends on the temperature. When air at a given temperature contains the maximum amount of water vapor, it is said to be saturated. The word *humidity* refers to the amount of water vapor in the air. The ratio of the amount of vapor actually in the air to the amount that would saturate the air at that same temperature is called the *relative humidity*. Expressed as a percent,

$$\text{relative humidity} = \frac{\text{actual vapor content}}{\text{saturated content}} \times 100\%.$$

The evaporation rate depends on the relative humidity. Relative humidity can be measured by determining the difference in the dry- and wet-bulb tempera-

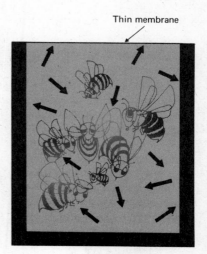

Figure 7.6 A box covered with a thin membrane and containing bees is analogous to a container of liquid.

Figure 7.7 Surface tension at the surface of water is sufficient to support small bugs. (Drawings by Lynne Coakley.)

◀ **Figure 7.8** Schematic illustration of forces on a molecule in the interior and on the surface of a liquid. The net force is zero on an interior molecule but there is a net force on a surface molecule that pulls it toward the interior.

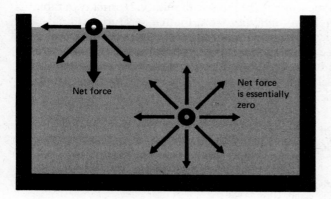

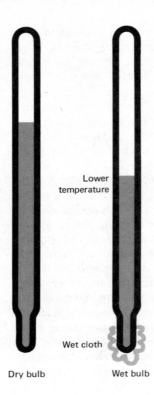

Figure 7.9 A wet- and a dry-bulb thermometer. The relative humidity is obtained from the difference in the readings of the two thermometers.

Lower temperature

Wet cloth

Dry bulb

Wet bulb

tures. The dry-bulb temperature is just the reading of an ordinary thermometer in the environment in question. If the bulb of a similar thermometer is surrounded by a cloth which is kept moist, it is referred to as a wet-bulb thermometer (Fig. 7.9). The wet-bulb thermometer will have a lower reading because the evaporation of water molecules will lower the temperature. The amount of lowering is a measure of the relative humidity. Normally, a table* is used to determine the relative humidity from the temperature readings. The wet-bulb temperature is the lower limit for cooling processes utilizing an evaporative process. The dry-bulb temperature is the lower limit for a cooling process based on a conduction process. This turns out to be quite important for two of the current schemes used to cool the water returning from the condenser of a steam turbine.

7.6 METHODS OF DISPOSING OF HEAT

The common methods of disposing of heat from the condensers of steam turbines are shown in Table 7.2. We will now proceed to discuss these methods.

---

* See, for example, *Handbook of Chemistry and Physics,* Chemical Rubber Publishing Co., 18901 Cranwood Parkway, Cleveland, Ohio 44128.

Table 7.2 Common methods of heat disposal

Method	Physical method	Comments
Once-through cooling	Evaporation, radiation, conduction	Utilizes a natural (lake, river) or artificial (cooling pond) watercourse
Wet-type cooling tower	Evaporation, convection	Recirculates water to condenser after evaporative process
Dry-type cooling tower	Conduction, convection	Heat dissipated to air through heat exchanger (radiator)

Once-through Cooling

In this process, cooling water is drawn from a source at one location, passed through the condenser, and returned to the source at a different location. Rivers, lakes, reservoirs, ponds, estuaries, and oceans are possible sources. The heated water is dispersed in a variety of ways. In a river, the natural flow is utilized. In all cases, the heat is dissipated from the body of the water to the atmosphere primarily by evaporation. Since evaporation is a surface phenomenon, the idea is to disperse the heated water to as large a surface area as possible. The longer the water stays in the condenser, the more heat it absorbs and the more its temperature increases. About 0.01 gal/sec is required for each kilowatt of power for a temperature rise of 15°F. A 1000-MW station would require a water flow of about 10,000 gal/sec. If the temperature increase were to be lowered to 5°F, the water flow would increase to 30,000 gal/sec. Water pumped out of the river must be replenished by the natural river flow at a rate at least equal to the rate of flow through the condenser. Few rivers can meet these requirements and, as a result, rivers are not being considered so frequently as sources for dissipating heat. The sheer magnitude of these flow rates is difficult to appreciate until we realize that such rates approach those which would supply the water needs of a city like Los Angeles.

Cooling ponds are artificial bodies of water used to dissipate heat by evaporation. They are sometimes used when a natural supply is not available and land is inexpensive. As we might expect, large areas are required. A 1000-MW plant would require a pond with a surface area of 1000 to 2000 acres (the playing area of a baseball field is about 3 acres). Most of the cooling ponds in this country are found in the the southwestern states.

Because certain areas of the United States have an abundance of large and deep lakes, they naturally come to mind as sources of cooling water. As far as predicting the thermal effects, they might seem at first to present an ideal situation. But unlike rivers for which the water flows are fairly well defined, lakes are subject to a variety of flow patterns and conditions which are strongly dependent on the weather. One feature, however, known as thermal stratifica-

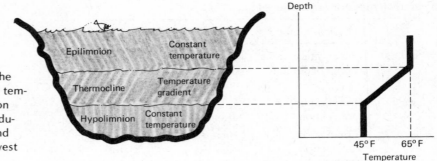

Figure 7.10 Schematic diagram of the thermal stratification of a lake. The temperature is constant in the epilimnion (upper region). The temperature gradually decreases in the thermocline and levels again in the hypolimnion (lowest region).

tion, makes lakes highly attractive sources. In winter, lakes tend to achieve a nearly uniform temperature throughout. However in the summer, when the upper portion is warmed, lakes tend to stratify into two layers of nearly constant temperature which are separated by a layer in which the temperature changes gradually (Fig. 7.10). The stratification results from another ex-

Figure 7.11 Density–temperature relationship for water. The density increases as the temperature decreases until a temperature of 4°C is attained. Thereafter the density decreases until the water freezes.

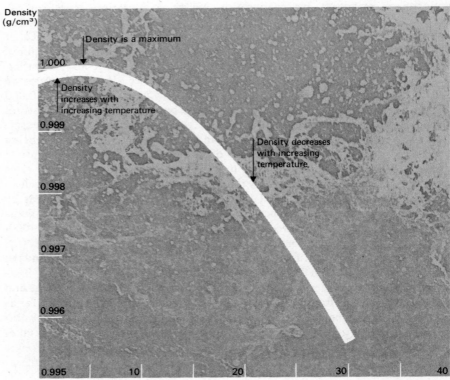

tremely interesting property of water, namely the dependence of its density on temperature.

Knowing that the size of an object changes with temperature, due to a contraction or expansion, it follows that density can vary with temperature. What may be surprising to learn is that, in the case of water, the density will either increase or decrease with an increase of temperature depending on the initial temperature (Fig. 7.11). The condition prerequisite to the floating of something in a fluid is that its density be less than the density of the fluid. A fluid having density less than water will float on the surface of the water. This is why an oil slick floats on water. Lakes in winter receive relatively little energy from the sun. As a result, the temperature tends to be low, say about 40°F in the central United States, and nearly constant throughout. In the summer months the lakes are warmed to a depth equal to the penetration of the sun's rays and their water is mixed by the wind. As a result of the warmed layers, which are of lower density, tending to stay on the surface, lakes stratify. From the standpoint of the cooling condenser of a steam turbine, it is desirable to extract water from the cool hypolimnion and disperse the heated water from the condenser over the surface of the warm epilimnion. Done properly, little or no increase in temperature of the surface water results. Although this is an ideal engineering feature, there is an undesirable environmental feature. Oxygen-consuming biological processes lower the dissolved oxygen in the hypolimnion. When this water, after passing through the condenser, is discharged to the epilimnion, the dissolved oxygen vital for plant and animal life is reduced. Caution must be exercised to see that this does not in fact happen.

The oceans constitute a nearly inexhaustible source of cooling water and surely they will be tapped. There are unique problems, such as corrosion, that result from the salt in the water, and these remain expensive propositions.

Cooling Towers

Wet- or dry-type cooling towers are normally used when a natural or artificial watercourse is unfeasible. The wet and dry nomenclature follows from whether water vapor is or is not released to the atmosphere in the cooling process. If a natural source with the proportion of a river or lake is required to achieve adequate cooling, we could well imagine that these tower structures are enormous. Typical cooling towers (Fig. 7.12) are some 400 feet high and as much as 400 feet in diameter at the base. Merely constructing these towers presents some challenging engineering and environmental problems.

The wet-type tower utilizes evaporative cooling (Fig. 7.13). The lowest cooling temperature is therefore the wet-bulb temperature (Fig. 7.9). As in any evaporative method, the idea is to disperse the water in such a way as to maximize the surface area. This is achieved by either breaking the water into droplets or forming it into thin films. The evaporated water molecules must be removed and this is achieved by either a natural or forced-air draft through the tower. The natural draft works by the ordinary chimney effect; the forced-air

Figure 7.12 Cooling towers for the Rancho Seco nuclear-electric generating station located near Sacramento, California. Note the relative size of the towers, the generating facility, and the automobiles. (Photograph courtesy of the Sacramento Municipal Utility District.)

draft functions by using a fan which either pulls the air through the top or forces it up from the bottom. About two percent of the water is lost through evaporation—a figure which is deceptively small. A 1000-MW facility will require evaporative losses of about 30 ft³/sec. This is equivalent to a daily rainfall of about one inch over an area of one square mile. Hence, severe fogging and icing could occur in the vicinity of the plant on a cold day.

The dry-type tower is, in principle, a huge radiator of the type found in most automobiles. The water to be cooled is pumped through a heat exchanger (Fig. 7.14) with a huge surface area that is in contact with air. Heat is transferred from the exchanger to the air by conduction and is carried away by either forced or natural convection currents.

The dry-type system seems ideal because it is closed and there is essentially no water loss. However, besides being the most expensive of the systems discussed, it produces an immense heat plume because the energy is transferred directly to the atmosphere. The heat plume is so pronounced that it could produce local alteration of the weather. Although systems of this type have never been used in the United States in conjunction with electric power plants, they have been used on a limited scale in Europe for some time. As the availability of cooling water declines, dry-type cooling towers are becoming increasingly attractive, and the federal government is funding investigative efforts.

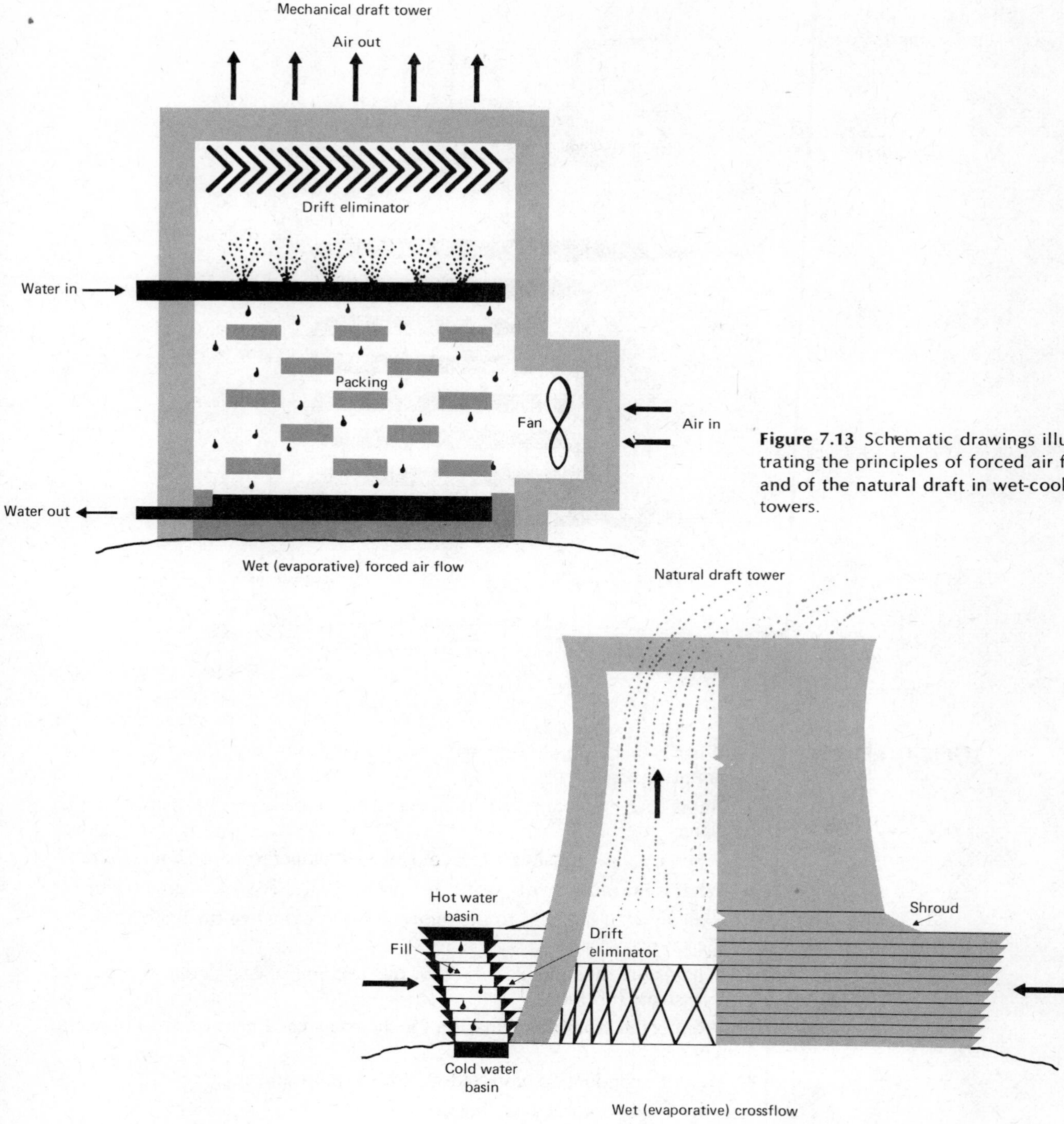

Figure 7.13 Schematic drawings illustrating the principles of forced air flow and of the natural draft in wet-cooling towers.

Figure 7.14 Schematic drawing illustrating the principle of the mechanical draft dry-cooling tower.

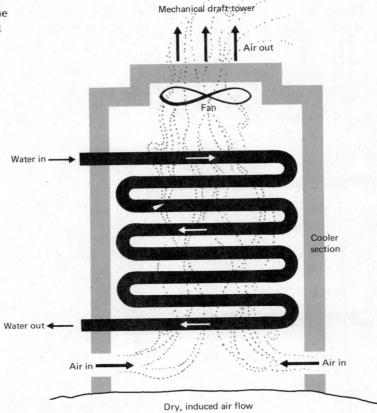

TOPICAL REVIEW

1. What is thermal pollution and how does it differ from the other forms of pollution discussed?
2. Name four physical effects that increased temperatures have on a body of water.
3. What are some possible effects of elevated temperatures on aquatic life?
4. What is the ultimate disposition of the heat rejected by a steam turbine?
5. Explain what happens to the molecules of a gas when the temperature of the gas decreases.
6. What is the fundamental origin of the mechanical energy that is produced by a steam turbine?
7. What is the atomic explanation for the transfer of heat from an object to a cooler object?
8. Describe conduction, convection, and evaporation.
9. Why is evaporation a cooling process?

10. Describe the concept of relative humidity.

11. Name three ways of disposing of heat to the atmosphere without first rejecting heat to a lake, river, or stream.

12. What is a cooling tower?

13. Distinguish between wet- and dry-type cooling towers.

14. Order once-through, wet-tower cooling, and dry-tower cooling according to their relative costs.

Atomic Description of Thermal Energy and Heat Transfer

Two texts that are useful for expansion of the material presented in Chapter 7 are

1. *Physics: A Descriptive Analysis,* by Albert J. Read, Addison-Wesley, Reading, Mass. (1970).

2. *Conceptual Physics,* by Paul G. Hewitt, Little, Brown, Boston (1977).

Disposal of Waste Heat

Very nice treatments of thermal pollution and the disposal of waste heat are presented in the following *Scientific American* articles:

1. "Thermal Pollution and Aquatic Life," by John R. Clark, *Scientific American* **220**, 19 (March 1969).

2. "The Calefaction of a River," by Daniel Merriman, *Scientific American* **222**, 42 (May 1970).

3. "Cooling Towers," by Riley D. Woodson, *Scientific American* **224**, 70 (May 1971).

7.2 Thermal Pollution

1. Conceivably, heat from the condenser of a steam turbine would be of more use if it were at a higher temperature. If the heat were extracted from the condenser at a higher temperature so that it would be more useful, what trade-off in efficiency would be made in the operation of the turbine?

7.3 Atomic Description of Internal Energy

2. Why does the size of a balloon inflated at room temperature change when it is immersed in a bath of ice water?

3. Formulate an argument in support of the following statement: "If there were a preferred direction of motion for the molecules of a gas in a container, then the container would experience a net force and could possibly move."

4. Work must be done on a car to increase its kinetic energy (or speed). How does this principle apply when work is done on a gas in a cylinder by compressing it with a piston?

5. The flow of heat from one object to another is similar to the flow of water from the top to the bottom of a waterfall. What are the energy aspects of both of these processes?

6. Use an energy argument to explain why heat will not flow from an object to another object at a higher temperature.

7. The average random kinetic energy of a molecule is directly proportional to the Kelvin temperature ($E = \frac{3}{2} kT$). Why is it that even if the temperature outside a house is fairly low, say $-20\,°C$, there is still an enormous amount of thermal energy in the ground and in the surrounding atmosphere?

8. Following an athletic event one often sees people milling around the area. Yet at an exit gate there is an orderly flow of people. How is this analogous to the conversion of random molecular motion into ordered motion in a steam turbine?

7.4 Conduction, Convection, and Radiation

9. Why will a pan of water on an electric stove heat faster if it is in firm contact with the heating coil rather than being as little as one sixteenth of an inch above the heating coil?

10. The conductive flow of heat is akin to the flow of charge in an electrical conductor. Some materials are better heat conductors than others. From your experience of touching the handles of spoons that are immersed in a hot liquid like coffee, judge which is a better heat conductor—metal or plastic.

11. The cross-sectional area and length of a wire affect the resistance to the flow of charge. How do the dimensions of a heat conductor affect the flow of heat? Where do you see this idea put into practice in a home?

12. Aluminum being a metal is a significantly better heat conductor than wood. Yet aluminum-frame storm windows are extremely popular. What, then, is the attraction to the aluminum storm windows?

13. From the standpoint of heat conduction, what is the advantage of wearing loose-fitting clothes in wintertime?

14. How does the heat flow compare through a brick wall and a concrete wall of the same thickness and cross-sectional area when the temperature difference across the two walls is the same?

15. Some windows are constructed of two thicknesses of glass with an air space in between. From a heat-loss standpoint, why is this arrangement better than a solid glass window of the same total thickness as the two layers and the air space?

16. Why is it common practice to wrap the metallic hot water pipes in a building with a glass wool material similar to the rock wool mentioned in Table 7.1?

17. What is the main heat transfer mechanism in the draft created in a fireplace?

18. When you warm yourself by standing near a fireplace, what is the main heat transfer mechanism? Why is it that, normally, most of the heat generated in a fireplace actually escapes up the chimney?

19. From the standpoint of energy conservation, why is it important to keep a chimney blocked off when an open fireplace is not being used?

7.5 Evaporation

20. Water often collects in puddles on a tennis court following a rain. To enhance the drying, the water is spread out over the surface of the court. Why does this accelerate the drying?

21. Sweating is a cooling process for the human body. What is the physical mechanism that is responsible for this cooling effect?

22. A simple type of household humidifier continually circulates a porous belt in and out of a bath of water. A fan then forces air through the belt when it is removed from the water. What is the purpose of the fan?

23. Why does an area of skin touched with perfume or any other volatile (easily vaporized) liquid feel cool?

24. Dehumidifiers are commonly used to reduce the water vapor content in rooms of a house. Manufacturers of these devices will state that dehumidifiers do not provide a cooling of the surroundings. Why is this so? However if you go into a room in which the humidity has been lowered, you feel cooler. Why?

25. Why are small bugs able to move around on the surface of a pond without sinking?

7.6 Methods of Disposing of Heat

26. What is the meaning of a relative humidity reading of 50 percent?

27. Describe the cooling system on an automobile and compare it with that used in a dry-type cooling tower.

7.3 Atomic Description of Internal Energy

NUMERICAL PROBLEMS

1. At some particular Kelvin temperature T a system of N molecules has a certain amount of random kinetic energy. How many molecules would be required in a system at one half the Kelvin temperature to have the same amount of random kinetic energy?

2. Ten trillion molecules (10^{13} molecules) from a container at 300 K are transferred to another container having 5 trillion molecules (5×10^{12} molecules) at 200 K. What is the equilibrium temperature in the combined system?

3. The average random kinetic energy of a molecule is given by $E = \frac{1}{2} kT$ where $k = 0.0000000000000000000000138$ joules/K • molecule and T is the Kelvin temperature.

   a) Calculate the average random kinetic energy of a molecule at 300 K (room temperature).

   b) Burning a pound of coal produces about 13,000 Btu of heat. How many molecules at 300 K are required to produce a total amount of average random kinetic energy equivalent to burning a pound of coal? (1 Btu = 1055 joules.)

4. a) A gram of water contains about 33 billion trillion (33 followed by twenty-one zeroes) molecules. How many joules of thermal energy are there in a gram of water at 300 K?

   b) There are 1055 joules in one Btu. How many Btu of thermal energy are there in a gram of water at 300 K?

   c) There are 454 grams in a pint of water. How many Btu of thermal energy are there in a pint of water?

   d) There are about a billion trillion (one followed by twenty-one zeroes) pints of water in the world's oceans. How many Btu of thermal energy are there in the world's oceans, assuming that the temperature is 300 K?

   e) The thermal energy content of a pound of coal is about 13,100 Btu. How much coal is required to match the thermal energy content of the oceans?

7.4 Conduction, Convection, and Radiation

5. A glass window is installed in a brick wall that is 4 inches thick. How thick would the window have to be in order that the heat loss through the window and an equal area of the brick be the same?

6. The inside of a house is maintained at 20°C when the outdoor temperature is 0°C. What is the rate of heat loss through a 0.5 cm thick pane of glass 60 cm wide and 120 cm long?

7.6 Methods of Disposing of Heat

7. A 1000-megawatt electric power plant using a wet-tower cooling method will disperse about 30 cubic feet of water to the atmosphere during each second of operation. Show that if this is uniformly distributed over a square mile of land it corresponds to a daily rainfall of about one inch?

8. In a day's time the amount of water that flows through the cooling condensers of a large electric power plant is comparable to the amount of

water handled by the entire water system of a large city. This may seem incredible but it is easy to figure out. Water for all uses amounts to about 150 gallons per day per person in a city. How much water would be used in a day in a city having five million people? Water flows through the cooling condensers of a large power plant at a rate of about 30,000 gallons per second. How much water flows through the condenser in a day? Now compare the two calculations.

9. If a steam turbine is cooled with water from a cooling pond, about an acre of pond is required for each megawatt of power produced. Thus the pond area required for a 1000-megawatt plant is about 1000 acres. To get some feel for the size of 1000 acres, determine how many football fields are needed to make 1000 acres. A football field has an area of about ¾ acre.

10. A cooling tower for a large electric power plant is about 400 feet in diameter. It is difficult to appreciate just how big these cooling units are. Compare the area occupied by a tower 400 feet in diameter with the area of a football field. The area of a circle of diameter d is $A = 0.785\ d^2$. A football field is 100 feet wide and 300 feet long.

11. One Btu of heat added to a pint of water will raise its temperature 1°F. The total world energy use in 1973 was about 230 trillion Btu. If all this energy were dumped into the world's oceans, how much would the temperature increase? Assume 10^{21} (one followed by twenty-one zeroes) pints of water in the world's oceans.

ATMOSPHERIC PROBLEMS

8

A dense midday haze shrouds this metropolitan area. Brought on by massive use of private automobiles, these hazy conditions are often confounded by weather conditions that prevent pollutant dispersal. Photograph by Gene Daniels. (Courtesy of the EPA.)

8.1 MOTIVATION

We have discussed the origin and the control methods for pollutants generated from energy conversion processes and the environmental standards that have been established to ensure the protection of public health and welfare. To meet the air quality standards we rely heavily on flushing of the air by natural air movement and by precipitation. A serious situation can result if nature rebels and does not provide the atmospheric circulation that is necessary to disperse the pollutants. The possibility of both local and worldwide climatic changes stemming from the accumulation of particles and gases in the atmosphere is an added problem. We are now in a position to develop an appreciation for the magnitude of these problems and to understand some of the basic mechanisms.

One of the most serious recorded incidences of air pollution occurred in 1948 in the small industrial city of Donora, Pennsylvania. While pollution problems in large industrial cities are common, why should Donora, hardly even of city status with a population of about 14,000 in 1948, suffer this fate? The reason is that a weather condition prevailed that essentially put a lid over Donora and prevented the dispersal of pollutants spewed out from energy conversion processes in the city's industries. Many cities are harassed by these atmospheric lids that contribute to stagnant air conditions. This situation prevails in Los Angeles for about 100 days of the year. Pollution problems from rapidly expanding industrialization in Denver, Colorado, are intensified by the existing atmospheric conditions. To a lesser but still significant extent the twenty-five easternmost states are victimized by intermittent stable weather conditions that occur for a variety of reasons. While many of the mechanisms that cause them are very complicated, some of the common ones are not difficult to understand. Because of their important role in the dispersal of by-products from energy production, we shall examine the common ones. Some good references on these problems are given at the end of the chapter should you desire further insights.

8.2 TEMPERATURE INVERSIONS

Anyone who has hiked up a mountain or ridden in a commercial jet airplane knows that the air temperature is lower at higher altitudes than it is on the ground. The variation of temperature with altitude is measured by a quantity called the lapse rate, expressed as the change in degrees Celsius per kilometer (°C/km) or as the change in degrees Fahrenheit per mile (°F/mi). If the temperature decreases 4°C for each kilometer increase in altitude, we would say the lapse rate is −4°C/km. The negative sign reminds us that the temperature decreases as altitude increases. Generally, the temperature decreases about 7°C for each kilometer increase in altitude up to a distance of about 13 kilometers (about 8 miles). This region is called the troposphere and contains about 80 percent of the entire mass of the atmosphere.

A condition whereby temperature decreases as altitude increases is necessary for a mass of warm air to rise.* This is no different than the conditions required for warm air to rise from the floor in a room. The dispersal of pollu-

---

* In the next section we will see that there are conditions under which warm air will not rise even though the temperature decreases with increasing altitude.

tants from a smokestack relies on rising warm air to carry the particles into the upper atmosphere. Now suppose that the temperature increases with altitude. This is called a temperature inversion because the condition is inverted from normal. The air from the smokestack no longer rises (Fig. 8.1). Pollutants become more concentrated, and serious effects can occur. Conditions inclining toward development of temperature inversions are further aggravated if fog is present as often happens in London. The word smog, meaning smoke and fog, was coined for these conditions.

Two common types of temperature inversions are prefixed radiation and subsidence. Radiation inversions develop in the following way. During sunny daylight hours, the temperature will decrease with altitude up to several thousand feet (Fig. 8.2a). At night, the ground and air in the first one or two thousand feet cool thereby producing an inversion in the temperature distribution (Fig. 8.2b). Pollutants released during the night will not rise and will tend to collect in this low "inverted" layer. Normally, when the sun comes up and the ground warms, the situation goes back to normal and the pollutants disperse. But if there is a cloud cover and no wind and the ground does not warm up, the pollutants remain and the situation worsens. A radiation inversion of this type prevailed at the time the photograph in Fig. 8.1 was taken. Temperature inversions of this type are common on cold, clear, winter mornings and they can be identified by observing smoke emanating from a chimney or smokestack.

Figure 8.1 An illustration of how smoke fails to rise because of a temperature inversion in the atmosphere. This type of inversion is often seen in the morning on a clear, cold day. After the sun is up for awhile, the inversion disappears and the smoke can be seen to rise.

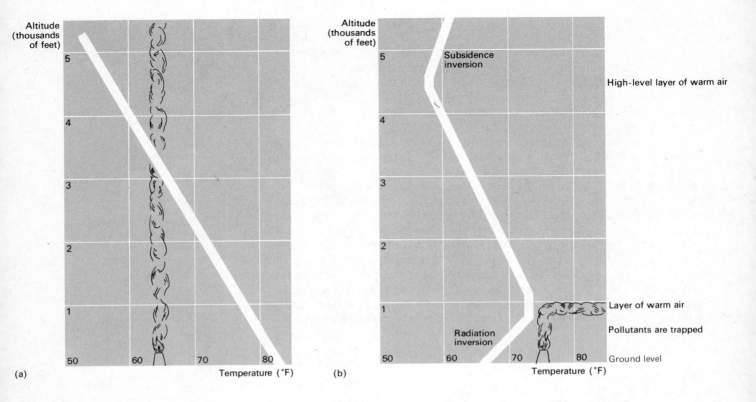

Altitude (thousands of feet)

50   60   70   80

Temperature (°F)

(a)

Altitude (thousands of feet)

Subsidence inversion

High-level layer of warm air

Layer of warm air

Pollutants are trapped

Radiation inversion

Ground level

50   60   70   80

Temperature (°F)

(b)

Figure 8.2 (a) Usually the temperature of the atmosphere decreases as the altitude increases, as shown in the left graph. These conditions allow warm air to rise and to be dispersed by air circulation. (b) The graph to the right illustrates a radiation inversion at ground level and a subsidence inversion starting at about 4500 feet. Both types of temperature inversion can prevent warm air from rising.

The second common type of temperature inversion results from the phenomenon of *subsidence.** Air pressure, as well as temperature, decreases with increasing altitude. Under certain conditions, cool, high-level air will sink to a higher pressure level. This happens, for example, when a high-altitude wind loses speed. Once in the high-pressure area, the air is compressed and it warms. (Remember, this principle is used in a diesel engine, Sec. 4.10.) As a result, the air temperature is higher than normal in a local area and a temperature inversion results. Inversions of this type tend to form at altitudes of 1000–10,000 ft. Subsidence inversions are common on the West Coast and, because of the enclosing mountainous terrain, produce the stagnant air conditions around Los Angeles. Such inversions are not as common in other parts of the United States, but they do happen. The Donora incident resulted from an inversion of this type and conditions there were aggravated by the surrounding mountains. Weather forecasters are often able to sense the onset of conditions which produce a subsidence inversion and are able to forewarn the public. Figure 8.3 shows how photochemical oxidant levels rose in Cincinnati, Ohio during a temperature inversion and a stagnant air condition that lingered on for nearly two

* From the word *subside,* meaning to sink to a lower level.

180   **Atmospheric Problems**

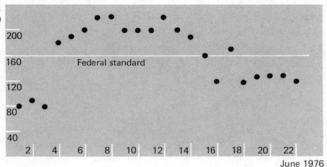

Photochemical oxidants
(micrograms/cubic meter)

Federal standard

June 1976

Figure 8.3 This plot is a record of the concentration of photochemical oxidants in Cincinnati, Ohio, during an extended stagnant air condition. The Federal standard for photochemical oxidants was exceeded for eleven consecutive days. Rain and wind eventually flushed the oxidants from the area. Conditions like these are common in the major cities east of the Mississippi River.

weeks. These conditions plague most major cities east of the Mississippi River during the summer months. Because most of the photochemical oxidants are a consequence of private automobiles, citizens are urged not to drive in the city until the weather changes and the pollutants are flushed.

A stable atmospheric condition that will prevent pollutant dispersal can occur without a temperature inversion. This happens when the prevailing temperature decrease with increasing altitude (the lapse rate) is less than a certain reference lapse rate. For example, if the reference lapse rate is $-4°C/km$ and the actual lapse rate in the atmosphere is $-2°C/km$, then conditions exist in which warm air will not rise even though there is no temperature inversion.

All of you have witnessed the behavior of a child's balloon filled with a gas "lighter than air." Sometimes it rises. Other times it sinks or perhaps remains at some position until a gust of air moves it. If it moves, then there must be a net force in the direction it is moving. If it is stationary, then the net force is zero. Since gravity is always pulling downward on the balloon then for the balloon to rise there must be a larger upward force. This upward force, called a buoyant force, is a consequence of the variation of air pressure in the atmosphere. Because the air pressure gets smaller as one moves up in the atmosphere the pressure is greater at the bottom of the balloon than on the top. Hence the force on the balloon is larger at the bottom than at the top thereby producing an upward force. If the buoyant force is larger than gravity the balloon rises; if it is not, the balloon sinks. Archimedes first deduced that the buoyant force is equal to the weight of the air displaced by the object. For example, if the object has a volume of 100 cubic centimeters then the upward buoyant force is equal to the weight of 100 cubic centimeters of air. So if the weight of a balloon is less than the weight of the air that it displaces, the balloon will rise. The volume of the displaced air is, of course, the same as the volume of the balloon. Hence, the balloon's weight will be less than the air that it displaces if its average density is less than the density of the air. Thus one can deduce whether a balloon will rise or fall by comparing its density with that of the air that it is placed in.

8.3 A STABLE ATMOSPHERIC CONDITION WITHOUT A TEMPERATURE INVERSION

Droplets of oil may be suspended in a mixture of isopropyl alcohol and water. The downward force of gravity is balanced by the upward buoyant force of the surrounding fluid, and the droplets remain motionless.

Figure 8.4 The white line represents a prevailing condition whereby the temperature decreases as one moves up into the atmosphere. The black line shows how the temperature of a gas might decrease with increasing altitude when the gas is released in the atmosphere. Initially, the gas is at a higher temperature than the air. However, it cools faster than the air as it rises. When the temperature of the gas and air equalize, a stable condition results. Forces on the gas always tend to restore it to the stable position.

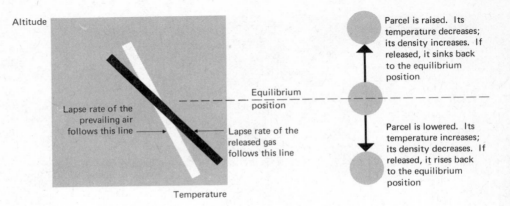

Altitude

Lapse rate of the prevailing air follows this line →

← Lapse rate of the released gas follows this line

Equilibrium position

Temperature

Parcel is raised. Its temperature decreases; its density increases. If released, it sinks back to the equilibrium position

Parcel is lowered. Its temperature increases; its density decreases. If released, it rises back to the equilibrium position

We can examine the behavior of a gas introduced into the air just as we examined the behavior of the balloon in air. We pick a certain volume of the gas released and compare its density with the density of the surrounding air. Three properties of a gas need to be recognized to understand its behavior in the atmosphere. The first is that a gas cools as it expands. You can observe this by letting gas escape rapidly from an inflated balloon. The second property is the change of density with temperature. Like most liquids, the density of a gas increases as its temperature decreases. This is why cool, dense air at the ceiling of a room falls to the warmer floor thereby creating convection currents. Finally, we must remember that a gas is a very poor conductor of heat (Sec. 7.4). When our parcel of gas is released into the air it essentially does not exchange heat with the surrounding air. It behaves as if it is confined within the boundaries of a balloon. As it rises and expands into the lower pressure surroundings, it cools at its own internal rate. Suppose that when the gas (for example, smoke) is released its temperature is higher than the surrounding air. Its density is less so it starts to rise, cooling as it ascends. If it cools faster than the surrounding air, its temperature eventually falls to the temperature of the surroundings, the densities become equal, and the gas (smoke) stops rising (Fig. 8.4). A stagnant condition then prevails.

8.4 ATMOSPHERIC EFFECTS DUE TO PARTICULATES

The natural beauty of a red sunset or a clear, blue sky is a treat for all to enjoy. And although the astronauts who explored the moon enjoyed many exciting views that we on earth cannot, they saw no red sunsets or blue skies. These effects are produced by the scattering of light by atoms and molecules in the atmosphere and there is no such atmosphere on the moon. White light is a mixture of the colors of the rainbow. The amount of light scattered depends very strongly on the color. Blue light is scattered by small particles much more than red light is. Sunsets are red because the "blues" are scattered out of the line of sight of the viewer by atoms and molecules. The sky is blue because the blues are scattered toward the earth by the atmosphere.

Particulates in the atmosphere will also scatter and absorb radiation. This is of concern because of the amount of particulates that are accumulating. The effects are often seen as dense hazes in areas of high particulate concentrations. The hazes limit the maximum distance from which objects can be distinguished from their background. This maximum distance is called the visual range.

A visual range of about 20 miles is needed for safe, optimum control of aircraft at a municipal airport. However, in an atmosphere containing a particulate concentration of 100 micrograms per cubic meter of air, which is not unusually high, the visual range is reduced to about eight miles. Visual ranges of this order begin to curtail air-traffic control.

The fact that visibility is reduced by scattering and absorption of radiation means also that energy associated with the radiation is diminished. The notion of turbidity* is introduced as a quantitative indicator of this energy loss. It accounts for the type of radiation and the size and concentration of the scattering and absorbing particles; it is a number which varies from zero to infinity. Zero means that no radiation is removed; infinity means that all

Figure 8.5 Two New York City scenes recorded one day apart. The thick smog is the result of heavy smoke concentrated by a stagnant air condition. (Photograph courtesy of the *New York Daily News*.)

---

* The word *turbidity* comes from *turbid* which implies cloudy, smoky, or hazy conditions.

radiation is removed. Solar radiation encounters a "thickness" of air (the atmosphere) in its path toward the earth. There will be a complex distribution of particulates mixed with the air in the atmosphere that will vary for different positions on the earth. So the turbidity and energy loss will also vary with position on the earth. There is, of course, some energy loss even if there are no particulates; particulates only enhance this loss.

Particulate concentrations have become high enough in some areas that sunlight has been significantly reduced at the earth's surface. Analyses of the radiation received in these areas have led to the following conclusion: For concentrations varying from $100 \mu g/m^3$ to $150 \mu g/m^3$, where large smoke turbidity factors persist, in middle and high latitudes direct sunlight is reduced up to *one-third* in summer and *two-thirds* in winter. Furthermore, it is estimated that the total sunlight is reduced five percent for every doubling of the particulate concentration.

An astronaut orbiting the earth outside the earth's atmosphere would be in a position to actually measure the amount of radiant solar energy proceeding to the earth. The astronaut would also observe that some of the radiation is returned back into space as a result of reflection and scattering in the atmosphere and on the earth. The energy reflected back into space divided by the incident energy is called the albedo. Measurements of the albedo average around 0.4, meaning that about 40 percent of the solar radiation returns to space. Particulates contribute to the albedo both through reflection and absorption. If particulate concentrations became significant and if reflection predominated over absorption, then the earth would be deprived of solar energy and the temperature of the earth would decrease. If absorption predominated, the earth would warm. There is no real evidence that either effect is occurring. However, the decrease in worldwide air temperature since 1940 may be due to reflection of solar radiation by particulates accumulating in the atmosphere.

8.5 CONCERNS OVER THE "GREENHOUSE" EFFECT

Carbon dioxide (CO_2) is not toxic and does not harm plants or property. Superficially it would seem that it should be of no concern even though a 1000-MW coal-burning electric power plant at full capacity dumps about 29,000 tons of carbon dioxide into the atmosphere each day. Yet the long-term effects of carbon dioxide accumulation in the atmosphere may be disastrous. The concern is for a warming of the earth by what is popularly called the "greenhouse" effect. It has been speculated that the earth could warm to the extent that the polar ice caps would melt and the resulting water would submerge the coastal cities.

As we proceed to understand the "greenhouse" effect and the future energy sources discussed in the remaining chapters, it will be necessary to grasp a deeper understanding of radiation. When finished, you may well conclude that the description is not very fundamental for we will find that solar radiation, for example, has a dual, or "split personality," behavior. Sometimes radiation behaves like particles similar to bullets fired from a gun. Other times it is wavelike akin to waves on the ocean or a lake. Although seemingly contradictory, these conclusions are well-founded both experimentally and theoretically. We will find the ideas to be essential not only for understanding the "greenhouse" effect but also for the study of many aspects of future energy sources.

When seeking a model for radiation, bear in mind that you never see the radiation as you see waves on the ocean. True, you see visible light radiation because of the nature of our eyes. But you cannot see infrared heat radiation nor can you see suntanning ultraviolet radiation from the sun. You see only the effects produced by the radiation when it interacts with matter or other radiation. From these observations one tries to present some physical explanation or model for the radiation. First, let us examine its wave aspects. A wave is described as "a disturbance or oscillation propagated with a definite velocity from point to point in a medium." A sharp clap of the hands produces a sound disturbance (wave) that propagates through air. A rock dropped onto the surface of a still body of water like a pond produces a disturbance that moves visibly across the surface of the water. If a float in the water is bobbed up and down continuously, then a continuous train of disturbances having peaks and valleys moves across the surface. Figure 8.6 is representative of a "slice" in the direction the disturbance is moving. The distance from crest to crest is called the wavelength. The number of crests that pass some point in the path of the wave each second is called the frequency. The speed of the wave is determined by following the motion of a crest. We find that speed (v) is related to the frequency (f) and wavelength (λ, Greek lambda):

speed = frequency × wavelength,

$$v = f\lambda. \tag{8.1}$$

Frequency is expressed in cycles (or oscillations) per second. A cycle per second is called a hertz, abbreviated Hz.

Figure 8.6 "Snapshot" of a slice of a water wave along the direction of propagation. A complete cycle corresponds to taking the float up to its maximum height above the equilibrium position, back down to an equal distance below the equilibrium position, and then back to the equilibrium position.

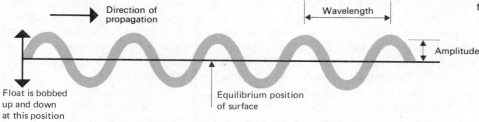

Waves can be observed to do several things. A water wave will bounce (reflect) off a shore. Sound echos are produced by sound waves bouncing off a building or a mountainside. Water and sound waves will progress (diffract) around a barrier placed in their paths. Waves will change their direction of propagation (refract) when they enter a medium of different density. One has

A stone dropped into a still pond creates a circular water wave that propagates outward from the initial disturbance.

only to get in the path of a water wave rushing to the shore to know that a wave possesses energy. All of these effects are well understood and can be described mathematically. Light radiation will also reflect, refract, and diffract precisely like water waves which we can actually see. When we read a book, we see light that is reflected from the pages. We see a ray of light refract (or bend) when it passes from air into water. We see light diffract around the edges of a very narrow opening placed in its path. If we get in the path of the sun's rays, we absorb energy and get a suntan. The speed of the radiation can be measured and a wavelength deduced from diffraction experiments. These measurements are consistent with the relation velocity equals frequency times wavelength, which holds for observable waves. For these reasons we say that radiation is a wave phenomenon and use a wave model to describe it in these situations. However, some radiation will release electrons when it impinges on the surface of many metals. This is called the photoelectric effect. (The photoelectric effect is utilized in some warning systems that are activated by crossing a beam of light.) The photoelectric effect cannot be explained by assuming a wave model. Quantum physics gives us a particle explanation.

Quantum physics assumes that the energy associated with radiation is comprised of fundamental energy units, or quanta, called photons. A photon behaves like a particle and has energy proportional to the frequency of the wave:

$$\text{photon energy} \quad \text{is proportional to} \quad \text{frequency,}$$
$$E \qquad\qquad \propto \qquad\qquad f.$$

$$E = hf, \tag{8.2}$$

where h is a fundamental number in nature and is called Planck's constant. Numerically Planck's constant is equal to 6.63×10^{-34} joule $\cdot$ seconds. As a decimal, $h = 0.000000000000000000000000000000000663$ joule · seconds. It is an extremely small number.

The wave aspects are apparent in the propagation of electromagnetic energy and when "waves" interact with "waves." The particle nature emerges when electromagnetic waves interact with matter. Both wave and particle aspects never occur in the same experiment.

There still remain the questions of "What is the disturbance associated with the waves?" and "How are the photons generated?" A water or sound wave possesses ordinary mechanical energy; that is, energy associated with particles in motion. Through actual collision, the waves can transmit energy. Radiation waves possess electric and magnetic energy by virtue of associated electric and magnetic fields which oscillate with time (Fig. 8.7). The magnitudes of the electric and magnetic fields at a particular position change with time much like the height of a water wave changes with time at a particular position. We call such a wave an electromagnetic wave. The electromagnetic wave transfers energy through the interaction of its associated electric and magnetic fields with the electric and magnetic fields that it encounters. For example, if a charge gets in the path of an electromagnetic wave, the electric field of the charge interacts with the wave, and the charge experiences a force which sets it into motion.

Photon emission is associated with the loss of energy that accompanies the deceleration of an electric charge. For example, the electromagnetic radiation from the antenna of a TV station is induced by accelerating electrons in the metallic conductors on the antenna. A free electron that is accelerated by entering the electromagnetic field of an atom will radiate photons. The radiation from an incandescent light bulb is due to random accelerations of electrons in the heated filament. As a result of this activity a broad, continuous spectrum of wavelengths is emitted. Photons are also emitted when atoms, molecules, and nuclei make transitions from higher energy states to lower energy states. This is analogous to an energy transformation that evolves when a ball makes a transition from the top of an inclined plane (higher energy state)

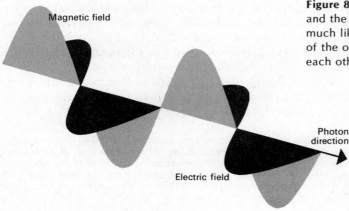

Magnetic field

Electric field

Photon direction

Figure 8.7 The oscillations of both the electric field (shown black) and the magnetic field (shown gray) are represented by a picture much like that for an ordinary water wave (Fig. 8.6). The directions of the oscillations of the two fields are always at right angles to each other.

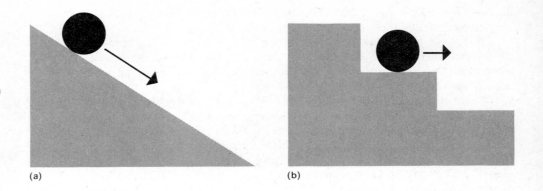

Figure 8.8 (a) A ball rolling down an inclined plane continuously loses potential energy and gains kinetic energy. (b) A ball rolling down a staircase loses potential energy in "jumps" each time it rolls over the edge of a step.

(a)   (b)

to the bottom (lower energy state). (See Fig. 8.8a.) But unlike the inclined plane example, the changes in energy occur discontinuously, more like those that occur with a ball rolling down a staircase (Fig. 8.8b). The radiation from a neon sign is not a continuous distribution of wavelengths (or colors) like that from an incandescent light bulb because the radiation is due to transitions between atomic states in neon atoms. Such a discontinuous change in energy is accompanied by the emission of a photon with energy equal to the difference in energy of the initial and final state.* The energy given up by the system is transformed into electromagnetic energy carried by the photon. The frequency of the radiation is given by the difference in energy divided by Planck's constant:

$$f = \frac{E_2 - E_1}{h}. \tag{8.3}$$

Thus, the greater the energy given up in the transition, the higher the frequency of the radiation. The range of energies covered by these radiations as well as the range of their associated frequencies (or wavelengths) is absolutely staggering (Fig. 8.9). Interestingly, the speed of all these waves in vacuum is 300 million meters per second *regardless of the frequency*. This speed is also called the speed of light. Infrared, ultraviolet, and visible radiation are all categories of electromagnetic radiation. They are distinguished by differences in wavelengths. X-rays and gamma rays, which we shall study in the next chapter, are also categories of electromagnetic radiation.

Several experiments verify the photon concept of electromagnetic radiation. One is the photoelectric effect mentioned earlier. Another is the study of thermal radiation from a heated object.

---

* When the atom emits a photon it is analogous to a person throwing a ball while standing on a skateboard. A reaction to the force that makes the ball move one direction causes the person and the skateboard to recoil in the opposite direction. The person and the skateboard acquire kinetic energy as a result of the recoil. When a photon is emitted by an atom, the atom recoils and acquires kinetic energy. We have neglected this recoil energy because it is generally small compared to the energy of the photon.

Figure 8.9 A portion of the spectrum of wavelengths of electromagnetic waves. An agreed upon range of wavelengths within the spectrum characterizes a certain class of radiation. For example, electromagnetic waves having wavelengths between 0.0000004 meters and 0.0000007 meters are classed as visible radiation. The boundaries between different types are not sharply defined.

Wavelength (meters)	Approximate wavelength range corresponding to a category of electromagnetic radiation
100,000	
10,000	
1,000	AM radio
100	
10	
1	FM, Shortwave
0.1	
0.01	
0.001	Microwave
0.0001	
0.00001	Infrared
0.000007	
0.000004	Visible
0.000001	
0.0000001	
0.00000001	Ultraviolet
0.000000001	
0.0000000001	X-ray
0.00000000001	
0.000000000001	Gamma
0.0000000000001	

Any object at a temperature above zero kelvins will emit electromagnetic radiation. This radiation is called thermal radiation. The amount and type of radiation depends critically on the temperature of the object. This is readily observed with the heating coils of an electric stove. In a dark room the coils are invisible if no electric current exists in them. They are invisible because the eyes are insensitive to the infrared radiation emitted by the coils. When charge flows, the coils warm and emit visible red radiation. The amount of radiation increases noticeably. To study the fundamental properties of thermal radiation, experimenters use a source called a cavity radiator. A cavity radiator is simply a hollow container of arbitrary shape with a tiny hole in the container wall (Fig. 8.10). The interior and walls are maintained at the same temperature and one measures only the radiation passing through the hole. The rate of energy flow (the power) associated with each wavelength (or color if visible) is measured. Figure 8.11 shows results of such a measurement. Carefully note that all wavelengths (colors) are not emitted with the same intensity. The position of the peak along the horizontal axis denotes the dominant type of radiation emitted. For example. the peak position in Fig. 8.11 is at 1.6 micrometers. This means that the dominant radiation for a temperature of 2000 K is infrared. If the measurement is repeated at different temperatures, similar results are obtained, but the peak in the curve shifts very systematically (Fig. 8.12). A study of results like those in Fig. 8.12 shows that if you multiply the wavelength corresponding to the peak position with the temperature expressed in kelvins, you always get the same answer. The result is

$$\lambda_m T = 2900 \text{ micrometer} \cdot \text{kelvins.} \tag{8.4}$$

8.7 THERMAL RADIATION

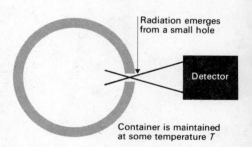

Figure 8.10 The cavity radiator shown here is spherical. However, the shape is unimportant. The only requirement is that it contain a tiny hole for the radiation to pass through.

Figure 8.11 As shown in this graph, the amount of radiant power from a cavity radiator associated with each wavelength (or color) depends on what wavelength is being considered. For a temperature of 2000 K, most of the power is associated with a wavelength of 1.6 micrometers, which is of the infrared type.

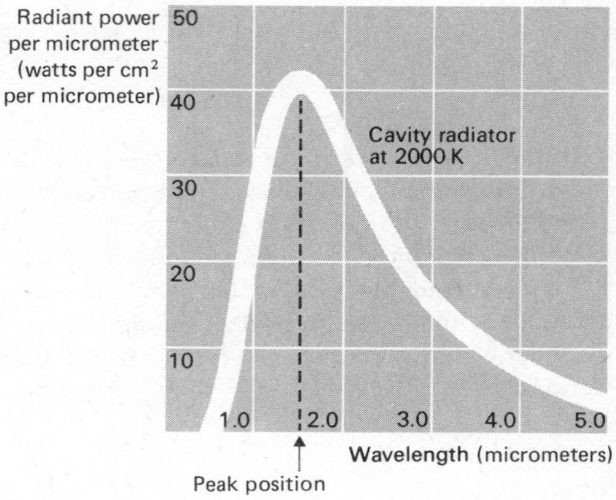

Radiant power per micrometer (watts per cm² per micrometer)

Cavity radiator at 2000 K

Wavelength (micrometers)

Peak position

Figure 8.12 Illustration of how radiant power changes with temperature. The wavelength of the dominant type of radiation (given by the peak position) gets smaller as the temperature of the radiator increases. Note that if you multiply the wavelength of the dominant type and the temperature, you always get a number around 3000 micrometer • kelvins.

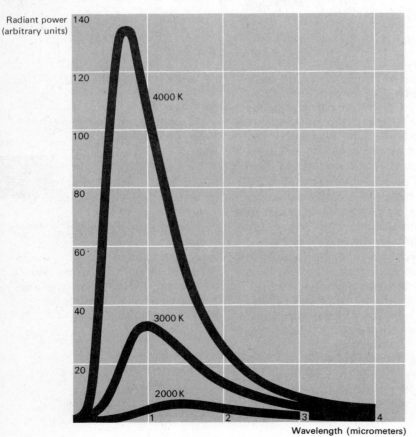

Radiant power (arbitrary units)

4000 K

3000 K

2000 K

Wavelength (micrometers)

Figure 8.13 Shown here is an aerial thermal photograph of a residential neighborhood in Hibbing, Minnesota. While it appears to be an ordinary black and white photograph taken with a conventional camera, the exposure is actually produced by infrared thermal radiation. The whiter the exposure, the higher the temperature of the source. A bright spot on a roof usually indicates active heating by a furnace or fireplace. The photograph was taken at 9 P.M. at an altitude of 1500 feet. Ground temperature was −20°F at the time of the exposure. Photograph courtesy of U.S. Environmental Protection Agency. ▶

Consequently, as the temperature increases, the position of the peak in the spectral radiancy curve shifts to smaller wavelengths. For example, if the temperature of two cavity radiators were 300 K (about the temperature of the earth) and 6000 K (about the surface temperature of the sun), the peaks would occur at about 10 and 0.5 micrometers, respectively. A wavelength of 10 micrometers is in the infrared region of the electromagnetic spectrum and one of 0.5 micrometers is in the visible region. The earth and sun are not constructed like cavity radiators. Nevertheless, they both radiate at fairly constant temperatures and there is a predominant type of radiation given off which is determined by the radiating temperature. It is this difference in emission characteristics that gives rise to the "greenhouse" effect that we will discuss in the next section.

The amount of radiated energy depends very strongly on the temperature. This strong temperature dependence allows us to determine differences in temperature from measurements of the radiant energy. If the temperatures of interest are around a temperature you might encounter in a house, then the radiation would be mostly infrared, and so we would use a camera sensitive to infrared radiation. This type of camera functions much like an ordinary photographic camera. In black and white imagery, hot areas produce greater exposure and produce white areas on the recording. Figure 8.13 shows a thermal photograph of a subdivision taken at night when the radiation is nearly all infrared. The very bright rectangular region near the center is due to a glass greenhouse from which the thermal radiation escapes much more easily than from a house. This infrared technique has become extremely useful for detecting heat leaks in buildings.

Max Planck's theoretical explanation of cavity radiation is a major triumph in physics. His interpretation provided one of the major links between classical and quantum physics. Planck assumed that the atoms in the walls of the radiator behaved as oscillators emitting electromagnetic radiation. An oscillator does not have just *any* energy, but only those that satisfy the relation $E = nhf$, where n is an integer $(0,1,2,3, \ldots)$, h is Planck's constant mentioned earlier, and f is the frequency of the oscillator. Secondly, the oscillators do not radiate continuously but only in "jumps" or quanta. He then proceeded to derive the theoretical curve shown in Fig. 8.14. The precise agreement between

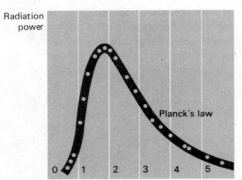

Figure 8.14 Planck's theoretical curve for radiant power superimposed on an experimental result. The precise agreement between theory and experiment is impressive.

his theory and experimental results gives enormous support to the quantum nature of electromagnetic radiation.

The warming of the earth by the "greenhouse" effect can now be explained using our understanding of thermal radiation. Radiation coming from the sun is mostly the visible type to which our eyes are sensitive. Visible radiation passes through carbon dioxide in the atmosphere with very little attenuation. After penetrating the earth's atmosphere, the radiation is absorbed and reflected by objects on the earth's surface. The earth radiates energy like the sun, but because the temperature of the earth's surface is much lower than the temperature of the surface of the sun, the type of radiation is mostly infrared. Carbon dioxide strongly absorbs infrared radiation at selected infrared wavelengths. Thus radiation comes freely through carbon dioxide in the atmosphere, but is prevented from escaping upon reradiation by the earth. The energy is trapped and the temperature of the atmosphere and the earth would obviously rise if significant amounts of carbon dioxide were to accumulate.

It is not particularly difficult to forecast the amounts of carbon dioxide produced by the combustion of fossil fuels. However, all the carbon dioxide does not remain in the atmosphere. Significant amounts are taken up by the oceans and some promotes plant growth via photosynthesis. These factors are difficult to estimate. So the only reliable measure is actual monitoring of carbon dioxide in the atmosphere. Sporadic measurements have been made for over a century. Systematic monitoring begun in 1959 at the Mauna Loa Observatory in Hawaii shows that global concentrations are increasing at an exponential rate of 4%/year, which corresponds to a doubling time of 17.5 years. The increase reflects the retention in the atmosphere of about half the carbon dioxide produced by combustion of fossil fuels.

Assessing the global impact of the accumulation of carbon dioxide in the atmosphere poses a unique problem to mankind. There is geological evidence that only a few degrees temperature variation differentiates an ice age and our present situation. Short term temperature variations occur naturally. From actual measurements, it is known that in the last 100 years the mean global temperature rose about 0.6 K to 1940 and has fallen about 0.3 K in the interim. While we have measurements of the accumulation of carbon dioxide, we do not have experimental evidence of its impact on global temperature. Any evidence is based on complex atmospheric models. Yet it seems clear that in a time period of the order of a century, the carbon dioxide "greenhouse" effect could produce temperature changes comparable to those between an ice age and the present. If we wait to see if the effect is occurring, then it may be too late. At the moment, there is insufficient evidence to warrant a halt to fossil fuel burning. But there is sufficient evidence to warrant concentrated studies to learn more about the carbon dioxide cycle and to develop more accurate climate models.

A suntan, admired and craved by many, is actually a mild form of skin damage produced by ultraviolet radiation from the sun. Overexposure to the sun's rays produces skin damage which in severe cases may lead to skin cancer.

Nature has provided a defense against the gross effects of ultraviolet radiation by creating a band of ozone (the ozone layer) at an altitude of 20,000 to 30,000 meters. The ozone layer removes the great bulk of the ultraviolet radiation emitted by the sun. Life as we know it would be drastically different without this ozone layer.

Scientists are becoming increasingly concerned about the possibility of depleting this ozone layer through the interjection of man-made products. The first of these anxieties was registered by the opponents of the American commercial supersonic transport (SST) plane. Since the SST would cruise at an altitude which puts it in the midst of the ozone layer, the fear was that nitric oxide (NO) from the engine exhausts would react with and deplete the ozone. Ozone can be broken down into normal oxygen (O_2) and atomic oxygen (O) by absorption of ultraviolet radiation:

$$O_3 + \text{sunlight} \longrightarrow O_2 + O.$$

This process and its inverse go on all the time. If nitric oxide (NO) were introduced into the ozone layer from the exhaust of an SST, then we could have

$$NO + O_3 \longrightarrow NO_2 + O_2,$$

followed by

$$NO_2 + O \longrightarrow NO + O_2.$$

Note that NO serves as a catalyst. It is regenerated in the process and O_3 is depleted. A small amount of NO could conceivably consume a large amount of ozone. If enough were consumed, serious consequences could result.

A more recent concern* involves the effect of fluorocarbons on the ozone layer. As the name suggests, fluorocarbons are molecules having fluorine and carbon atoms as constituents. Two fluorocarbons of interest are fluorocarbon 11 ($CFCl_3$) and fluorocarbon 12 (CF_2Cl_2). Fluorocarbons are inert and are widely used as propellants in aerosol spray cans. When these propellants are released they float up into the ozone layer and are irradiated with short wavelength ultraviolet radiation. Chlorine freed from the fluorocarbons then reacts with and depletes the ozone much as nitric oxide from an SST would. The overall effect is still speculative. However, scientists conceive of conditions which could produce a forty percent decrease in the ozone layer by 1995. Predictions such as these have stimulated vigorous theoretical and experimental research programs to learn more about the potential problems, and have instigated several federal investigations. Aerosol sprays may well be banned by federal legislation if the concerns for them prove to be true.

TOPICAL REVIEW

1. What is a temperature inversion?
2. Distinguish between radiation and subsidence inversions.
3. Why are temperature inversions of concern in the environment?

_____

* "The Ozone Layer: The Threat from Aerosol Cans is Real," *Science,* vol. 144 (8 October 1976): p. 170.

4. What is the meaning of lapse rate?

5. Explain how warm air rises in the atmosphere using the behavior of a balloon as an analogy.

6. How can a stagnant air condition result without a temperature inversion?

7. Why is there concern over accumulation of particulates in the upper atmosphere?

8. Describe the "greenhouse" effect. Why is it of concern?

9. Comment on the wave-particle duality of electromagnetic radiation.

10. What is a photon?

11. How does the dominant type of radiation from a heated source change when the temperature changes?

12. What definitive statements can you make about the present day accumulation of carbon dioxide and particulates in the atmosphere?

Atmospheric Problems

1. An elegant discussion of effects due to the accumulation of gases and particulates in the atmosphere is presented in *Fundamentals of Air Pollution,* by Samuel J. Williamson, Addison-Wesley, Reading, Mass. (1973).

2. Useful information on problems associated with the accumulation of pollutants in the atmosphere is presented in *Environmental Pollution*, 2d ed., by Laurent Hodges, Holt, Rinehart, New York (1977).

3. An assessment of the global atmospheric scene is given in *Man's Impact on the Global Environment,* the M.I.T. Press, Cambridge, Mass. (1971).

The Greenhouse Effect

1. The role of carbon dioxide in the environment is presented very nicely in "The Carbon Cycle," by Bert Bolin, *Scientific American* **223**, 124 (Sept. 1970). This same article appears in *The Biosphere,* W. H. Freeman and Co., San Francisco, Calif. (1970).

2. The "greenhouse" effect and its implications are discussed in some detail in "Carbon Dioxide and the Climate: The Uncontrolled Experiment," by C. F. Baes, Jr., H. E. Goeller, J. S. Olson, and R. M. Rotty, *American Scientist* **65**, 3 (May–June 1977).

3. "Carbon Dioxide and Climate: Carbon Budget Still Unbalanced," *Science* **197**, 1352 (30 September 1977).

Waves and Photons

The wave-particle duality of electromagnetic radiation is discussed in nearly all elementary physics and physical science textbooks. Two such texts that would be useful for the expansion of the material presented in Chapter 8 are

1. *Physics: A Descriptive Analysis,* by Albert J. Read, Addison-Wesley, Reading, Mass. (1970).

2. *Conceptual Physics,* 3d ed., by Paul G. Hewitt, Little, Brown, Boston, Mass. (1977).

SUGGESTIONS FOR FURTHER UNDERSTANDING OF CHAPTER 8

8.1 Motivation

1. The farther up in the atmosphere pollutants are emitted, the easier it is to disperse the pollutants. Why, then, aren't industries built on mountain tops rather than in valleys as they usually are?

2. What is meant by saying "the solution to pollution is dilution"?

3. Why should there be concern over particulate emissions from power plants and industrial sources if only ten percent of the total particulate emissions into the atmosphere come from man-made sources.

8.2 Temperature Inversions
8.3 A Stable Atmospheric Condition without a Temperature Inversion

4. Why does smoke from a burning cigarette generally rise up into the air?

5. Two identical balloons are inflated to the same size. One is filled with air, the other with helium. Why will the balloon filled with helium rise more readily in the atmosphere?

6. If you inflate a balloon by blowing your breath into it, seal it by tying a knot in the open end, and then release it in the atmosphere, is it possible for the balloon to rise?

7. Why do you sometimes have to exert considerable force to hold an inflated beach ball under water?

8. If a friend observes that the smoke emanating from the chimney of her fireplace floats toward the ground, what explanation could you offer for the reason?

9. Sometimes one sees the smoke rising from a cigarette seemingly hit a barrier and stop rising. What is the reason for this?

10. As a result of a severe radiation temperature inversion occurring in the nighttime, pollutants are often trapped in a low-lying layer. Then when the sun comes out, the pollutants will migrate toward the ground. What temperature conditions from the ground up would give rise to this effect?

11. A variation of temperature with altitude is shown in the accompanying graph. Pick out the portions of the plot that indicate temperature inversions.

Altitude

Temperature

8.4 Atmospheric Effects Due to Particulates

12. Why does a sunrise or sunset appear particularly red when viewed through a polluted atmosphere?

13. Why does smoke rising from a cigarette in an ash tray often have a bluish color?

8.6 Waves and Photons

14. What everyday experiences with sound can you think of that suggest that it is a wave phenomenon?

15. If the wavelength of electromagnetic radiation increases, how does the frequency change?

16. If the energy of a photon increases, how does the frequency change?

17. Why don't our eyes detect individual photons from a light source?

8.7 Thermal Radiation

18. All objects at a temperature greater than zero kelvins radiate electromagnetic waves. Why, then, do we not see objects in an unlighted room?

19. The human body, like any object at a temperature above zero kelvins, radiates energy. Why is it that we do not continually cool down as we sit in a room of a house?

20. An ordinary 100-watt incandescent light bulb when lit is much too hot to handle with bare hands. Knowing that the dominant radiation is visible, explain why the temperature is necessarily high.

21. Why does the temperature of an auditorium tend to rise when filled with people?

22. If you were given the task of developing a camera that would delineate a warm area in a body of water, what special property is needed for the camera film?

23. A homeowner suspects that the builder of his home has neglected to insulate selected regions of the walls. Without disassembling the walls, how could he determine if the builder has cheated him?

24. As an energy conservation measure, it is important that the seals around the doors of a refrigerator fit properly. How would a thermal photograph of a refrigerator be a useful energy diagnostic tool? What areas of a refrigerator would you want to show up dark and light in the thermal photograph?

25. How could a thermal photograph of an automobile help a police officer determine if a car of interest had been involved recently in a high-speed chase by police?

26. Light from an incandescent light bulb is due to thermal radiation from a metallic filament heated by electric charges flowing through it. What physical restriction limits severely the number of materials that can be used for the filament?

27. Why does an iron ingot in a foundry first turn red and then become a more "whitish" color when heated to the melting temperature?

8.8 The Greenhouse Effect

28. If the accumulation of carbon dioxide in the atmosphere proves to be a serious problem, what alternatives would there be to burning fossil fuels?

29. Nitrogen dioxide (NO_2) is a strong absorber of ultraviolet and visible blue radiation. If you view solar radiation after passing through an atmosphere containing nitrogen dioxide, what overall color would you expect to see?

8.9 Depletion of the Ozone Layer

30. Why does it take energy to separate chlorine atoms from a chlorinated fluorocarbon molecule like $CFCl_3$? What is the source of energy that causes a chlorinated fluorocarbon to break up in the ozone layer?

31. The chlorinated fluorocarbons used in aerosol sprays are extremely inert, i.e., they do not react easily with other chemicals. Why is this property extremely useful in aerosol sprays? Why does it lead to the problems associated with depletion of the ozone layer?

32. Nature generates the same gaseous pollutants as human beings. Why, then, is there concern about depletion of the ozone layer by nitrogen oxides produced from human activities?

33. On a trans-Atlantic flight an SST is in the air about three hours less than a conventional jet transport. Using the rate concept (amount = rate × time), explain why this does not necessarily mean that the SST pollutes the atmosphere less than a jet transport.

NUMERICAL PROBLEMS

8.2 Temperature Inversions
8.3 A Stable Atmospheric Condition Without a Temperature Inversion

1. A certain inflated balloon has a volume of 5000 cm³ and a mass of 3 grams. The balloon is released in air where the density is 0.0011 grams per cm³. Will the balloon rise or fall?

2. A small toy balloon has a mass of 2 grams. It is inflated to form a sphere of diameter 20 cm and is filled with ¼ gram of helium. When it is placed in air having a density of 0.001 g/cm³, will the balloon rise or fall? The formula for the volume of a sphere of diameter d is $V = 0.52d^3$.

3. If the temperature is 27 °C at ground level and the lapse rate in the atmosphere is −5 °C/km, what is the temperature at an altitude of 1600 meters?

4. A commercial jet plane often cruises at an altitude of about 35,000 feet (about 7 miles). If the lapse rate is −20 °F/mile, what is the temperature outside the plane if it is 75 °F on the ground?

5. On a certain day the temperature at ground level is 21.5 °C. At the top of a building 400 meters high, the temperature is 19.7 °C. What is the average lapse rate?

6. If the temperature at the base of a 10,000 feet high mountain is 80 °F, what is the approximate temperature at the top under ordinary weather conditions?

7. On a certain day the temperature of the air near the ground is 30 °C and the temperature decreases 4 °C for each kilometer of altitude. If the temperature of gas released from a smokestack is 36 °C and its temperature decreases 7 °C for each kilometer as it rises, how far will it rise before stabilizing? One way of doing this is to make a plot of temperature versus altitude as is done in Fig. 8.4 and seeing where the two lines cross.

8. In a day's time a large coal-burning electric power plant may burn 3000 tons of coal. If the coal contains 3 percent sulfur, 180 tons of sulfur oxides will be produced. Suppose that because of a temperature inversion all the sulfur oxides are uniformly distributed in a volume 80 km × 80 km × 0.5 km. What is the concentration of sulfur oxides in micrograms per cubic meter? How does this concentration compare with the primary standard given in Section 4.4? Useful information: 1 ton = 2000 pounds, 1 pound = 453 grams.

8.4 Atmospheric Effects Due to Particulates

9. To a reasonable approximation the visual range multiplied by the particulate concentration always produces the same number,

$$750 \text{ miles} \cdot \left(\frac{\text{micrograms}}{\text{cubic meter}}\right).$$

Determine the visual range for particulate concentrations of 50, 100, 150, and 200 micrograms per cubic meter.

10. In 1883 the volcano Krakatoa in the East Indies erupted and spewed about 10 cubic kilometers of particulates into the atmosphere. The particulates were spread worldwide by atmospheric circulations and produced spectacular sunsets for more than two years. In 1974 all the coal-burning electric power plants in the United States released about one million cubic meters of particulates into the atmosphere. At the 1974 rate, how many years of operation by these power plants would be required to produce the particulate output of the Krakatoa explosion?

8.6 Waves and Photons

11. A certain radio station broadcasts electromagnetic radio waves with a frequency of 700,000 hertz. What is the distance between crests of these waves? The speed of the waves is 300 million meters per second.

12. Regardless of frequency, all electromagnetic waves travel with the same speed in vacuum ($v = 300,000,000$ meters/second). Using Eq. 8.1, which relates speed, frequency, and wavelength of a wave, compare the frequency of an AM radio wave and a gamma wave with a light wave. Values of the wavelengths are presented in Fig. 8.9.

13. Observing crests of a wave moving past some position is like watching evenly spaced cars move by some location on a highway. What wave property would be analogous to the distance between cars? If T is the time between the passing of two cars and V is the speed of the cars, how far will a car travel in time T? Point out the similarities between this result and Eq. 8.1 that relates speed, frequency and wavelength of a wave.

14. If the energy of a photon is 0.00000000000000000002 joules, what is its type of radiation?

15. The energy of a photon is related to frequency by $E = hf$, where

$$h = 6.63 \times 10^{-34} \text{ joule} \cdot \text{sec}$$
$$= 0.000000000000000000000000000000000663 \text{ joule} \cdot \text{sec.}$$

Burning a pound of coal liberates about 10,000,000 joules of heat energy. Knowing that the frequency of visible light is 600,000,000,000,000 Hz, or 6×10^{14} Hz, how many visible photons are needed to have an energy equivalent to the heat energy liberated from burning a pound of coal?

16. A 100-watt light bulb produces about 10 joules of radiant energy each second. Assuming that this radiation is green with a wave frequency of 6 trillion hertz (6×10^{12} Hz), how many photons are produced each second that the bulb is lit?

8.7 Thermal Radiation

17. The tip of a cigarette often becomes "red hot" when a smoker inhales. Using the relation between dominant wavelength and temperature for a cavity radiator, estimate the cigarette tip temperature knowing that red color has a wavelength of about 0.65 micrometers. Is it any wonder that an errant cigarette burns a hole in a carpet?

18. An object at room temperature radiates energy at a rate of about 0.3 watts per square inch of surface. The human body has about 2800 square inches of surface area. (You might like to verify this by assuming the body is a cylinder of your height (h) and diameter (d) of 12 inches. The formula needed is $A = 1.57d^2 + 3.14dh$). What is the rate of emission of energy by the human body at room temperature? How does this rate of emission compare with a typical light bulb?

8.9 Depletion of the Ozone Layer

19. Chlorine oxide (ClO) molecules, formed from chlorine atoms separated from chlorinated fluorocarbons in the ozone layer, enter into chemical reactions with ozone much as does nitric oxide (NO). Following the analysis presented in Sec. 8.9, fill in the steps in the equations below:

$$ClO + O_3 \longrightarrow \underline{\hspace{2cm}} + O_2,$$
$$\underline{\hspace{2cm}} + O \longrightarrow ClO + \underline{\hspace{2cm}}.$$

What constitutes the catalyst in these reactions?

NUCLEAR PHYSICS PRINCIPLES

9

A view into the core of a research-type nuclear reactor. The glow near the bottom of the reactor is called Cerenkov radiation. It occurs when charged particles enter a transparent medium with a speed in excess of the speed of light in that medium. The particles slow down (lose energy) by emitting visible (mostly blue) radiation. Photograph courtesy of DOE.

9.1 MOTIVATION

The strength of our modern society is founded on our ability to procure energy sources and convert them to our needs. For more than three decades after the turn of this century, no source challenged the superiority of the fossil fuels. But as oil and natural gas became premium and environmental problems mounted from burning coal, a need emerged for new forms of energy. Nuclear energy began making inroads into the supremacy of fossil fuels around 1970. If the trend continues to the year 2000, then one fourth of our energy will come from nuclear sources. But nuclear energy is no panacea. Nuclear fuel will generate electricity for our homes, but it will not replace gasoline as a fuel for our cars. It is completely devoid of all the emissions from the burning of fossil fuels, but it brings radioactivity to the forefront as a pollutant. Public fear shrouds nuclear energy because of the possibility of accidental holocausts from the large-scale release of nuclear radiations. Nevertheless, the nuclear energy machine is underway. Whether it is only an interim source until a more acceptable alternative is developed remains to be seen. Our goal now is to elucidate the mysteries of this fascinating subject.

No single equation in physics has the universal prestige of the Einstein relation $E = mc^2$. In a strikingly simple way, it relates mass (m) to energy (E). The letter c denotes the speed of light or electromagnetic radiation in a vacuum (300,000,000 meters per second). If energy is to be in joules, then m must be expressed in kilograms and c in meters per second. To illustrate, let us calculate the energy equivalent of a penny, which has a mass of about 3 grams (0.003 kg).

Energy = mass in kilograms multiplied by the square of the speed
of light in meters per second,

$$= 0.003 \, (300,000,000)^2,$$

$$= 270 \text{ trillion joules.}$$

Figure 9.1 Atomic nuclei are a vast source of energy. A single truckload of nuclear fuel can supply the total electric power needs of a city of 200,000 people for a year. (Photograph courtesy of the Department of Energy.)

Because heat from burning one ton of coal is about 27 billion joules, the intrinsic energy in one gram of matter is nearly 3000 times the heat produced from burning a ton of coal. In fact, the bombs exploded over Hiroshima and Nagasaki in World War II released energy equivalent to about one gram of matter. Knowing that a potential energy source exists is one thing; exploiting it is something else. However, our knowledge of the physics of the nucleus of the atom has led to a scheme involving reactions between atomic nuclei which converts mass to useful energy. Let us now look into these important principles.

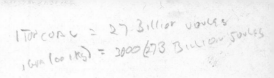

Our discussions thus far have required a model of the atom consisting only of a positively charged nucleus surrounded by negatively charged electrons. This model is adequate because those things discussed—molecules, chemical reactions, etc.—involved only interactions of the electrons. Nuclear reactions involve the details of atomic nuclei and this requires that the model be improved. Because both the atom and the nucleus are *bound* systems, many of the concepts used in the atomic model are also useful in the refinement. The atom is bound together by attractive *electric forces* between the positively charged nucleus and negatively charged electrons. The nucleus is bound together by "strong" *nuclear forces*. In the hierarchy of fundamental forces, the strong nuclear force has the same status as the gravitational and the electromagnetic force. It is the strongest force of all and exists between protons and neutrons. A proton possesses the same amount of charge as an electron but it is positive. The neutron is electrically neutral and its mass is slightly larger than the mass of the proton. The proton and neutron are respectively, 1836 and 1839 times as massive as the electron. (It is useful to remember that both a neutron and a proton are about 2000 times more massive than an electron.) Because the atom is electrically neutral and only protons in the nucleus have charge, an atom has equal numbers of electrons and protons. The nuclear force has characteristics radically different from the gravitational and electric forces. First, it is enormously stronger. Consequently, nuclear energies tend to be much larger than gravitational and electric energies. For example, the total energy, kinetic plus potential, of an electron about a nucleus is about 10 electron volts whereas the total energy of a proton in the nucleus is about 50,000,000 electron volts (or 50 MeV).* Secondly, the nuclear force acts only over a distance about the "size" of a nucleus. This distance is roughly 0.0000000000001 cm and is about 1/100,000 the diameter of an electron orbit about the nucleus.

The mathematical form of the electric force between two charges is quite simple. It is this simplicity which allows fairly elementary calculations of the structure of atoms. The Bohr model† of the atom evolves from such a calcula-

9.2 PROPERTIES OF THE ATOMIC NUCLEUS

---

* The electron volt (eV) is a favorite energy unit in atomic and nuclear physics. One eV is the energy acquired by an electron (or a proton) accelerated through a potential difference of one volt. Using the definition of potential difference ($W = QV$) it follows that $1\text{eV} = 1.602 \times 10^{-19}$ joules. It is not terribly important to be able to do calculations in terms of electron volts. It is important to remember the relative energy differences between atomic and nuclear systems.

† Named for Niels Bohr, Danish physicist, 1885–1962.

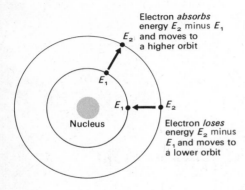

Electron *absorbs* energy E_2 minus E_1 and moves to a higher orbit

E_2
E_1
E_1
E_2
Nucleus

Electron *loses* energy E_2 minus E_1 and moves to a lower orbit

Figure 9.2 An atom in the energy state E_1 absorbs energy and the atom "moves" to a state with larger energy E_2. In the Bohr model of the atom, this corresponds to an electron moving to an orbit of larger diameter. An atom in an energy state E_2 loses energy and the atom "moves" to a state with less energy E_1. In the Bohr model of an atom, this corresponds to an electron moving to an orbit of smaller diameter.

tion and gives rise to the idea that electrons revolve about the nucleus in orbits much as planets revolve about the sun in our own solar system. The electrons exist in certain well-defined orbits, each characterized by a well-defined (quantized) energy. An electron can move to a higher orbit by absorbing just the right amount of energy. Conversely, an electron can descend to a lower orbit by releasing a precise amount of energy. (See Fig. 9.2.) The concept of orbits is not particularly important. In fact, this notion does not even arise in more sophisticated calculations. However, the idea of well-defined energies is a characteristic of all atomic models. Although it is known that the nuclear force is very strong and acts over a very short distance, its precise mathematical form is still not completely understood. The search for a detailed understanding constitutes one of the challenges and major efforts of current research. There is no such thing as an *elementary* calculation for a model of the nucleus. However, like electrons around the nucleus, protons and neutrons (or nucleons as they are termed collectively) have kinetic energy due to their motion and potential energy associated with forces operating within the nucleus. Therefore, it is probably not surprising that the total energy of a nucleus, like the atom, is quantized. The language for absorption and emission of energy still applies, but the energy scale is characteristic of the strong nuclear force (Fig. 9.3).

A nucleus is identified by the number of protons and neutrons that it possesses. The same chemical symbol is used to denote an atom or its nucleus. When referring to the nucleus, it is customary to also affix the proton number (Z), the neutron number (N), and the nucleon number (A) to the chemical symbol (X) in the manner $_Z^A X_N$. The nucleon number is just the sum of the pro-

Figure 9.3 Energy diagrams for systems in which the energies take on only discrete values.

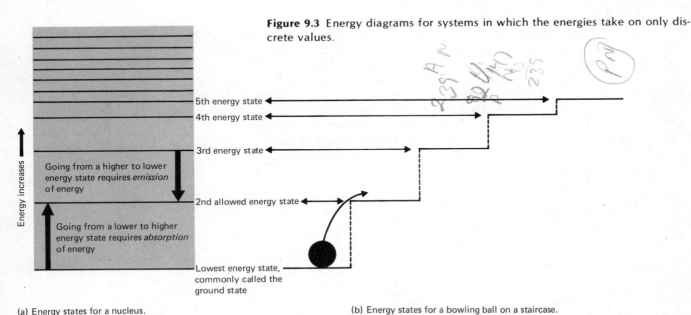

Energy increases

Going from a higher to lower energy state requires *emission* of energy

Going from a lower to higher energy state requires *absorption* of energy

5th energy state
4th energy state
3rd energy state
2nd allowed energy state
Lowest energy state, commonly called the ground state

(a) Energy states for a nucleus.

(b) Energy states for a bowling ball on a staircase. Energy is required to move the ball up the staircase.

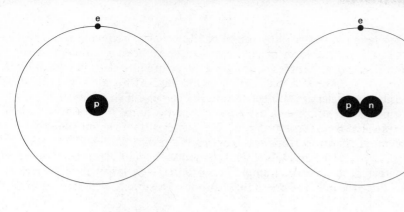

Hydrogen atom—
single proton with
one orbiting electron

Deuterium atom—bound
combination of one proton
and one neutron with one
orbiting electron

Tritium atom—bound
combination of one proton
and two neutrons with one
orbiting electron

Hydrogen nucleus—
single proton

Notation for the
hydrogen nucleus
(proton)
$_{1}^{1}H_{0}$

Deuterium nucleus—
bound combination of one
proton and one neutron

Notation for the
deuterium nucleus
(deuteron)
$_{1}^{2}H_{1}$

Tritium nucleus—
bound combination of one
proton and two neutrons

Notation for the
tritium nucleus
(triton)
$_{1}^{3}H_{2}$

Figure 9.4 Schematic illustration of the three isotopes of hydrogen.

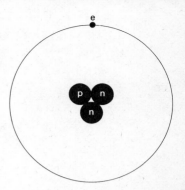

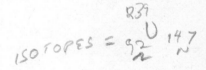

ton and neutron numbers, $A = N + Z$. The nucleus of the hydrogen atom would be denoted $_{1}^{1}H_{0}$. This complete description is necessary in order to keep track of neutrons and protons in nuclear transformations. Those atoms with nuclei having the same number of protons but a different number of neutrons are called isotopes. The nuclei of the hydrogen, deuterium, and tritium atoms all have one proton but 0, 1, and 2 neutrons, respectively (Fig. 9.4). The most abundant isotope of carbon has 6 protons and 6 neutrons. It is this isotope that is assigned a mass of exactly 12 atomic mass units (amu) on the atomic mass scale. All other nuclear and atomic masses are measured relative to this isotope of carbon. The most abundant isotope of oxygen, $_{8}^{16}O_{8}$, has a mass of 15.994915 on this scale.

9.3 NUCLEAR STABILITY

A system that tends to revert to an initial condition after receiving some perturbation is said to be stable. A marble resting inside a bowl is stable because, if moved and released, the forces acting tend to restore it to its stable position at the bottom of the bowl. A ball sitting at the edge of a stair step is in an unstable position. Nuclei are subject to internal forces which promote breakup. Some nuclei are stable against these "provocations." Others will be induced to nuclear transformations. Stable nuclei with few nucleons have nearly equal numbers of neutrons and protons. For example, $_1^1H_1$, $_2^3He_1$, $_2^4He_2$, $_3^6Li_3$, $_4^9Be_5$, $_5^{10}B_5$, $_6^{12}C_6$, $_8^{16}O_8$, and $_8^{17}O_9$ are all stable. As the number of nucleons in a nucleus increases, the neutron number gets progressively larger than the proton number. For example, the one stable isotope of cesium has 55 protons and 78 neutrons.

A plot of neutron number (N) versus proton number (Z) for the stable nuclei defines a "region" of stability as shown in Fig. 9.5. Most nuclei are

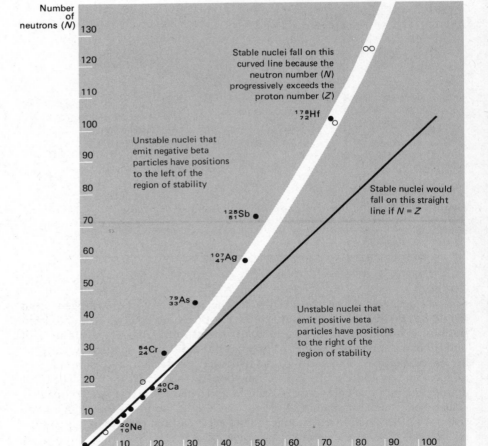

Figure 9.5 The stability "region" for nuclei. Some stable nuclei are shown for reference.

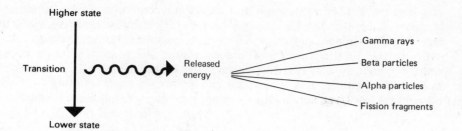

Higher state

Transition

Lower state

Released energy

Gamma rays

Beta particles

Alpha particles

Fission fragments

Figure 9.6 A nucleus in a state of excess energy may give off energy and "drop" to a lower state. It is analogous to a ball on top of an inclined plane. By rolling down the plane, the ball goes to a lower potential energy state, but kinetic energy is imparted to the ball in the process.

unstable and may release energy and descend to a lower energy state (Fig. 9.6). Ultimately the nucleus proceeds to a stable state in which no more spontaneous energy transformations are possible. These energy transitions are accompanied by emission of a particle and/or radiation. A nucleus which is unstable and, therefore, capable of emission of energy is said to be radioactive. The process of energy emission is called radioactivity. These terms evolved from early studies and really say little about the actual mechanism. Several different mechanisms are possible. However, only the emission of alpha, beta, and gamma radiation and the nuclear fission process are of interest now.

Electromagnetic radiation (refer to Chapter 8) is released from an atom when an electron moves from one energy state to a state of lower energy. The frequency of this radiation is given by the Planck formula (Eq. 8.3):

$$f = \frac{E_{higher} - E_{lower}}{h}, \tag{9.1}$$

where h is Planck's constant. Exactly the same mechanism can take place in a nucleus. However, the energies involved are characteristic of nuclear processes and therefore the frequency of the radiation is some 1 million times greater than for an atomic process. This nuclear radiation is called gamma radiation. It is in the form of photons with energies greater than about 100,000 eV (or 100 keV). X-rays (which are nothing more than photons with energies 1 to 100 keV) are often emitted from nuclei by the same mechanism.

Frequency, wavelength, and speed of electromagnetic radiation are related by $v = f\lambda$. The speed in vacuum is independent of wavelength and frequency. So if the frequency goes up, the wavelength goes down. As the wavelength goes down, the ability of electromagnetic radiation to penetrate matter increases. Although the wavelength of visible light (about 0.00005 cm) is comparatively short, it is not short enough to enable the light to penetrate even a few sheets of paper or the surface layer of skin. However, the wavelength of an X-ray (about 0.00000001 cm) or a gamma ray (about 0.00000000001 cm) is short enough that either can easily penetrate matter. This explains why an X-ray can be used to photograph the bone structure of a person and also why it is difficult to protect oneself against the biologically damaging effects of such radiations.

9.4 TYPES OF NUCLEAR RADIATION OF ENVIRONMENTAL INTEREST

Beta particles emanate from the nucleus and come in two types which are identical except that one has negative charge, (β^-), the other positive, (β^+). In fact, β^- differs in no way from an electron. A beta particle does not exist as an entity in the nucleus for this would contradict the proton–neutron model which provides for only two discrete particles. Rather, a beta particle is created at the time of emission. A β^- particle can be thought of as resulting from the transformation of a neutron into a proton, a β^- particle, and an antineutrino:

$$n \longrightarrow p + \beta^- + \bar{\nu}.$$

The proton remains in the nucleus and the β^- and antineutrino escape. A β^+ particle can be thought of as resulting from the transformation of a proton into a neutron, a β^+ particle, and a neutrino:

$$p \longrightarrow n + \beta^+ + \nu.$$

The neutron remains in the nucleus and the β^+ and neutrino escape. With "the stroke of the pen," a neutrino and antineutrino are introduced. Although not important as an environmental danger, they are extremely important for the physics of the process. Like photons, they possess no charge and no mass. Their interaction with matter is extremely weak. A neutrino impinging on the earth would likely whiz through unscathed. But, importantly, they have energy and, if for no other reason, are necessary to conserve energy in the transformation. The equations for the transformations are merely convenient ways

Figure 9.7 Schematic illustration of beta decay. The charge of the residual nucleus always changes by one unit in beta decay. The charge goes down one unit in positive beta particle decay; up one unit in negative particle decay. As a result, the chemical species always changes.

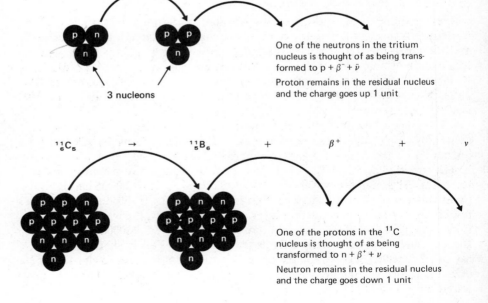

$^3_1\text{H}_2 \quad \rightarrow \quad ^3_2\text{He}_1 \quad + \quad \beta^- \quad + \quad \bar{\nu}$

3 nucleons

One of the neutrons in the tritium nucleus is thought of as being transformed to $p + \beta^- + \bar{\nu}$

Proton remains in the residual nucleus and the charge goes up 1 unit

$^{11}_6\text{C}_5 \quad \rightarrow \quad ^{11}_5\text{B}_6 \quad + \quad \beta^+ \quad + \quad \nu$

One of the protons in the ^{11}C nucleus is thought of as being transformed to $n + \beta^+ + \nu$

Neutron remains in the residual nucleus and the charge goes down 1 unit of thinking about the reactions. Actually, all the nucleons in a nucleus are involved in a transformation leading to the emission of a beta particle.

It is important to note that charge and number of nucleons are conserved in the process of beta decay. However, because the newly formed nucleon remains in the nucleus, the net charge of the nucleus will always change, thereby changing the chemical species. To illustrate, consider the beta decay of a tritium nucleus and a $^{11}_{6}C_5$ nucleus (see Fig. 9.7). Helium-3 ($^3_2He_1$) is actually produced commercially from the radioactive decay of tritium nuclei.

β^- emission occurs in nuclei that have more than enough neutrons for stability and β^+ emission occurs in nuclei that have more than enough protons for stability. Thus β^- and β^+ decay would occur for nuclei to the left and right, respectively, of the stability region shown in Fig. 9.5.

Given a nucleus containing Z protons and N neutrons, it is not hard to envision that some of these nucleons might get together and form their own nuclear state and split off from the main nucleus. Alpha particle emission is just such a process. An alpha particle is a bound nucleus of 2 protons and 2 neutrons. It is identical to the nucleus of one of the helium isotopes. The nuclei for which alpha particle emission is possible are usually very massive. The alpha particle decay of $^{238}_{92}U_{146}$ is an example:

$$^{238}_{92}U_{146} \longrightarrow {}^{234}_{90}Th_{144} + {}^4_2He_2.$$

Note that the process is simply the removal of 2 protons and 2 neutrons from the original nucleus. One might suspect that there is really nothing special about a helium nucleus and therefore one should see nuclei spontaneously emitting all sorts of other nuclei. However, most of these processes are prohibited when energy balances are considered.

In principle, nuclear fission is a process like alpha emission. However, it constitutes a gross fracturing or fissuring of a nucleus. A nucleus splits into two or more (but usually two) fragments of comparable mass plus some lighter fragments which are usually neutrons. While spontaneous alpha particle emission is common, spontaneous fission is very rare. It competes with alpha emission in the very heavy nuclei, but is typically 10 to 100 million times less probable. However, nuclear fission can be induced by an appropriate projectile and it is this process that is exploited as a source of energy.

A neutron is a good nuclear "axe" for splitting a nucleus. If $^{235}_{92}U_{143}$ reacts with a neutron, then the following reaction *may occur* with a release of about 200 MeV energy:

$$^{235}_{92}U_{143} + {}^1_0n_1 \longrightarrow \underbrace{{}^{141}_{56}Ba_{85} + {}^{92}_{36}Kr_{56}} + 3\,{}^1_0n_1 + \text{energy.}$$

Two relatively massive fission fragments

The "may occur" is emphasized for two reasons.

1. There are a variety of possible pairs of fission fragments. The only restriction is that the reaction be consistent with conservation laws for nucleons.

2. Even though the conservation laws are satisfied, there is still no guarantee that the reaction will proceed. The conservation laws are a necessary condition

but not a sufficient condition. There is only some chance that the reaction will proceed and this depends on the nature of the nuclear force between projectile and target and the structure of the interacting nuclei. $^{238}_{92}U_{146}$ is an isotope of uranium with only 3 more neutrons than $^{235}_{92}U_{143}$. Yet their nuclear fission characteristics are so different that ^{235}U can be used as a fuel in contemporary nuclear reactors and ^{238}U cannot.

The energy aspect of the fission reaction is impressive, but an even more intriguing prospect is the use of the neutrons emitted in a reaction to instigate other reactions. This is the mechanism of a chain reaction which gives rise to self-sustaining, energy-producing nuclear reactions. It is this process which is exploited in nuclear reactors discussed in the next chapter.

9.5 HALF-LIFE OF RADIOACTIVE NUCLEI

A nucleus is said to disintegrate (or decay) when it emits a particle or a photon. From the standpoint of protection against nuclear radiations, it is important to know the rate at which a sample of nuclei is disintegrating. The curie, symbolized Ci, is used as a measuring unit for disintegration rate:

1 curie = 37 billion disintegrations per second.

Most radioactive sources encountered in environmental situations have disintegration rates much less than a curie. Therefore, these rates are often expressed as thousandths of a curie, called millicuries (mCi); millionths of a curie, called microcuries (μCi); or trillionths of a curie, called picocuries (pCi).

Suppose that you have collections of equal numbers of tritium and ^{11}C atoms at some given time. Will the decay rate of these two samples at this given time be the same? Interestingly, they will not. The ^{11}C decay rate will initially be more than 100,000 times larger than the tritium decay rate. This may seem surprising at first. However, it is much like taking two equal numbers of people having average ages of 65 and 25 and asking if the death rates are the same. Both of these groups are people but their physical characteristics are quite different. Tritium and ^{11}C are both radioactive but each is bound together quite differently from the other and, therefore, their decay characteristics are different. Given a collection of people, we can say with absolute certainty that everyone in the group will die eventually. If the collection is large enough and there is enough information about the previous history of the types of people involved, we can determine with considerable accuracy how many will die in a certain time period. (Life insurance companies survive on this principle.) But we can never say when a given individual will die. Similarly, given a large collection of tritium atoms, we cannot predict when a given one will decay, but we can accurately determine how many will decay in a certain time interval and what the chance is that a given tritium nucleus will decay. This type of information is extremely important from the standpoint of safety at nuclear power plants.

A purist would prefer to calculate the disintegration rate from first principles, which would mean starting with the nuclear force. But like all nuclear calculations this is extremely difficult. Rather, let us look at what

Table 9.1 Data for the decay rate of ^{137}Ba.

Time (min)	Particles detected/sec
0	4617
0.6	3963
1.2	3332
1.8	2904
2.4	2542
3.0	2114
3.6	1820
4.2	1541
4.8	1279
5.4	1151
6.0	960
6.6	789
7.2	728
7.8	603

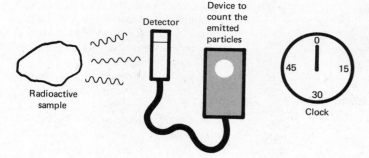

Detector

Device to count the emitted particles

Radioactive sample

0

45   15

30

Clock

Figure 9.8 An elementary setup for measuring the decay rate of a radioactive sample. One records the number of particles or photons interacting with the detector in a measured time interval. The measurement is repeated for several points in time.

happens experimentally and then deduce a model for explanation. While we will use the language of a nuclear process, always keep in mind that the analysis may apply to other systems with different characters. We start with a sample of radioactive atoms, a clock, and some instrument for counting the number of products emitted (Fig. 9.8). The measurement involves the rate at which nuclei decay. This simply means counting the number of particles detected in a certain time interval:

$$\text{decay rate} = \frac{\text{number of particles detected}}{\text{time interval}},$$

$$R = \frac{N}{t}.$$

R can be expressed in particles detected per second that could be related to the curie unit mentioned earlier. This measurement is taken at regular intervals of time. Table 9.1 shows data from such an experiment using ^{137}Ba that emits gamma rays. Figure 9.9 shows a plot of the data in Table 9.1. Note that the counting rate decreases by a factor of two at regular intervals of time. This is just what was found for electric energy production in the United States, except that the progression represented an increase, indicating growth, instead of a decrease. It is appropriate to talk about a time required for the sample to decrease by a factor of two. This is called the half-life. If you start with a sample of 1 million radioactive nuclei, then there will remain 500,000 nuclei after one half-life has elapsed. After two, three, and four half-lives have elapsed, 250,000, 125,000 and 62,500 nuclei will remain. This is shown as a generalized plot in Fig. 9.10.

The range of half-lives of radioactive nuclei is impressive—less than 10^{-20} seconds to more than 10^{15} years.* The half-lives of some nuclei of interest in nuclear power plants are shown in Table 9.2.

Table 9.2 Half-lives of some nuclei of concern in nuclear power plants.

Nucleus		Half-life
^{3}H	(tritium)	12.26 years
^{90}Sr	(strontium)	28.8 years
^{137}Cs	(cesium)	30.2 years
^{131}I	(iodine)	8.05 days
^{85}Kr	(krypton)	10.76 years
^{133}Xe	(xenon)	5.27 days

* The half-lives of radioactive nuclei can be found in nearly all nuclear physics texts, for example, *Introduction to Nuclear Physics* by H. A. Enge, Addison-Wesley, Reading, Mass. (1966).

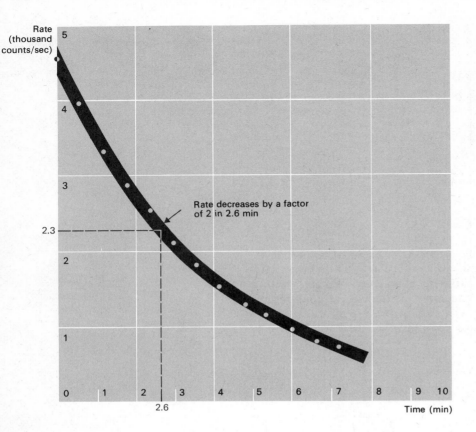

Figure 9.9 Plot of the ^{137}Ba decay rate data shown in Table 9.1. Initially, the decay rate was 4617 decays/sec. After an elapse of 2.6 minutes the decay rate decreased to 2300 decays/sec—a decrease of a factor of two. After an elapse of 5.2 minutes the decay rate decreased to 1190 decays/sec—a decrease of a factor of four. Regardless of what time you pinpoint, the decay rate will be half as much 2.6 minutes after that time.

Figure 9.10 A generalized plot for radioactive decay. N denotes the number of radioactive nuclei at the beginning. N was one million in the example given in the text. If one starts out with N radioactive nuclei, then after one half-life (designated T) has elapsed there will remain N/2 nuclei that have not decayed. After two half-lives (designated 2T) there will remain N/4 nuclei, and so on.

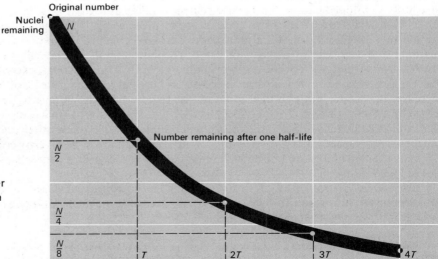

The fundamental origin of energy from either fossil or nuclear fuel is from a reaction which may be symbolized:

$$A + B \longrightarrow C + D + \text{Energy}.$$

reacting   reaction
particles   products

The reacting particles in a chemical reaction are atoms and molecules. In a nuclear reaction, they are the nuclei of atoms. The principles of analysis are very much the same, but some of the bookkeeping language is different.

In a nuclear reaction the number and types of nuclei may be different after the reaction. (Recall that the number and types of atoms are unchanged in a chemical reaction.) But the number of nucleons and the total charge are unchanged. The nucleons may be bound together to form different species. For example, consider the spontaneous fission of uranium-233 ($^{233}_{92}U_{141}$) into two nuclei, one of which is krypton-90 ($^{90}_{36}K_{54}$). Because the initial nucleus, $^{233}_{92}U_{141}$, contains 92 protons there must be the same number of protons in the combined total of the two fission fragments. Because there are 36 protons in $^{90}_{36}Kr_{54}$ this means that there must be 56 protons in the other fragment. This identifies the second fragment as an isotope of barium (Ba). The same philosophy applies to the number of neutrons. You might like to go through the analysis. The complete process can be summarized as follows:

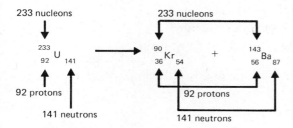

The analysis of the energetics of a nuclear reaction is precisely the same as for a chemical reaction. However, the magnitude of the energies involved is usually more than a million times larger. For comparison, the energy released in a chemical reaction is of the order of electron volts while in a nuclear reaction the energy released is of the order of millions of electron volts.

The inclusion of mass energy is crucial in a nuclear reaction. To illustrate consider the fission reaction just discussed. The combined mass of ^{90}Kr and ^{143}Ba after the fission process is 0.2 amu less than the mass of ^{233}U. This loss in mass is converted to energy in accordance with the Einstein relation $E = mc^2$. Since one amu is equivalent to 931 MeV, this means that the 0.2 amu mass difference is equivalent to 186 MeV of energy. Remember that the energy released in a chemical reaction is typically a millionth or a ten-millionth of this amount. So the energy release that occurs during a nuclear fission reaction is very significant.

We have used the spontaneous fission reaction for illustration. The analysis applies to the energetics of all nuclear reactions, however. Sometimes you will find that the combined mass of the reaction products exceeds the combined mass of the products before the reaction. In that case energy must be added to the reacting products to make up the mass difference.

TOPICAL REVIEW

1. What is the significance of the Einstein equation $E = mc^2$?
2. Describe the Bohr model of the atom and discuss the forces that play a role in binding the atom and its nucleus.
3. What does the symbol ${}^3_2\text{He}_1$ mean?
4. What is radioactivity?
5. What are the emissions of interest from unstable nuclei and how are they produced?
6. Distinguish between X-ray and gamma ray.
7. What is an isotope? Name two isotopes of hydrogen.
8. What is an atomic mass unit (amu)? How is it defined?
9. Describe nuclear fission.
10. What is the nuclear meaning of disintegration?
11. What is a curie (Ci)?
12. What is meant by half-life?
13. How is energy produced in a nuclear reaction?
14. Describe the fission of a heavy nucleus by a neutron.
15. What is a chain reaction?

REFERENCES

Properties of Atoms and Nuclei

1. A readable extension of the ideas presented in this chapter are presented in Chapter 3, "Atoms, Molecules, and Nuclei," in the paperback book *Nuclear Energy: Its Physics and Its Social Challenge,* by David R. Inglis, Addison-Wesley, Reading, Mass. (1973).

2. A less sophisticated but enjoyably readable description of atoms, nuclei and nuclear reactions is given in Chapter 4, "The Age of the Bomb," in *Physics for Society,* by W. B. Phillips, Addison-Wesley, Reading, Mass. (1971).

SUGGESTIONS FOR FURTHER UNDERSTANDING OF CHAPTER 9

9.2 Properties of the Atomic Nucleus

1. In a newspaper article concerning nuclear energy, reference is made to Pu-238. What does this notation mean?

2. Neutrons are sometimes referred to as the "glue" needed to hold the nucleus together. Explain the origin of this expression.

3. A person's weight is usually expressed in pounds, but there is no reason why it could not just as well be expressed in ounces. Nuclear energies are usually expressed in millions of electron volts, but there is no reason why they could not be expressed in joules, which are more basic. Why do we find it convenient to express our weight in pounds and nuclear energies in millions of electron volts? Since a pound is equal to 4.45 newtons, do you see why people might be reluctant to adopt the metric unit of newton for their weight?

4. An electron volt (eV) is equivalent to 0.00000000000000000016 joules (1.6×10^{-19} J). What advantage is there in doing numerical calculations involving atomic and nuclear energies in eV rather than in J?

5. An elevator in rising or descending stops only at certain levels. Could you associate an energy with each stopping position? How is this situation like the energy states of an atom or a nucleus?

6. To lift a box from the floor to a tabletop requires that the lifter do work. If the box falls back to the floor, it converts its potential energy into some other form of energy. How are these energy transformations like those that take place between energy states of an atom or a nucleus?

7. The electric force between two charges and the gravitational force between two massess decreases as the separation of the objects increases. However, in principle they have to be separated an infinite distance before there is no force. The nuclear force between two nuclei is essentially zero beyond some separation distance which is about the size of a nuclear diameter. Suppose that the gravitational force between two masses were zero beyond a distance of 100 miles; what effect would this have on our space program?

9.3 Nuclear Stability

8. Experiment shows (Chapter 5) that two positive (or two negative) charges will tend to push each other apart. If *only* electric forces operated within an atomic nucleus, could a nucleus ever be a bound system?

9. While it is possible to rest a small sphere like a marble on top of a round ball, the arrangement is unstable. Discuss (a) the stability aspects in terms of the forces on the sphere, and (b) the energy transformations when the sphere rolls off the ball.

10. The radioactive isotopes of strontium (Sr) are among those of environmental concern. Two of these isotopes are $^{83}_{38}\text{Sr}_{45}$ and $^{90}_{38}\text{Sr}_{52}$. One of these is a β^- emitter; the other is a β^+ emitter. Make the identification using Fig. 9.5 as a guide.

9.4 Types of Nuclear Radiation of Environmental Interest

11. In what ways are alpha and beta particles different? In what ways are gamma rays different from beta particles?

12. If a nucleus emits some type of radiation and the chemical nature of the emitter does not change, what is the type of radiation?

13. In beta decay, why is it not possible to ever produce an isotope of the disintegrating nucleus?

14. Knowing that an alpha particle is identical to the nucleus of a helium atom ($^4_2He_2$) and that many minerals contain radioactive atoms that emit alpha particles, explain why helium gas is often retrieved from oil wells.

15. Taking into account mass-energy considerations, explain why an isolated proton cannot spontaneously disintegrate into a neutron, a β^+ particle, and a neutrino.

9.5 Half-life of Radioactive Nuclei

16. The half-life was defined as the time required for one half of a sample of radioactive atoms to disintegrate. In principle, how long does it take for any sample of a radioactive material to disintegrate completely?

17. Does the concept of half-life as presented in the text have any meaning if the sample is very small? Suppose, for example, that you started with 15 radioactive nuclei.

9.6 Energy Aspects of Nuclear Reactions

18. Water behind a dam has potential energy by virtue of an advantageous position. The potential energy is converted to kinetic energy when the water is rearranged as it plummets to the bottom of the dam. What is being rearranged when the potential energy of a nucleus is converted to kinetic energy?

19. $^8_4Be_4$ is unstable and disintegrates by emitting two identical particles. What are the particles?

20. A deuteron ($^2_1H_1$) is composed of one neutron and one proton. An alpha particle ($^4_2He_2$) is composed of two neutrons and two protons. One might suspect that an alpha particle might spontaneously break up into two deuterons. This cannot possibly happen, though. What conservation principle forbids this?

21. The accompanying sketch shows a set of weighing scales with before and after products of the fission of ^{235}U by a neutron. Why would you expect the scales to be tipped as shown?

22. How is a nuclear fission chain reaction similar to a collision chain reaction of closely spaced cars on a highway?

23. The nuclear fission reaction illustrated in Sec. 9.4 shows isotopes of barium and krypton being formed. Why should you *not* conclude that only isotopes of barium and krypton can be produced when ^{235}U fissions?

24. A nucleus at rest has only mass energy, $E = mc^2$. If this nucleus disintegrates by emitting an alpha particle, for example, then the alpha particle clearly has kinetic energy. The principle of conservation of energy requires that the total energy before the transformation equal the total energy after the transformation. What is the origin of the kinetic energy of the alpha particle?

25. What would you conclude if the nuclear fission reaction illustrated in Section 9.4 showed $4_0^1 n_1$ rather than $3_0^1 n_1$?

26. A nuclear reaction will not proceed unless there is energy, including mass energy, available for the reaction. However, even if the energy is available, why can you not say anything about when the reaction will happen?

NUMERICAL PROBLEMS

9.1 Motivation

1. Using the Einstein equation, $E = mc^2$, show that the intrinsic energy of one gram of matter is 90 trillion (i.e., 9×10^{13}) joules.

2. The heat from burning one pound of coal is 13,100 Btu. Knowing that one Btu is equivalent to 1055 joules, show that the heat from burning one ton of coal is 27.6 billion (i.e., 27.6×10^9) joules.

3. An average house in Central United States requires annually about 100,000,000 Btu of energy for space heating.

   a) If this energy were obtained from coal having a heat content of 10,000 Btu per pound, how many pounds of coal are required?

   b) The complete fissioning of 1 pound of ^{235}U produces 50,000,000,000 Btu of energy. If all the fission energy could be converted to heat, how many pounds of ^{235}U would be required to provide the 100,000,000 Btu for space heating?

4. The complete fissioning of 1 kg of pure ^{235}U produces the energy equivalent of burning 6 million pounds of coal. Prove this using the logic outlined below.

   > 235 grams of ^{235}U contains 6×10^{23} atoms.
   > Hence, 1 kg (1000 grams) contains_____atoms.
   > The fissioning of one ^{235}U nucleus produces about 3×10^{-11} joules.
   > Burning one pound of coal produces about (see Table 3.2) 1.3×10^7 joules.
   > Hence the complete fissioning of 1 kg of ^{235}U produces energy equivalent to burning_____pounds of coal.

9.2 Properties of the Atomic Nucleus

5. If a nucleus of an atom were the size of an orange (about three inches in diameter), about how many lengths of a football field away would an electron be? A football field is 300 feet long.

6. The plutonium nucleus, chemical symbol Pu, having 239 nucleons, 94 of which are protons, has important implications in nuclear technology. Write the nuclear formula for this plutonium nucleus.

7. A proton has a mass 1836 times greater than the mass of an electron. What percentage of the mass of the hydrogen atom is contained in its nucleus?

9.4 Types of Nuclear Radiation of Environmental Interest

8. $^3_2\text{He}_1$ is an isotope of helium of considerable interest in physics but its natural abundance is very small. However, it is an end product of the beta decay of tritium. Fill in the blanks below to show how ^3He can be made starting from the capture of a neutron by deuterium.

$$^2_1\text{H}_1 + \text{n} \longrightarrow \underline{\hspace{2cm}}$$
$$\underline{\hspace{2cm}} \longrightarrow \underline{\hspace{2cm}} + \beta^- + \overline{\nu}$$

9.5 Half-life of Radioactive Nuclei

9. Carbon-14 ($^{14}_6\text{C}_8$) is a β^- emitter with a half-life of 5730 years. It is commonly used to radioactively date carbon-based relics in our environment. Write down the reaction for the β^- decay of $^{14}_6\text{C}_8$.

10. A certain radioactive nucleus has a half-life of ten days. What fraction of the sample will have decayed in 30 days? What fraction of the sample will remain after 30 days?

11. Suppose that a radioactive material has a half-life of ten years and that at some given time it had a strength of 640 Ci. If it can be safely disposed of when its strength has diminished to 10 Ci, how many years will you have to wait?

12. After 15.2 minutes the disintegration rate of a radioactive sample is only 1/1024 of its initial value. What is the half-life of this radioactive sample?

13. A measurement of the disintegration rate of a radioactive sample was taken at 2 P.M. on Monday. At 2 P.M. on Friday of the same week, the disintegration rate has dropped to one-fourth the Monday value. What is the half-life of this radioactive sample?

14. Suppose that you had 50 pennies in your hands and, after thoroughly mixing them by shaking, you tossed them on a table. Theoretically, how many pennies will fall "heads up"? Although you can state your answer with some confidence can you predict which pennies will fall "heads up"? How is this situation analogous to the disintegration of radioactive atoms?

15. Picocurie radioactive sources are commonly used for demonstration purposes in classrooms and student laboratories. How many disintegrations occur in one minute with a one picocurie source?

16. Student measurements of the rate of decay of ^{131}I yielded the data shown at the right. Show that these measurements are consistent with the known half-life of 8.05 days for ^{131}I.

Disintegration rate (disintegrations per second)	Time (days)
3400	0
2600	3
2000	6
1600	9
1200	12
950	15
720	18

9.6 Energy Aspects of Nuclear Reactions

17. If the spontaneous fission of $^{235}_{92}U_{143}$ produces only two nuclei, one of which is $^{91}_{36}Kr_{55}$, what is the other nucleus?

18. Two nuclei interact to form a single nucleus having a mass 0.01 amu less than the combined masses of the two interacting particles. What happens to this 0.01 amu of mass?

19. Each of the nuclear reactions listed below cannot happen because a fundamental physics principle is violated. Name the principle for each reaction.

$$^{235}_{92}U_{141} \longrightarrow {}^{90}_{36}Kr_{54} + {}^{142}_{56}Ba_{86}$$

$$^{3}_{1}H_{2} \longrightarrow {}^{3}_{2}He_{1} + \beta^{+} + \nu$$

$$^{11}_{5}B_{6} \longrightarrow {}^{11}_{6}C_{5} + \beta^{-} + \bar{\nu}$$

20. The masses of ^{233}U, ^{143}Ba, and ^{90}Kr are 233.03965, 142.92055, and 89.91960 amu, respectively. Show that 186 MeV of energy are released when ^{233}U spontaneously fissions into ^{90}Kr and ^{143}Ba. (One amu is equivalent to 931 MeV of energy.)

21. One possible way that ^{235}U can be fissioned by a neutron is

$$^{235}U + n \longrightarrow {}^{141}Ba + {}^{90}Kr + 5n.$$

Using the masses given below for the nuclei involved in this reaction, show that 163 MeV are released in the fission. Remember, one amu is equivalent to 931 MeV of energy.

Nucleus	Mass (amu)
^{235}U	235.0439
n	1.0087
^{141}Ba	140.9141
^{90}Kr	89.9198

NUCLEAR-FUELED ELECTRIC POWER PLANTS

10

A nuclear reactor vessel being installed in its protective containment structure. It is in the reactor vessel that energy is derived from nuclear reactions. The comparative size of the workmen and the vessel gives an indication of the magnitude of the engineering aspects. Photograph courtesy of the General Electric Company.

10.1 MOTIVATION

An electric current can be produced in a number of ways. But to date there is only one commercial method and that is with an electric generator in which a massive coil of wire is rotated in a strong magnetic field. Additionally, there are many engines that produce rotational motion but the steam turbine is the most widely used for supplying mechanical energy to a commercial electric generator. A contemporary nuclear electric power plant retains these same conventional features, but through the use of a nuclear reactor provides a different mechanism for supplying the heat to the steam turbine (Fig. 10.1). The nuclear reactor converts the intrinsic potential energy of an appropriate nuclear source into heat. While it is easy to depict these systems operationally, one should never lose sight of the magnitude and engineering aspects of these systems. The reactor vessel alone is typically 70 feet high and 20 feet in diameter and is constructed from steel six inches thick. The development time for commercial application of nuclear power should be firmly entrenched in our minds because it is typical of the time required for any large-scale commercial

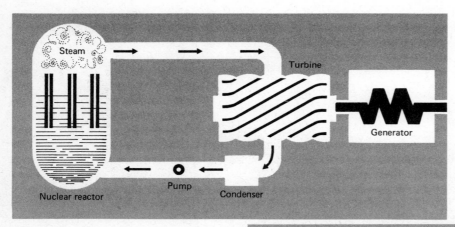

Figure 10.1 Schematic illustrations of the nuclear-fuel and fossil-fuel electric generating systems. Both use a steam turbine to provide power for an electric generator. Heat for vaporizing the water is provided by a nuclear reactor in the nuclear system and by a fossil-fuel-fired boiler in the fossil-fuel system.

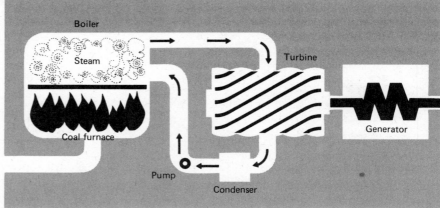

energy system. The feasibility of nuclear energy as a practical source was demonstrated at the University of Chicago in 1942. Yet in 1976 only 2.6 percent of our total energy came from nuclear sources. Many of the problems are technical, but there are also economic, political, and environmental aspects as well. The purpose of this chapter is to provide some insight into the hopes for and promises of contemporary nuclear power plants.

There is very little physical similarity between an automobile engine and a nuclear reactor. Certainly the size of a reactor dwarfs an engine. Yet from the standpoint of energy conversion and control they are quite similar in principle. Both require sources of energy (gasoline and fissionable nuclei), mechanisms for converting the energy (engine and reactor), control mechanisms (accelerator and control rods), and various auxiliaries (carburetor and moderator, for examples). The coupling of these elements is shown schematically in Fig. 10.2. The schematic layout of the essential elements of a reactor shown in Fig. 10.3 provides a basis for our subsequent discussion of the function of each of the components.

10.2 THE NUCLEAR REACTOR PRINCIPLE

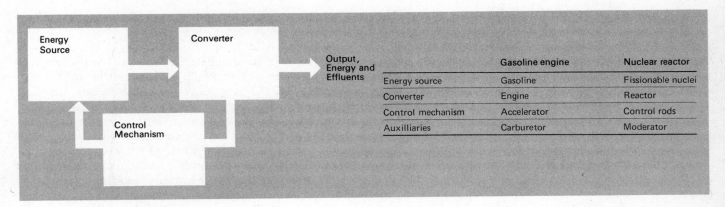

	Gasoline engine	Nuclear reactor
Energy source	Gasoline	Fissionable nuclei
Converter	Engine	Reactor
Control mechanism	Accelerator	Control rods
Auxilliaries	Carburetor	Moderator

Figure 10.2 The basic elements of a gasoline engine and a nuclear reactor. Although the components differ in size and physical function, the energy and control functions are similar in principle.

Fuel

The fission of a ^{235}U nucleus by a neutron is illustrative of the neutron-induced nuclear-fission process:

$$^{235}U + n \longrightarrow {}^{92}Kr + {}^{141}Ba + 3n + Energy.$$

The basic idea in a nuclear reactor is to instigate a fission reaction with a neutron from some source. Neutrons produced in the fission process are used to bombard other fissionable nuclei and thereby produce a self-sustaining chain reaction (Fig. 10.4). Energetically this is similar to starting a fire with a match to produce a self-sustaining combustion process. It would appear that any nucleus comparable in size with ^{235}U would work. However, the chance of fission taking place depends critically on the nucleus chosen and on the energy of the neutron. Figure 10.5 shows the relative fission probability for ^{235}U and

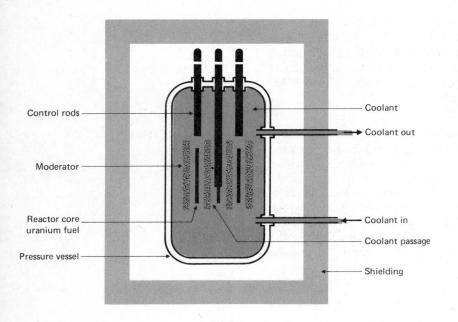

Figure 10.3 Schematic cutout view of a nuclear reactor.

Control rods

Moderator

Reactor core
uranium fuel

Pressure vessel

Coolant

Coolant out

Coolant in

Coolant passage

Shielding

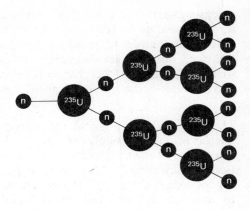

Figure 10.4 Schematic illustration of a nuclear chain reaction. The neutrons released when ^{235}U fissions as a result of interacting with a neutron are used to instigate other fission reactions and keep the process going.

^{238}U as a function of the energy of the bombarding neutron. The probabilities are roughly the same for energies greater than a million electron volts. However, the fission probability for ^{235}U progressively rises for energies less than one million electron volts while that for ^{238}U drops essentially to zero. A sustained chain fission reaction is impossible unless the bombarding neutrons have the proper energy. It turns out that the neutrons are emitted with several millions of electron volts of energy. Still, from Fig. 10.5 it would appear that either ^{235}U or ^{238}U would suffice as a fuel. However, there are many processes other than fission that compete for the neutron. The neutron may be absorbed, scattered, or involved in another type of reaction. It happens that those neutrons that are not removed are rapidly degraded in energy, especially by collisions with hydrogen nuclei in the coolant, to the point that the probability of their inducing fission of ^{238}U is significantly reduced. Hence, given ^{235}U and ^{238}U as possible sources, only ^{235}U has a chance. Unfortunately, natural uranium contains only 0.7 percent ^{235}U. The remaining 99.3 percent is ^{238}U.

Present day reactors require a fuel in which two to three percent of all uranium atoms are of the ^{235}U variety. Thus, the proportion of ^{235}U must be enhanced from its natural abundance. Separating ^{238}U from ^{235}U to achieve the desired enrichment is a nontrivial task because their only distinguishing feature is a slight (about 1%) mass difference. The most widely used method is called gaseous diffusion. It is based on the principle that the ability of a molecule to diffuse through a semiporous barrier depends on the mass of the molecule. Uranium hexafluoride (UF_6) gas is allowed to pass through thousands of these barriers. The gas at the end of the process is richer in ^{235}U. Huge plants at Oak

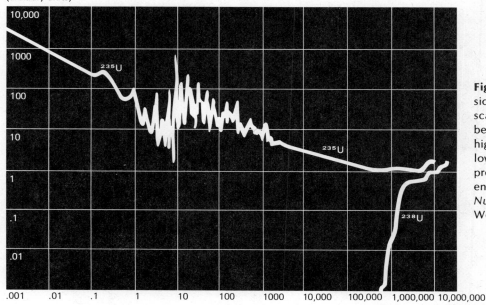

Fission probability
(arbitrary units)

Neutron energy (eV)

Figure 10.5 Relative probability of fission for ^{235}U and ^{238}U. Note that both the scales change by a factor of ten at the beginning of each small block. There is high probability for fission of ^{235}U by low-energy neutrons and essentially zero probability for fission of ^{238}U by low-energy neutrons. (From *Introduction to Nuclear Physics,* by H. A. Enge, Addison-Wesley, Reading, Mass., 1966, p. 444.)

Ridge, Tennessee; Portsmouth, Ohio; and Paducah, Kentucky were built to effect this separation.

Considerable interest has developed for another separation scheme called the centrifuge process. In principle, it operates like the cyclone particulate separator discussed in Chapter 4. A closed cylindrical container of a gas having differing molecular species is set into rapid rotational motion. As a result of the rotational motion, the molecules move away from the center of the cylinder and the gas tends to be compressed near the walls of the container. The thermal motion of the molecules tends to keep the molecules uniformly distributed throughout the volume occupied by the gas. However, the lighter molecules tend to separate from the bulk of the molecules because of their greater speed. The greater speed follows from the fact that although all molecules at the same temperature have the same kinetic energy ($E = \frac{3}{2} kT$) their speed depends on mass ($E = \frac{1}{2} mv^2$). Thus the more massive molecules tend to collect near the walls of the cylinder. A "gas-scooping" device is used to collect the more massive molecules. The separation of ^{235}U from ^{238}U is done by converting uranium to a gas composed of uranium hexafluoride (UF_6) molecules. Those molecules having a ^{238}U atom are slightly more massive than those containing a ^{235}U atom. Three pilot centrifuge separation plants are operated by a consortium consisting of the United Kingdom, West Germany, and the Netherlands. Expansion of these centrifuge separation facilities is underway. A demonstration plant in the United States is operating at the Oak Ridge National Laboratory in Tennessee.

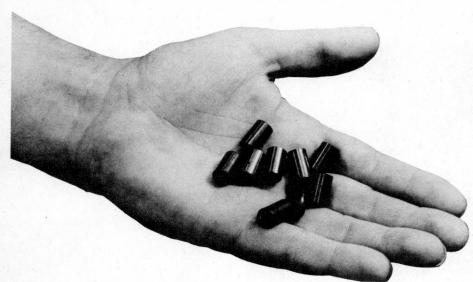

The basic fuel element (Fig. 10.6) is made of a compressed, ¾-inch long, ⅜-inch diameter, cylindrical pellet of uranium oxide (UO_2). The pellets are loaded into tubes 12 feet long made of a zirconium-based metal called zircaloy. These tubes are called cladding. The tubes are assembled into 8 inch by 8 inch by 12 feet long fuel assemblies for use in the reactor (Fig. 10.7). A typical commercial reactor will use about 175 fuel assemblies, each made up of 208 fuel rods.

The energy-release process becomes self-sustaining when more neutrons are generated than are lost in processes that compete with the fission. In this sense, it is somewhat like a profit-oriented business that becomes self-sustaining when more dollars are returned than invested. This condition is achieved only if the resources of the business are properly managed. There are many things such as taxes, overhead, and labor that compete for the invested dollar and often the best management can make a profit only if there is a critical size for the business. Many small stores fail to survive because they cannot compete with the giants of the industry. An analogous situation occurs in a nuclear reactor. The nuclear chain reaction is self-sustaining (profitable) only when there is a critical size or critical mass. It is fairly easy to see why this should be. Figure 10.8 shows schematically a neutron released somewhere within a spherical volume. This volume contains fissionable material as well as material which simply removes neutrons. The neutron will bounce around, but eventually it either enters into a nuclear reaction or escapes from the reactor. Clearly, if the size of the reactor is too small, the probability of escape is large;

hence the probability of fission is small. Therefore, a critical size and arrangement is required which depends on such things as

1. the geometry of the fuel elements,
2. the type of moderator,
3. the purity of the fissionable material,
4. the neutron energy used for the fission process, and
5. the type of material used to reflect neutrons back into the system.

When the fission process is self-sustaining, the reactor is said to be critical. This concept of criticality is extremely important both in the making of a nuclear reactor and a nuclear bomb.

Moderator

Because the fission probability of ^{235}U is higher for low-energy neutrons, an effort is made to degrade the fission neutrons as rapidly as possible. This is the purpose of the moderator. The moderator is made of a material which will degrade the neutron energy by collisions with its atomic constituents. A specific particle will lose the maximum amount of energy in a collision with another particle of the same mass. If, for example, a nonrotating billiard ball makes a head-on collision with a stationary billiard ball, the moving ball will come to rest after the collision. In a neutron-nucleus collision, the neutron will lose the most energy if its mass matches that of the nucleus. Hence, the most efficient moderators are made of materials whose atomic constitutents have masses comparable to the mass of the neutron. The original reactor built by Enrico Fermi and his coworkers used pure carbon (graphite). Because a water molecule contains two hydrogen atoms and the masses of a hydrogen atom and a neutron are nearly equal, present-day reactors use water as the moderator. The water serves as both the moderator and coolant.

Figure 10.7 A drawing of a fuel assembly for a nuclear reactor. The tubes into which the fuel pellets are packed are shown in the cutaway view in the forefront. (Drawing courtesy of General Electric Company.)

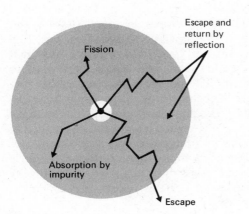

Figure 10.8 There are several fates of a neutron in a nuclear reactor. This figure schematically illustrates a few of the important ones.

As noted in the previous chapter, deuterium is an isotope of hydrogen having about twice the mass of ordinary hydrogen. Because the chemical properties of hydrogen and deuterium are the same, a water molecule can be made using either hydrogen or deuterium. About one of every 7000 hydrogen isotopes in natural water is deuterium.

Some reactors, in particular those made in Canada, use water made with deuterium atoms for the moderator. Although deuterium is a somewhat poorer moderator than hydrogen it is less likely to actually capture a neutron than is hydrogen. Consequently, using this "heavy water" as a moderator it is possible to make a self-sustaining chain reaction using natural uranium (i.e., not enriched in ^{235}U) as a fuel. These reactors are designated heavy water reactors (HWR). Reactors, such as those manufactured in the United States, using ordinary water for the moderator are termed light water reactors (LWR).

Control Mechanism

A nuclear chain reaction is something like a series of collisions which sometimes occurs in a stream of traffic on a crowded highway. That both of these processes can get out of control is obvious. The nuclear chain reaction is controlled by introducing a material that is a more efficient absorber of neutrons than the fissionable material. Cadmium and boron function well in this capacity.

At first thought, it might appear that the chain reaction happens so fast that any mechanical method of control would be ineffective. However, all neutrons from the fission processes are not emitted instantaneously. About 0.7 percent of the neutrons are delayed in emission by times of the order of minutes. This delay in emission is enough to allow mechanical devices to operate.

Converter

The energy of the fission products ultimately produces heat in the reactor core. This heat is usually extracted by circulating water around the fuel elements but gases such as helium and carbon dioxide are sometimes used. Two basic types of water-cooled reactors are called boiling water reactors (BWR) and pressurized water reactors (PWR). These are shown schematically in Figs. 10.9 and 10.10. In the PWR system, the water is kept under pressure so that it doesn't come to a boil. This constraint allows the temperature of the water to be made greater than in a BWR system. The pressure and temperature in a BWR are about 1000 pounds per square inch and 545 °F, respectively, whereas, in a PWR, they are about 2250 pounds per square inch and 600 °F. The steam that ultimately drives the turbine in a PWR comes from water brought to a boil in a heat transfer system. Water is in intimate contact with the reactor cores in both systems, although the contact with the fuel is not direct because the fuel is clad in a tight container. Nonetheless there is always the possibility that water may penetrate the fuel elements. For this reason, the coolant is always contained in a closed cycle system.

Figure 10.9 The water-steam circuit in a boiling water reactor. Water is vaporized in the reactor and the steam is channeled directly to the turbine. Water condensed after passing through the turbine is pumped back into the reactor. The cycle is closed and neither the steam nor the water come into contact with anything exterior to the system. The condenser cooling water never comes into contact with the water circulating in the reactor. ▶

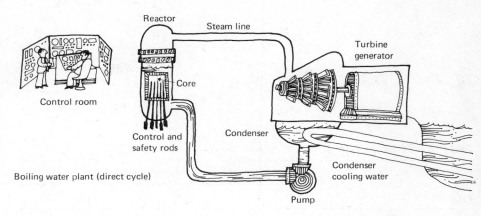

Boiling water plant (direct cycle)

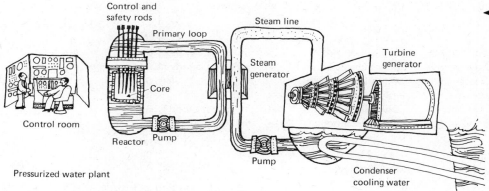

Pressurized water plant

◀ **Figure 10.10** The water-steam circuit in a pressurized water reactor. Water circulating through the reactor in the primary loop is kept under pressure and never comes to a boil. Heat is transferred from this loop to a steam generator that provides steam for the turbine. Water in the primary loop never comes into direct contact with water in the steam turbine circuit. In principle, the steam turbine circuit is identical to that in a boiling water reactor.

Output Energy and Effluents

Table 10.1 gives comparative operational data for electric power stations using fossil fuel and those using nuclear fuel. The lower efficiency for the nuclear plant is a slight disadvantage because the heat disposal problems are compounded. However, its efficiency (based on present reactor technology) is expected to surpass that for fossil-fueled plants in the next generation of reactors (see Chapter 13). Table 10.1 also shows that the problematic emissions from fossil-fueled plants are completely eliminated in the nuclear systems. However, a different type of effluent, radioactive atoms, is released from the nuclear power plants. It is appropriate now to discuss the consequences. It is interesting to note that a certain amount of radioactivity is released in the fossil fuel system. This radioactivity is not a product of the basic energy-conversion mechanism. Radioactive atoms exist as unwanted by-products in the fuel and are released when the fuel is burned.

Table 10.1 Effluents from a 1000-megawatt electric power station.

Type of fuel	Fossil-fueled plant			Nuclear-fueled plant
	Coal	Oil	Gas	Uranium
Annual fuel consumption	2.3 million tons	460 million barrels	6800 million ft^3	2500 pounds
Annual release of pollutants				
Oxides of sulfur	306 million pounds	116 million pounds	0.03 million pounds	0
Oxides of nitrogen	46 million pounds	48 million pounds	27 million pounds	0
Carbon monoxide	1.15 million pounds	0.02 million pounds		0
Hydrocarbons	0.46 million pounds	1.47 million pounds		0
Aldehydes	0.12 million pounds	0.26 million pounds	0.07 million pounds	0
Fly ash (97.5% removed)	9.9 million pounds	1.6 million pounds	1.0 million pounds	0
Annual release of nuclides				
Radium-226	0.0172 curies	0.00015 curies		0
Radium-228	0.0108 curies	0.00035 curies		0
Radioactive noble gases Krypton-85 and Xenon-135	0	0	0	600 curies (PWR) 1.11 million curies (BWR)
Overall efficiency	38%	38%	38%	32%

SOURCE: "Radiation in Perspective: Some Comparisons of the Environmental Risks from Nuclear- and Fossil-Fueled Power Plants" by Andrew P. Hull in *Nuclear Safety*, vol. 12, no. 3 (May–June 1971). The data for fossil-fueled plants are published by J. G. Terrill, E. D. Harward, and I. P. Leggett in *Ind. Med. Surg.*, vol. 36 (1967): p. 412. The data for nuclear-fueled plants are from "Hearings Before the Joint Committee on Atomic Energy, Congress of the United States, Ninety-First Congress, Second Session on Environmental Effects of Producing Electric Power, January 27, 28, 29, 30; February 24, 25, and 26, 1970, Part 2 (Vol. II), Superintendent of Documents, U.S. Government Printing Office, Washington, D.C. 1970. The data for radioactive noble gases are from "Radioactivity in the Atmospheric Effluents of Power Plants That Use Fossil Fuels" by M. Eisenbud and H. Petrow, *Science*, vol. 144 (1964): p. 288.

Table 10.2 Principal radioactive fission products and actinides in high-level radioactive wastes. Half-lives were taken from Nuclear Data Sheets, published by Academic Press, Inc., New York.

Fission fragments			Actinides		
Isotope	Half-life (years)	Radiation	Isotope	Half-life (years)	Radiation
Strontium-90	28.9	beta	Radium-226	1600	alpha, gamma
Zirconium-93	950,000	beta	Thorium-229	7340	alpha, gamma
Technetium-99	210,000	beta, gamma	Plutonium-238	87.8	alpha, gamma
Cesium-135	2,300,000	beta	Plutonium-239	24,390	alpha, gamma
Cesium-137	30.1	beta, gamma	Americium-241	433	alpha, gamma
			Curium-244	17.9	alpha, gamma
			Curium-245	8700	alpha, gamma

Radioactive wastes are broadly categorized as low activity and high activity. The first comes from routine operation of the reactor and fuel processing plants. Contributions come from such things as leaks in the cladding of the fuel elements and irradiation of the coolant and air by neutrons. This radioactive material is collected by various filtering systems and released to the environment under controlled conditions. Gaseous radioactive wastes are normally vented to the atmosphere. Solid and liquid radioactive wastes are buried. In either case, the disposal is done within a framework of regulations administered by the Nuclear Regulatory Commission (NRC). Disposal of the wastes is considered a low-hazard process.

The high-level activity comes from the spent fuel. There are two primary sources. One is from radioactive fission fragments. The other, called actinides, is due to radioactive nuclear species produced from the bombardment of ^{235}U and ^{238}U. For example, plutonium-239 (^{239}Pu) with a half-life of 24,390 years can be made by bombarding ^{238}U with neutrons. There are about 35 fission fragments and 18 actinide radioactive species in the spent fuel element. The most important of these from the hazard standpoint are presented in Table 10.2. This waste is analogous to the hot ashes that are left over in a coal-burning stove or furnace. Hot ashes can be cooled by dousing with water. However, the radioactive "hot ashes" cannot be cooled by any external process. They rid themselves of the excess energy on a time scale determined by the half-lives of the nuclear constituents. Many of these nuclei have half-lives greater than 10 years (Table 10.2). The levels of radioactivity involved are so large that hundreds of years must elapse before the leftover radioactive by-products can be considered safe. The long-term stability required of an appropriate disposal system places an unprecedented burden on our society. Several alternatives have been proposed. We will examine them after discussing the concerns for radioactivity.

10.3 BIOLOGICAL DAMAGE DUE TO RADIATION

Fundamental Mechanisms for Biological Damage

Many of our discussions have used the notion of bound systems built from some basic constituent(s). Living tissue can be viewed within this same framework with cells as the basic units (Fig. 10.11). Cells are responsible for both the metabolic processes (movement, respiration, growth, and reaction to environmental changes) and reproduction. Those cells that control metabolism are termed somatic; those that control reproduction are termed genetic. Control of these cell functions is exercised by the nucleus of the cell. Cells reproduce in tissue through a process of division called mitosis. The hereditary aspects of life are passed on from the parent cell to the divided ones (daughters) in chromosomes that are rich in a substance called deoxyribonucleic acid (DNA). The growth of a cell and its reproduction is governed by biochemical reactions involving many complex molecules. Damage to the molecules involved in these processes *may* alter the function of a cell to the extent that

1. it dies outright and is removed from life processes,
2. it is damaged so that it cannot reproduce and therefore it eventually dies, or
3. it is still able to divide but the functioning of the new cells is altered (mutations).

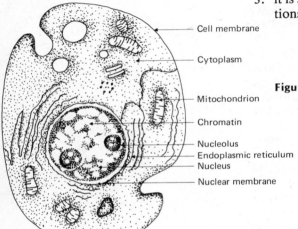

Cell membrane
Cytoplasm
Mitochondrion
Chromatin
Nucleolus
Endoplasmic reticulum
Nucleus
Nuclear membrane

Figure 10.11 Schematic depiction of a cell showing its main components.

Some mutations may be inconsequential. However, some can produce somatic effects such as uncontrolled cell growth which results in cancerous growths. Mutations in genetic cells can be passed on to succeeding generations producing defects in the newborn.

Any mechanism that can supply the energy necessary to break a chemical bond, ionize an atom, or alter the chemistry of cells is capable of producing biological damage. Roughly it takes 10–30 eV to damage a molecule. The alpha, beta, gamma, and neutron radiations from radioactive nuclei typically have energies in the MeV range. Thus they are capable of damaging hundreds of thousands of cells. The actual damage produced depends on the physics of the interaction that, in turn, depends on the type and energy of the particle involved.

A particle loses energy by some appropriate force. In the process, its energy is transformed or converted. For example, the brakes of an automobile exert a force on the wheels which decelerates the automobile. The kinetic energy of the car is converted into heat. A charged particle, like a beta or an alpha particle, loses its energy primarily through the electric interaction with the charged particles associated with atoms. Atoms may absorb energy and move to higher energy states or, if enough energy is transferred, electrons may be completely removed from the atoms. The amount of energy lost depends on the energy, charge, and mass of the radioactive particle. An alpha particle, being about 7000 times as massive and having twice the charge of an electron, loses energy at a much greater rate. For example, a one MeV alpha particle cannot penetrate a sheet of paper, but it takes many sheets of paper to stop a one MeV beta particle. Thus, an alpha particle can produce much more local damage than a beta particle of the same energy.

A neutron, by virtue of having no charge, is very difficult to stop. It loses energy by collisions with nuclei the same way billiard balls lose energy in collisions. If a neutron does make a collision and loses its energy, it can do considerable damage. Because a neutron does not lose energy continuously as a charged particle does, it is not meaningful to talk about a thickness required to stop a neutron of given energy. Rather one talks about a thickness required to reduce the intensity of a neutron source by some given amount. Concrete is often used as a material to attenuate neutrons. It takes ten inches of concrete to reduce the intensity of a 10 MeV neutron source by 90 percent.

A gamma ray has no mass and no charge. It does have electromagnetic energy and does experience electromagnetic forces when it traverses matter. A gamma ray loses energy by three mechanisms:

1. It may scatter from an electron and impart some energy to it. This is called the Compton effect.

2. It may be absorbed by an atom whereupon energy is relinquished to one of the electrons surrounding the nucleus. This is called the photoelectric effect.

3. It may interact with an atom and transform its energy into a negative beta particle (β^-) and a positive beta particle (β^+). This is called pair production. Pair production requires a minimum of 1.022 MeV of energy to compensate for the masses of the β^- and β^+ particles.

Like neutrons, gamma rays do not lose energy continuously and again one talks about a thickness required to reduce the intensity of a beam of gamma rays. Lead is often used to attenuate a source of gamma rays. It takes about 4 cm of lead to reduce the intensity of a beam of 10-MeV gamma rays by 90 percent.

Radiation Units

In terms of energy, a 9-ounce baseball and a 9-ounce chunk of glass moving 60 miles per hour are identical. To a physician, who has to repair the damage inflicted on a person who has had the misfortune to intercept these particular objects, they produce quite different effects. The situation for nuclear radiation is quite similar, with different effects produced by the various types of radiation.

For some situations, one needs to know only how much energy is deposited in some type of matter. This is referred to as a *dose*. The earliest unit for quantifying this concept was the roentgen (abbreviated R.) It is a useful measure for X-rays and gamma rays but is not a unit appropriate for radiations in general. A more appropriate unit is the rad, meaning radiation absorbed dose. One rad is defined as 0.00001 joules of energy absorbed per gram of substance. When the substance is human tissue, the rad and roentgen are essentially the same and they are often used interchangeably. The rem, meaning roentgen equivalent man, is a biological unit which accounts for the biological damage produced. It is much less precise than a physical unit like a rad because of the many factors entering into biological damage. The energy of the particle is a major consideration. For example, both visible light and gamma rays are electromagnetic radiation. However, a photon of visible light is not biologically dangerous because of insufficient energy. Intense local tissue damage can be more serious than diffuse damage. For this reason, an energetic fission fragment which travels a very short distance in tissue is more dangerous than 250-keV X-ray photons, which distribute their energy over a much longer distance. Also, some body organs are much more susceptible to radiation damage than others. Because the amount of radiation damage is directly proportional to the dose received, the rem and rad are related to each other. But, because some radiation is more effective than others in producing biological damage, the proportionality factor, called the relative biological effectiveness (RBE), depends on the type and energy of the particle. Stated formally,

$$\text{dose equivalent in rems} = \text{RBE times dose in rads,}$$
$$DE = \text{RBE} \cdot D.$$

It is customary to use the damage done by a whole body irradiation of 250-keV X-rays as the norm and assign an RBE of unity to this radiation. The RBE for any other particle is then measured relative to this standard. Table 10.3 lists some representative RBE's for various radiations of interest.

Background Radiation

Knowing that radiation can produce very undesirable biological damage, it is only natural to seek protection by avoiding radiation exposure. It is, however, impossible to avoid all radiation because everyone is routinely exposed to a variety of natural and man-made radiation sources. As inhabitants of the earth we are continually bombarded with nuclear radiations coming from radionuclides in the earth (Table 10.4) and from radiations produced by the interaction of cosmic rays with elements of the atmosphere. Cosmic rays are primarily very high-energy protons and gamma rays of extraterrestrial origin.

Table 10.3 Some representative RBE's.

Radiation-type examples	Biological effect	Recommended RBE
X, gamma, and beta rays (photons and electrons) of all energies above 50 keV	Whole-body irradiation, hematopoietic system critical	1
Photons and electrons 10–50 keV	Whole-body irradiation, hematopoietic system critical	2
Photons and electrons below 10 keV, low-energy neutrons and protons	Whole body irradiation, outer surface critical	5
Fast neutrons and protons, 0.5–10 MeV	Whole-body irradiation, cataracts critical	10
Natural alpha particles	Cancer induction	10
Heavy nuclei, fission particles	Cataract formation	20

There is considerable variation in the intensities of these radiations depending on geographic location and altitude. The cosmic background radiation at Boulder, Colorado, may be a factor of two larger than it is in an eastern city that is at a substantially lower elevation. The water from wells in Maine has some 3000 times more radium content than water from the Potomac River. On the average, a person in the U.S. receives an annual exposure of 50 mrem* each from cosmic radiation and from the earth and building materials. An additional 25 mrem is received internally from inhalation of air (5 mrem) and isotopes formed naturally in human tissue (20 mrem). From diagnostic X-rays and radiotherapy, a person will receive about 60 mrem and from the nuclear industry, television, radioactive fallout, etc., 5 mrem. Thus the grand total comes to about 190 mrem/year.†

Table 10.4 Primary radionuclides from the earth that contribute to background radiation.

Isotope	Half-life (years)	Type of radiation
Radium-226 (^{226}Ra)	1622	alpha, gamma
Uranium-238 (^{238}U)	4,500,000,000	alpha
Thorium-232 (^{232}Th)	14,000,000,000	alpha, gamma
Potassium-40 (^{40}K)	1,300,000,000	beta, gamma
Tritium-(^{3}H)	12.3	beta
Carbon-14 (^{14}C)	5730	beta

---

* mrem is the symbol for millirem, which is one-thousandth of a rem.

† See, for example, "Ionizing-Radiation Standards for Population Exposure," by Joseph A. Lieberman, *Physics Today*, vol. 24, no. 11 (November 1971): p. 32.

Radiation Doses for Various Somatic Effects

It is clear that human beings will always be exposed to some radiation over which we have no control and that some risk will always be taken when there is exposure to additional amounts. It is only natural to want to know how much radiation is required to produce a given effect. This is a very difficult question to answer because

1. controlled experiments cannot be done on human subjects, and
2. damage to genetic cells shows up only in succeeding generations.

Information must be obtained from controlled experiments on animals, accidental (and unfortunate) exposures to human subjects, and scientific judgment. The most intense radiation exposures to large numbers of people occurred in the World War II nuclear bombing of Japan and in the Marshall Islands following the hydrogen bomb testing at Bikini Atoll in 1954. Studies of these and other victims have yielded considerable information on somatic effects. Some of these results are shown in Table 10.5. Most of the effects given in Table 10.5 are for one time, whole-body exposures. Larger exposures are required to produce the same effects if the net dose is accumulated over a period of time. This is because somatic cell damage is repairable to some extent. A sunburn, for example, produces considerable cell damage that is normally repaired by metabolic processes.

Table 10.5 X-ray and gamma ray doses required to produce various somatic effects.

Dose (rads)	Effect
0.3 weekly	Probably no observable effect
60 (whole body)	Reduction of lymphocytes (white blood cells formed in lymphoid tissues as in the lymph nodes, spleen, thymus, and tonsils)
100 (whole body)	Nausea, vomiting, fatigue
200 (whole body)	Reduction of all blood elements
400 (whole body)	50% of an exposed group will probably die
500 (gonads)	Sterilization
1000 (skin)	Erythema (reddening of the skin)

Radiation Standards

From our knowledge that risks are involved in exposure to nuclear radiations and that a certain amount of radiation will be released to the environment in the routine operation of a nuclear reactor, the question arises, "Who will decide the maximum allowable exposure for the general public?" In the U.S. this job is given to the Environmental Protection Agency (EPA). It is the function of the Nuclear Regulatory Commission (NRC) to set emission standards consistent with the guidelines of the EPA. These standards have changed throughout the years but since the mid-1950s the maximum allowable *average* exposure for the general population has been set at 170 mrem/year above background. This figure is based on the recommendation that a person should receive no more than 5 rems of radiation above background in a 30-year period. The recommendation is based on studies of the genetic effects of radiation by a committee established by the U.S. National Academy of Sciences–National Research Council and the U.K. Medical Research Council in the mid-1950s. If we add this value of 170 mrem/year to that of the approximate 190 mrem/year that a person normally receives from natural and man-made sources, we find that his total could mount to about 360 mrem/year.

Now that we have a guideline, the question becomes, "Will the emissions from a nuclear reactor fall within these standards?" Table 10.6 gives data for a 1000-megawatt nuclear plant. Although these data are for a single plant, it is clear that the emissions are considerably lower than the maximum EPA levels and, therefore, a high level of confidence should be placed in the ability of nuclear power plants operating routinely to stay within the established radiation levels. There is, however, controversy on whether the 170 mrem/year exposure for the general population is or is not acceptable. That radiation induces human cell damage and forms of cancer such as leukemia is unquestioned. In fact, for exposures above 100 rads, experimental evidence indicates a direct linear relationship between incidences of cancer and radiation exposure (Fig. 10.12). Reliable data for exposures below 100 rads is lacking. Data for these low exposure levels are of extreme importance because they

10.4 EXPECTED EMISSIONS FROM A NUCLEAR POWER PLANT

Table 10.6 Anticipated radiation exposure of a person located 5 and 20 miles from a nuclear power plant. Those labeled A are based on a one percent fuel leak. The actual fuel leak is expected to be somewhat less than one percent, which would make the exposures lower. The actual expected exposures are labeled B.

		Low population zone 5 miles from site	General population zone 20 miles from site	
A.	Air	0.1040	0.0156	Reactor designed
	Water	0.0103	0.0093	for one percent
		0.1143 mrem/year	0.0249 mrem/year	fuel leak
B.	Air	0.0001	0.0000	Reactor with less
	Water	0.0022	0.0020	than one percent
		0.0023 mrem/year	0.0020 mrem/year	fuel leak

SOURCE: "Electric Power Generation and the Environment," by James H. Wright, *Westinghouse Engineer*, May 1970. Reprinted by permission.

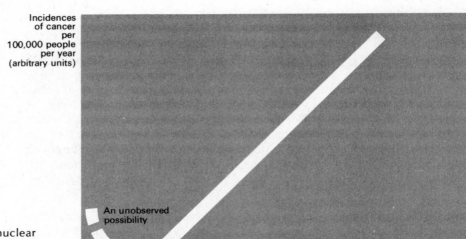

Incidences
of cancer
per
100,000 people
per year
(arbitrary units)

An unobserved
possibility

Linear hypothesis

Threshold hypothesis

0   100   200   300   400   500

Radiation exposure   (mrads)

Figure 10.12 The induction of cancer by nuclear radiation is proportional to the dose received for doses greater than about 100 millirads. Experimental information is lacking for smaller doses. The graph illustrates three possible ways of extrapolating the cancer incidence–radiation behavior for low-dose exposures.

apply to levels of radiation in the realm of background exposures and those which could be achieved under present EPA guidelines. Any estimate of cancer incidences at low dosages must be based on an extrapolation of this curve. The feeling is that if the incidence rate increased as the exposure decreased, then it would be observed; it has not been. There are two other alternatives commonly offered. The first is that no effect occurs until some level of exposure, say around 100 mrad, is obtained. Then the cancer incidence rate increases with exposure as observed. This is referred to as the threshold hypothesis. If this were true, it would be the best alternative. The second proposal is that the observed linear relationship continues down to zero exposure. This is the linear hypothesis. In the absence of conclusive contrary evidence, the linear hypothesis is the safest assumption. Based on this assumption, it has been estimated* that an added exposure of 170 mrem per year to the entire population for 30 years would produce 16,000 cancer cases annually. These estimates are probably realistic *if* the entire population did, in fact, get 170 mrem/year. The NRC maintained that by the year 2000 the radiation from nuclear facilities will be only about 3 mrem/year. This seems entirely reasonable in light of the maximum radiation levels (Table 10.7) to be achieved by light-water-cooled nuclear power reactors (the only type that is widely used at present).

---

* See, for example, "Radiation Risk: A Scientific Problem?" *Science*, vol. 167 (6 February 1970): p. 853.

Table 10.7 Nuclear radiation standards for light-water-cooled reactors.

1. For radioactive material above background in liquid effluents to be released to unrestricted areas by *all* light-water-cooled nuclear power reactors *at a site,* the proposed higher quantities or concentrations will not result in *annual* exposures to the whole body or any organ of an *individual in excess of 5 millirems;* and

2. For radioactive noble gases and iodines and radioactive material in particulate form above background in gaseous effluents to be released to unrestricted areas by *all* light-water-cooled nuclear power reactors *at a site,* the proposed higher quantities and concentrations will not result in *annual* exposures to the whole body or any organ of an *individual in excess of 5 millirems.*

SOURCE: *Federal Register,* vol. 36, no. 111 (June 9, 1971).

Much has been written about the radiation effects controversy. A list of references is included at the end of the chapter. You are strongly encouraged to delve further into this matter.

The fuel rods in a reactor must be replaced before all the fissionable fuel is depleted. Thus, the used fuel rod contains some valuable unspent fuel intermixed with a variety of radioactive products. This unspent fuel is recovered when the rods are returned to the reprocessing center. During reprocessing, the waste is converted to a liquid. It is stored as liquid for a short time so that the short-lived radioactive isotopes can decay. It is converted to a solid within five years and shipped to a repository within ten years after the liquids were produced. The choice of repository presents major concerns. It is not that the volumes involved are so enormous since a site covering 10 acres (a football field is about ¾ of an acre) could handle all solid wastes through the year 2010. Rather it is the potential public hazard associated with the radioactivity and the fact that the radiations are very long-lived (Table 10.2). It is estimated that by the year 2000 there will have accumulated about 150 trillion curies of high-level radioactive wastes (Table 10.8). The thermal power generated by these wastes amounts to about 700 million watts. A relatively simple aboveground repository in a remote area would suffice *if* the site could be isolated from both intruders and nature for several hundred years. If the wastes are stored elsewhere, then the repository must be stable for similar periods of time. Five storage schemes that warrant some consideration are storage in geologic formations, storage in ocean trenches, storage in polar ice caps, projection of wastes into outer space, and nuclear transmutation of the radioactive materials into less harmful products.

Salt formations have received the most attention of the geologic formations considered. The United States is underlaid with about 400,000 square miles of salt beds. This is roughly 10% of the entire area of the country. The beds are located in areas not subject to active geological processes. The areas are devoid of water; otherwise the salt would not be there since it is water soluble. Salt is easily deformed, has good thermal properties, and is easily

10.5 DISPOSAL OF THE HIGH-LEVEL RADIOACTIVE WASTES

Table 10.8 Projected radioactivity levels for high-level radioactive wastes for the year 2000.

	Fission fragments		Actinides	
Isotope	Radioactivity (millions of curies)	Isotope	Radioactivity (millions of curies)	
Strontium-90	13,200	Plutonium-238	214	
Zirconium-93	0.392	Plutonium-239	0.739	
Technetium-99	2.92	Americium-241	64.9	
Cesium-135	0.0786	Curium-244	891	
Cesium-137	19,000	Curium-245	0.318	

SOURCE: "Managing Radioactive Wastes," *Physics Today,* vol. 26, no. 6 (August 1973): p. 36.

mined. Thus, salt formations appear to be highly suitable as radioactive waste disposal sites. The Atomic Energy Commission (AEC) conducted a test project in abandoned salt mines in Lyons, Kansas (Fig. 10.13). The commission deemed the project successful technologically and was prepared to make the mines into a permanent disposal site. However, it was discovered that the area was vulnerable to potential problems caused by water emanating from nearby improperly plugged gas and oil wells and from a nearby mining company.

Figure 10.13 The Carey Salt Company Mine in Lyons, Kansas, where the AEC conducted an experiment in storage of nuclear wastes. Although this particular site was abandoned, salt formations still are potential sites for storing radioactive wastes. (Courtesy of the DOE.)

Although the AEC maintained that the problems were solvable, the state of Kansas objected and the project was abandoned. The NRC has moved the salt repository project to southeastern New Mexico, hoping to have pilot projects built by the early 1980s. There are other geologic formations like shales, clays, limestones, and granites that are prospective disposal sites but they are somewhat less attractive than the salt formations.

Deep trenches exist at various places on the ocean floor. These trenches are being closed by geologic activity in the earth. It has been suggested that radioactive wastes could be dropped into these trenches and then sealed by natural geological processes. This has the advantage of removing the wastes from the earth's crust. However, they would be difficult to deposit, virtually impossible to retrieve, would require risky shipment by ship, and would be in an area subject to geological instability.

The polar ice caps are attractive sites because of their remoteness. The disposal would be simple in principle. Using heat generated by the decaying products, the wastes would simply melt through the ice and sink from their own weight. The disadvantages are that the deposit would have to be made in an extremely harsh climate, shipping would be risky, the ice caps are transient geological features, and their geological structures are not completely known.

The ideal disposal site would be extraterrestrial. The wastes could, for example, be rocketed into the sun or orbited around the sun. They would thus be completely removed from the earth. The disadvantages are obvious. The technology is unproven and would surely be expensive. Nevertheless, as the space shuttle program develops this method will continue to receive attention.

We discussed the process of changing one nuclear species to another by a nuclear transformation. The fission of ^{235}U by a neutron produces new nuclear species. It is conceivable that the radioactive products could be converted to nonradioactive, or perhaps short-lived, products through nuclear transmutations. This would become attractive if vast sources of neutrons became available to effect the transmutations.

Summarizing, it is fair to say that the salt formations appear to be best suited for the permanent storage of radioactive wastes at this time. It is essential, though, that all avenues continue to be explored and that we proceed forcibly to obtain an acceptable storage method for the rapidly expanding nuclear industry.

The transportation of radioactive wastes both from the reactor site to the reprocessing plant and from the reprocessing facility to the repository presents problems no less difficult than those in the permanent disposal process. It is estimated that by the year 2000 between 20 and 50 thousand annual trips will be required to transport radioactive wastes to the reprocessing plant. Transporting solidified wastes to the repository will require between 200 and 500 trips. The potential for vehicular accident is significant. Thus the containment of the radioactivity is essential in the event of a vehicular accident. There is also concern over the theft of potentially useful fissionable materials for bombs or other terrorist activities. These concerns mount if we proceed toward the next generation of nuclear reactors that will produce plutonium-239 in prodigious quantities.

10.6 POSSIBILITY OF THE REACTOR BEHAVING LIKE A BOMB

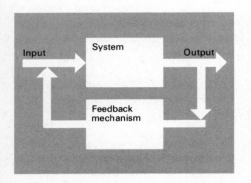

Figure 10.14 Schematic representation of a system with a feedback mechanism. If the system is the earth, the input might be carbon dioxide from burning fossil fuels and the output might be heat from warming by the "greenhouse" effect. Warming might release carbon dioxide trapped in polar ice—accelerating the "greenhouse" effect. This is an example of positive feedback.

When a person driving a car spots a stalled vehicle in his lane, information is fed to his brain which causes him to respond by applying brakes to the car. This is an example of the notion of *feedback*. Feedback is a widely used concept in many disciplines, but especially in physics and engineering. Basically, it means that something from the output of a system is fed back to its input (Fig. 10.14). The "something" can be information, as in the example cited, or such things as energy and electrical signals (voltage or current). For instance, some of the electric energy from coal-burning power plant is fed back to the plant to run the electrostatic precipitators. The feedback may tend to either increase or decrease the output and thus is referred to as being either positive or negative feedback. The feedback in the car example is negative because it results in a decrease in output, namely the speed of the car. In a nuclear bomb the idea is to have a positive feedback mechanism so that the fission reaction proceeds uncontrolled. This is achieved by

1. using no control mechanisms such as cadmium rods,
2. using nearly 100 percent pure ^{235}U or other fissionable materials, and
3. forcing and holding together two pieces of the fissionable material to form a critical mass.

Because a nuclear reactor employs the fission process, the question naturally arises, "Is it possible that these conditions could be achieved accidentally?" The answer is no for the slow neutron reactors considered here. The reasons are as follows. Although the ^{235}U is enriched above its natural 0.7 percent abundance, it is still only about three percent of the uranium used. Thus it is some 30 times *less* concentrated than the fuel needed for bombs. Also, every design precaution is taken to see that a bomb condition cannot develop. If an uncontrolled reaction does start, the intense heat developed will melt the uranium containers and the uranium will separate, tending to decrease the fission reactions. Finally, there is a multitude of fast-acting, negative feedback mechanisms which control the fission process. The control rods are the only moving parts in the reactor. If an "excursion" from normal is sensed, these rods are slammed into a position in which the reactor cannot possibly operate. Again, the time involved in delayed neutron emission is crucial for success of the control operation. Irregular excursions are indicated by such things as abnormal variations in the pressure and temperature of the reactor and in the flow rate of the coolant water. If the water supply fails or is purposely shut off, the neutron-moderating mechanism is removed. This drastically reduces the reaction probability for fission.

10.7 POSSIBILITY OF RELEASE OF LARGE AMOUNTS OF RADIATION

At the beginning of this chapter it was emphasized that the functional diagrams used to discuss the operating principle of a nuclear reactor are exaggerated simplifications. In reality, the reactor and its auxiliaries involve, among a multitude of other things, a complex array of pumps, pipes, and valves used in the process of circulating water into and steam out of the reactor core. Minor malfunctions of the components are inevitable. Routine maintenance of this system is inescapable. Apart from this there is particular concern over the remote possibility of the complete rupturing of a main line that feeds water into the reactor to cool the core. This is referred to as a loss of cooling accident (LOCA). Such an accident could conceivably occur from faulty construction, improper maintenance, natural disaster, sabotage, etc. Although the energy-producing fission process shuts down once the coolant stops flowing, the reactor core still produces heat at a level of five to seven percent of its rated power because of the high-level radioactive fission fragments and actinides in the core. The first effect of a shutdown of this type would be to release steam and water containing radioactivity to the exterior of the reactor vessel through the fracture. This material would be somewhat radioactive because it would have been in contact with the fuel rods which are not sealed perfectly. This radiation would be contained by the surrounding structures. There are two containment procedures currently used. One employs a large spherical or cylindrical steel vessel which essentially surrounds the entire reactor. This is the function of the spherical structure seen on many older reactor installations (Fig. 10.15). In the second type, the reactor vessel is located in a steel containment tank which is surrounded by thick layers of high-density concrete (Fig. 10.16). This structure is also located, at least par-

Figure 10.15 Photograph of the Yankee Atomic Electric Power Plant in Rowe, Massachusetts, which was built in 1964. This reactor uses a spherical radiation containment structure. (Photograph courtesy of Yankee Atomic Electric Company.)

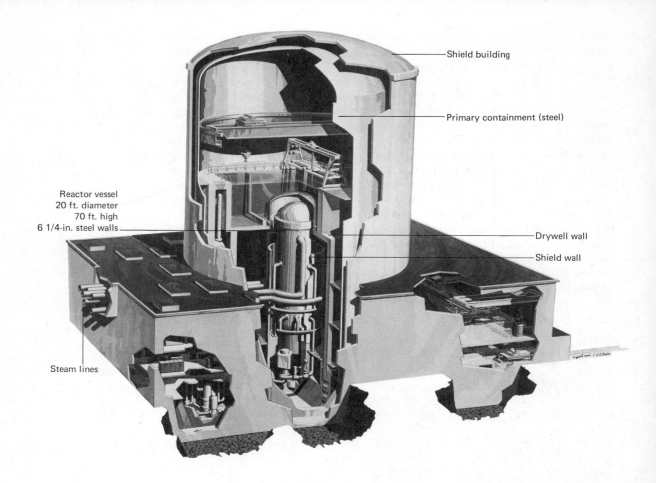

Shield building

Primary containment (steel)

Reactor vessel
20 ft. diameter
70 ft. high
6 1/4-in. steel walls

Drywell wall

Shield wall

Steam lines

Figure 10.16 Drawing of a boiling water nuclear reactor containment structure using steel, reinforced concrete, and earth barriers to contain radiation in the unlikely event of an accident. (Courtesy of the General Electric Company.)

tially, underground. Because a contemporary reactor generates about 3000 megawatts of thermal power when operating normally, the radioactive products will produce about 150–200 megawatts of power after the reactor has shut down. Although the power level drops following shutdown because many of the radioactive products decay rapidly, there is sufficient power to melt the core in about 30 seconds. Unless checked, this molten radioactive mass could melt its way through the bottom of the containment facilities. A set of conditions could be imagined that would release enormous quantities of radiation into the environment. Thus, an emergency core-cooling system (ECCS) must operate within this 30-second time period to cool the core and prevent its melting. Emergency systems operating from completely independent power facilities are built into the reactor system to pump cooling water into the deactivated reactor. A question then naturally arises. How reliable are the emergency core-cooling systems? The answer to this is controversial because it cannot be given definitively and the consequences of a future accident are staggering. These systems have never been tested in the sense that a main water line in a 1000-megawatt reactor has fractured and the emergency system activated. Rather, the reliability is based on computer simulations. Testing on scaled-down versions is in progress at this time. The nuclear industry has maintained that this procedure is reliable. Opponents are skeptical. We must accept and evaluate the reality of risks and then ask ourselves, "Are the benefits of nuclear energy worth the risks?"

An attempt (1974) to evaluate the risks was performed by a group of 60 specialists headed by Dr. Norman C. Rasmussen from the Massachusetts Institute of Technology.* The basic idea in the study was to use the "best engineering judgment" to determine the likelihood that a reactor could have a major accident in an operating year. They tried to envision all possible sequences of failures and from the probabilities of failure of individual components determine the overall probability. They concluded that there is one chance in a million of a given water-cooled reactor failing to the extent that 1000 people would be killed. This is comparable to the possibility of a meteor striking a population center in the United States and killing 1000 people. Less serious accidents are more probable. For example, an accident which might kill 10 people has a probability of one chance in 250,000. Still this is extremely low.

In the event of a major nuclear accident, a nuclear power industry could sustain significant financial losses through liability claims. In order to protect the nuclear industry against liability claims and to encourage the development of nuclear energy, the Price–Anderson Act was enacted in 1957. The act requires a nuclear power industry to purchase the maximum liability insurance available from private insurance companies. In 1957, this maximum amount was $60 million. It is now $125 million. In addition to the private insurance, the Government guaranteed an indemnity equal to the difference between $560 million and the amount obtained privately. A utility pays a premium for the Government's liability coverage. In any accident, no more than $560 million can be paid out in liabilities. Thus, if a given accident produced $1120 million in liabilities, claims would be settled on the basis of 50 cents for each dollar claim. The 1957 act covered a ten-year period. The act was extended for a third ten-year period beginning in August 1977. Amendments to the 1977 renewal provide for a gradual phaseout of Government indemnity through a system of private insurance. Opponents of nuclear power claim that the act proves the "unsafety" of nuclear power plants. Proponents argue that the act provides nothing more than what is done for other industries where lack of insurance experience prevails.

10.8 THE PRICE-ANDERSON ACT

---

* "Reactor Safety Study: An Assessment of Accident Risks in U.S. Commercial Nuclear Power Plants," WASH-1400, August 1974, United States Atomic Energy Commission, Washington, D.C.

10.9 NUCLEAR PARKS AND OFFSHORE SITING

Figure 10.17 An economic advantage accrues from conglomerating nuclear electric power plants. The Calvert Cliffs Nuclear Power Plant located in Lusby, Maryland, utilizes two pressurized water reactor systems, each producing 850 megawatts of electric power. (Photograph courtesy of the Baltimore Gas and Electric Company.)

A bright future has been forecast for nuclear electric power since its feasibility was recognized in the early 1940s. Yet the actual development has consistently lagged behind the forecasts. One reason for this is that every nuclear power station has been designed to accommodate a particular site. Environmental considerations have been justifiably strict and this has increased the delays. The time from "decision to build" to "production of electricity" is often eight to ten years. The industry is striving for standardization so that the construction time can be reduced. One way to standardize is to cluster several plants at the same site. Presently, it is fairly common to build two or three 1000-megawatt plants at the same site (Fig. 10.17). Nuclear parks envisioned for the future would have a combined output of 40,000–50,000 megawatts. With a capacity of this magnitude, nuclear parks could have on-site chemical reprocessing and fuel fabrication and possibly even waste disposal. Nuclear parks would avoid most of the problems associated with transportation of radioactive materials. The isolation of the parks from urban centers would give the public added protection in the event of a nuclear accident.

A still more advanced concept calls for siting nuclear plants in the coastal waters offshore from the mainland (Fig. 10.18). There are several attractive features to this idea. About half of the United States electrical demand is within 300 miles of the shores of the country. Nuclear stations located about three miles offshore could transmit electric power to the bulk of the users

(a)

(b)

Figure 10.18 (a) An artist's conception of a platform-mounted offshore nuclear electric power plant. (b) The power plants would be assembled in a construction facility like that shown here and towed to the installation site. A facility like that illustrated in this conceptual drawing is now under construction. (Drawings courtesy of Offshore Power Systems, Jacksonville, Florida.)

within reasonable transmission distances. Advantage could be taken of the ability of the oceans to assimilate heat and wastes. Offshore siting would also isolate the plants from populous areas. The much sought after standardization would be relatively easy to obtain because of the uniformity of the siting areas. This would tend to reduce the initial costs of the power plants. The safety risks in an offshore plant would be much the same as those in a land-sited plant, but there would be added concern for the possibility of releasing radioactive products into the ocean streams in the event of a large-scale accident.

There are several possible designs for offshore plants. One advanced design requires that plants be built on land in the form of barges that are then towed to and anchored at the site. After anchoring, the facility would be surrounded with a protective breakwater.

10.10 LIFETIME OF URANIUM FUEL SUPPLIES

When discussing energy sources of the magnitude of the reserves of a country, it is appropriate to define an energy unit equal to a billion-billion (10^{18}) Btu. This is symbolized by Q, which denotes both quintillion and a common symbol for heat. The total quantity of the energy derivable from fossil fuels in the United States amounts to about six to eight Q. Depending on projected estimates of consumption rates (Chapter 2), this supply may last for 100–150 years. The estimated fission energy available from thorium and uranium in the United States is about $900Q$. It would appear that if nuclear energy becomes an acceptable energy source, our energy resources would be extended by hundreds of years. And it can, but not with light-water reactors. The reason is that although the $900Q$ is available, it includes ^{238}U and ^{232}Th that will not work in a light-water reactor because of the low fission probability for slow neutrons. Because the useful material, ^{235}U, constitutes only about 0.7 percent of natural uranium, the amount of energy available using nuclear reactor technology like that discussed in this chapter would be comparable to that available from fossil fuels. The ^{235}U resources could last, perhaps, 25 years. These facts create somewhat of a misunderstanding between the public and the promoters of nuclear energy who sometime lead us to believe that the energy available from nuclear sources is nearly limitless. What may not have been explicitly stated is that although the source is nearly limitless, the present technology is not available to exploit it. There are, indeed, genuine prospects, but they are possibly 10 to 15 years from fruition. This and other advanced nuclear technologies are the order of business in Chapter 13.

TOPICAL REVIEW

1. List the energy conversion steps in the production of electricity in a nuclear power plant. How does the process differ from a coal-burning power plant?
2. Name the four principal parts of a nuclear reactor. What are their functions?
3. What is a "chain reaction"?
4. What percentage of the uranium in a nuclear reactor is of the ^{235}U variety?
5. How is the process of gaseous diffusion utilized in the nuclear energy field?

6. What is meant by "critical mass"? Why is it important in nuclear reactors?

7. Distinguish between BWR and PWR reactors.

8. Compare the types of effluents from coal-burning and nuclear power plants.

9. Distinguish between low-level and high-level radioactive wastes.

10. Distinguish between fission fragment and actinide. Name some that are of concern in radioactive wastes.

11. List three possible damaging effects on living cells that can be induced by nuclear radiations.

12. Distinguish between somatic and genetic radiation effects.

13. Distinguish among rem, rad, and RBE.

14. What is background radiation?

15. What is the concern over disposal of high-level radioactive wastes?

16. Name three proposed schemes for disposing of high-level radioactive wastes.

17. What prevents a nuclear reactor from exploding like a bomb?

18. Distinguish between LOCA and ECCS.

19. What is the Price–Anderson Act?

20. How plentiful are the reserves of ^{235}U?

21. What is a nuclear park?

22. Name some attractive features of offshore siting of nuclear power plants.

Nuclear Energy (General Description)

The subjects treated in Chapter 10 are covered in much more detail in the following books.

1. *Nuclear Energy: Its Physics and Its Social Challenge* (paperback), David R. Inglis, Addison-Wesley, Reading, Mass. (1973).

2. *Nuclear Power and the Public,* Harry Foreman (ed.), University of Minnesota Press, Minneapolis, Minn. (1970).

   The following articles and pamphlets are pertinent and well-written.

3. The DOE publishes a series of booklets about nuclear reactors, several of which are particularly relevant to material covered in Chapter 10. These booklets may be obtained from USDOE Technical Information Center, P.O. Box 62, Oak Ridge, Tenn. 37830. The cost is 40¢ (1 booklet), 30¢ (2–10 booklets), 25¢ (11–30 booklets), and 20¢ (31 or more booklets).

Atomic Fuel	Radioactive Wastes
Atomic Power Safety	Sources of Nuclear Fuel
Genetic Effects of Radiation	The First Reactor
Nuclear Power Plants	The Natural Radiation Environment
Nuclear Reactors	Your Body and Radiation

REFERENCES

4. "Fission: The Pros and Cons of Nuclear Power," in *Energy and the Future,* American Association for the Advancement of Science, Washington, D.C. (1973).

5. "Nuclear Energy: Benefits Versus Risks," Walter H. Jordan, *Physics Today* **23**, No. 5, 32 (May 1970).

Disposal of Radioactive Wastes

The following review articles give excellent accounts of the status of the radioactive waste disposal methods.

1. "Managing Radioactive Wastes," John O. Blomeke, Jere P. Nichols, and William C. McClain, *Physics Today* **26**, No. 8, 36 (August 1973).

2. "Disposal of Nuclear Wastes," A. S. Kubo and D. J. Rose, *Science* **182**, No. 4118, 1205 (21 December 1973).

3. "High-Level and Long-Lived Radioactive Waste Disposal," Ernest E. Angino, *Science* **198**, 885 (2 December 1977).

4. "The Disposal of Radioactive Wastes from Fission Reactors," Bernard L. Cohen, *Scientific American* **236**, 21 (June 1977).

Nuclear Reactor Safety

1. "How Safe Are Reactor Emergency Cooling Systems?" Charles K. Leeper, *Physics Today* **26**, No. 8, 30 (August 1973).

2. "Reactor Safety," Herbert J. C. Kouts in *Physics and the Energy Problem,* M. D. Fiske and W. W. Havens, Jr. (eds.), American Institute of Physics, New York (1974).

3. "Nuclear Salvation or Nuclear Folly?" Ralph E. Lapp, *New York Times Magazine,* February 10, 1974.

4. "Reactor Safety Study," WASH-1400, United States Atomic Energy Commission, Washington, D.C., August 1974.

Enrichment of Uranium Fuel

1. "Supplying Enriched Uranium," Vincent V. Abajian and Alan M. Fishman, *Physics Today* **26**, No. 8, 23 (August 1973).

2. "The Reprocessing of Nuclear Fuels," William P. Bebbington, *Scientific American* **235**, 30 (December 1976).

Radiation Effects Controversy

1. "Radiation Risk: A Scientific Problem?" Robert W. Holcomb, *Science* **167**, 853 (6 February 1970).

2. "Tamplin-Gofman, Pauling and the AEC," *Bulletin of the Atomic Scientists* (September 1970).

3. Dr. Ernest J. Sternglass, Department of Radiology and Division of Radiation Health, University of Pittsburgh, has tried to associate infant mortality and nuclear tests. This is also a subject of controversy. "Infant Mortality and Nuclear Tests," Ernest J. Sternglass, *Bulletin of the Atomic Scientists* (April 1969).

4. "Infant Mortality Controversy," Leonard A. Sagan and "A Reply" by Ernest J. Sternglass, *Bulletin of the Atomic Scientists* (October 1969).

5. "Fetal and Infant Mortality and the Environment," Arthur R. Tamplin, *Bulletin of the Atomic Scientists* (December 1969).

Price–Anderson Act

"Price–Anderson: Exploring the Alternatives," Hubert H. Nexon, *Nuclear News* **17**, No. 4, 56 (March 1974).

Offshore Siting of Nuclear Power Plants

1. "Offshore Nuclear Power Stations," Michael W. Golay, *Oceanus* **17**, 46 (Summer 1974).

2. "Floating Nuclear Power Plants for Offshore Siting," Alan R. Collier and Roger C. Nichols, *Westinghouse Engineer* **32**, No. 6, 162 (November 1972).

10.2 The Nuclear Reactor Principle

SUGGESTIONS FOR FURTHER
UNDERSTANDING OF CHAPTER 10

1. On the average, a certain neutron-induced nuclear fission reaction produces 0.9 neutrons. Why can't this reaction be used as a fuel in a nuclear reactor?

2. If a number of dominoes are standing on end in a line and the first is tipped over, then a chain reaction of falling dominoes may develop. Why is a critical arrangement required so that the chain reaction can proceed?

3. Fuel for a nuclear reactor is packed into metallic cylindrical pellets. Why is it important that the material holding the fuel have a high thermal conductivity? What are some other desirable characteristics for the material holding the fuel?

4. Uranium has very few uses other than as a fuel in nuclear reactors. Do fossil fuels have very few uses other than for burning?

5. Why does a pressurized water reactor (PWR) release considerably less radioactivity (see Table 10.1, for example) during its operation than does a boiling water reactor (BWR)?

6. The overall efficiency for generating electric energy is nearly the same for a coal-burning and a nuclear-fueled electric power. Yet a nuclear power plant releases 12–15 percent more heat to the cooling water for the condenser than does a coal burning power plant. Why is this so?

7. Although it is possible to separate ^{235}U from ^{238}U by the gaseous diffusion process, it is difficult because there is relatively little difference in mass. Deuterium ($^2_1H_1$) can be also separated from hydrogen ($^1_1H_0$) by gaseous diffusion. Why would you expect this to be a simpler process?

8. The nuclear equation for the capture of a neutron by a proton is

$$n + p \longrightarrow {}^2_1H_1 + \gamma$$

where γ represents a gamma photon. From mass energy considerations, why is the inclusion of the gamma photon reasonable?

9. How does the structure of ^{238}U differ from that of ^{235}U? Does this slight difference affect the use of either one in a nuclear reactor?

10. There are some neutrons in the "rain" of subnuclear particles from cosmic sources. What prevents a cosmic neutron from striking a ^{235}U nucleus that exists naturally in the ground and triggering a fission chain reaction?

11. Scientific jargon often gets carried over to nonscientific arenas. What do you think is implied by the title "Critical Mass on Capitol Hill," by Anna Mayo (*Bulletin of the Atomic Scientists,* Vol. 30, no. 4 (April 1974): p. 8)?

12. On the average, there are 2.5 neutrons produced in each fission of ^{235}U. These neutrons are used to produce a self-sustaining chain reaction. The reaction $n + {}^9Be \longrightarrow {}^8Be + 2n$ produces enough neutrons for a self-sustaining chain reaction, but it will not work. What mass conservation principle prevents it?

13. Consider the following game. A circle is laid out on a flat piece of ground. You are placed within the circle and your movement is confined to the area of the circle. Another person, preferably of the opposite sex, runs through the circle and you try to capture him (her). Give an argument as to why your probability of success is inversely proportional to the speed of the person running through the circle. In what ways is this similar to the capture of a neutron by a nucleus?

14. All molecules in a gas have the same average kinetic energy regardless of mass. Given that kinetic energy is $\frac{1}{2}mv^2$ where m is mass and v is speed, why do the less massive molecules have the greater average speed? If a container of gas contains two molecular species of the same size but of slightly different mass, which type will make the greater number of collisions with a wall of the container? If the wall is slightly porous so that a molecule might possibly pass through when it collides with the wall, why are the lighter molecules more likely to pass through? This is the principle of the gaseous diffusion process.

10.3 Biological Damage Due to Radiation

15. Order X-rays, neutrons, and fission products in increasing relative biological effectiveness (RBE).

16. Suppose that someone persisted in drinking a radioactive solution for some unknown reason. Even though you know the biological hazards of nuclear radiation, you might argue that it is none of your business and not interfere. But suppose that person were a sloppy drinker and some of the solution spilled off on you. Would your attitude change? Discuss the analogy between this and smoking. (See "Health: Air Pollution and Smoking," Theodor D. Sterling, *Environment,* Vol. 15 no. 6, (July/August 1973): p. 3.)

17. Much like light bulbs that emit visible radiation, many radioactive sources emit nuclear radiation uniformly in all directions. Knowing this, explain why getting away from a radioactive source is an effective form of protection from its damaging radiation.

18. Strontium has many of the physical and chemical properties of calcium. Structures that can be made from calcium can also be made from strontium. Can you see why it is especially important to minimize the ingestion of radioactive strontium by young people?

19. On a yearly average, an American receives about 100 millirems of nuclear radiation from cosmic sources and minerals in buildings and the ground. A resident in Denver, Colorado may receive twice this amount in a year's time. As the public learns more about nuclear radiation, do you think that there will be a migration away from the Denver area?

10.4 Expected Emissions from a Nuclear Reactor

20. How does the routine annual radiation exposure that one might receive from a nuclear power plant compare with the maximum allowable average annual exposure?

21. Some people argue that the only safe level of radioactivity emissions is zero. On this basis, what would be a safe speed for an automobile? How many deaths by automobiles are recorded in the United States on a three-day holiday weekend by deviating from this "safe" speed? (Actually the number of deaths on a three-day holiday weekend is not much greater from any other three-day period.)

10.5 Disposal of the High-level Radioactive Wastes

22. Two radioactive isotopes of strontium are ^{90}Sr and ^{89}Sr. Both are produced in nuclear reactors and nuclear bombs and both emit only beta radiation. Why, then, wasn't ^{89}Sr included in the list of important radioactive nuclei given in Table 10.2?

23. The rate (disintegrations per second) at which a radioactive sample decays depends on how many atoms are present and on the half-life. If the half-life is very long, then the rate will be small even if there are many atoms present.

    a) Calling N the number of atoms and T the half-life, explain why N divided by T has the form of a rate of disintegration.

b) Does this form for disintegration rate agree with the idea that very long half-life means low disintegration rates?

c) Does it follow that radioactive wastes with very long half-lives are necessarily the most problematical?

24. Tables 10.2 and 10.8 give the half-lives and amounts of the radioactive atoms of particular concern in nuclear reactor wastes. Note that for two isotopes such as cesium-135 and cesium-137 the one with the longer half-life has the smaller amount of radioactivity. Give an explanation of this based on the idea that half-life is related to the chance that a radioactive atom will decay in some time period.

10.6 Possibility of the Reactor Behaving Like a Bomb

25. How would you convince a concerned citizen that a light-water-cooled reactor cannot explode like a nuclear bomb?

26. The electric power system in the United States can be thought of as a system which is being driven by positive feedback. The desires of the public for more electric energy have been fed back to the electric power industry, which has responded accordingly. The electric power industry has also encouraged the public to use more electric energy. This driving of the system has led to some forms of pollution and concern for depletion of fuel supplies. What sort of negative feedback do you think is being or could be used to effect more control of the system?

27. Is the loss of cooling water in a light-water reactor a negative feedback or a positive feedback mechanism?

28. About 1200 people are electrocuted each year in the United States. Since there are about 200,000,000 people in the United States, the individual chance of electrocution is about 1 in 160,000. With 100 nuclear power plants in operation the Rassmussen study indicated the individual chance of being killed by a nuclear accident as 1 in 300,000,000. What differences are there in the way these chances are assessed?

29. A failure of the steering system *or* the failure of the brakes may cause a serious accident with an automobile. The simultaneous failure of both the brakes and the steering system could lead to a much more serious accident. Is it possible in a practical sense for a car manufacturer to ascertain by actual experimentation the likelihood of the simultaneous failure of brakes and steering? How is this like trying to determine the likelihood of nuclear accidents by experimental testing?

10.7 Possibility of Release of Large Amounts of Radiation

30. If you were going to establish a home downstream from a large dam, what assurances would you desire that the dam would not break? If you were to build a home near a nuclear power plant, what assurances would you desire that a large release of radioactivity would not occur? In your opinion, are these assurances generally fulfilled?

10.9 Nuclear Parks and Offshore Siting

31. How does the idea of "critical size" pertain to the justification for nuclear parks?

32. Salt is routinely placed on ice-laden highways to induce the melting of ice. However, the salt–water mixture attacks automobile bodies. Knowing this, can you suggest some related problem with offshore siting of nuclear power plants?

10.1 Motivation

NUMERICAL PROBLEMS

1. The chemical reaction $C + O_2 \rightarrow CO_2$ releases 4.1 eV of energy. A nuclear fission reaction liberates about 200 MeV of energy. How many of these chemical reactions does it take to deliver the same amount of energy as a nuclear fission reaction?

2. On the average a city of 1 million people requires 2000 million watts of electric power.

   a) How many joules of electric energy would be required in an average day?

   b) The complete fissioning of 1 kilogram of ^{235}U produces about 82 trillion (82×10^{12}) joules of energy. If the fission energy is converted to electric energy with an efficiency of 30 percent, show that about 7 kilograms of ^{235}U would provide the daily electric energy needs of a city of 1 million people.

   c) A cube of pure ^{235}U one centimeter on a side has a mass of 0.0187 kilograms. What would the lengths of the sides of a cube of ^{235}U be if its mass were 7 kilograms?

10.2 The Nuclear Reactor Principle

3. Boron-10 (^{10}B) is commonly used as a control rod material in a nuclear reactor because of its high probability of capturing neutrons. Complete the reaction below for this process:

   $$^{10}_{5}B_5 + n \rightarrow {}^{7}_{3}Li_4 + \underline{\hspace{2cm}}.$$

4. If a hard sphere of mass M makes a head-on collision with another hard sphere of mass m that is initially at rest, then the kinetic energy of the mass M before and after the collision is related as follows:

   $$KE \text{ (after collision)} = \left(\frac{M-m}{M+m}\right)^2 \times KE \text{ (before collision)}.$$

   What condition prevails if the two masses are the same? A neutron and a proton have nearly equal mass; this is why water with many hydrogen atoms is used to reduce the speed of neutrons in a light-water reactor.

10.3 Biological Damage Due to Radiation

5. For identical doses of radiation, X-rays will produce about 20 times less biological damage than fission fragments. Determine the relative biological effectiveness (RBE) of fission fragments.

6. How many rads of fission fragments would be biologically equivalent to five rads of X-rays?

7. If it takes 10 electron volts of energy to break an important biological molecule, how many of these molecules is a one million electron volt gamma ray capable of disrupting?

8. If a beam of gamma photons each having an energy of 2 MeV are incident on a 6 centimeter thick piece of lead, then the intensity of the radiation after penetrating the lead is only 10 percent (or 0.10) of the original intensity. If this beam of reduced intensity impinges on a second piece of lead 6 centimeters thick, by how much will the intensity be reduced by this additional 6 centimeter thickness of lead? By what percentage has the initial intensity been reduced by the total 12 centimeters of lead? Explain why you can never insert a sufficiently thick piece of lead to reduce the intensity completely to zero.

10.4 Expected Emissions from a Nuclear Power Plant

9. The General Electric Company states that the nearest neighbor radiation exposure from gaseous effluents from its BWR/6 reactor is less than 0.01 millirems per year. How many years of exposure would be required for a person to receive 30 rems of radiation at 0.01 millirems per year?

10.5 Disposal of the High-level Radioactive Wastes

10. The total volume of solidified radioactive wastes from nuclear power plants will be about 470,000 cubic feet in the year 2000. How thick would this be if it were placed uniformly over a football field that is 100 feet wide and 300 feet long? In volume, does this seem like a prohibitively large amount of material to deal with?

10.6 Possibility of the Reactor Behaving Like a Bomb

11. Neutrons in a reactor (or a bomb) are produced in generations. When the number in some generation just equals the number produced in the previous generation, the reaction is just self-sustaining. If the number produced in some generation is only slightly larger than the number produced in the previous generation, the neutron population increases rapidly. We can talk about the number of generations it takes for the population to double. The number produced in one generation divided by the number in the previous generation is called the production ratio. Complete the table at the side until the initial neutron number has doubled, assuming a production ratio of 1.1.

The time between generations in a reactor is about 0.00001 to 0.001 seconds but is typically about 0.00000001 seconds in a bomb. This is another reason why a reactor cannot be made to explode like a bomb.

Generation	Number of neutrons produced
1	100,000
2	110,000
3	121,000
4	
5	
6	
7	
8	
9	
10	

10.7 Possibility of Release of Large Amounts of Radiation

12. About 7 percent of the thermal energy produced in the core of a nuclear reactor comes from the emissions from radioactive atoms in the core. If a reactor is producing 3000 megawatts of thermal energy, how much of this is due to radioactivity?

13. If the cooling water for a nuclear reactor is lost suddenly, then the chain reaction mechanism terminates, but for a reactor producing 3000 megawatts of thermal power, about 200 megawatts remains because of the radioactive atoms present.

    a) Show that in one minute, a power of 200 megawatts produces thermal energy amounting to 12 billion (12×10^9) joules.

    b) To raise the temperature of iron from 500°C to its melting point requires about 670,000 joules per kilogram of iron. How many tons of iron initially at 500°C could be melted by 12 billion joules of thermal energy? Remember, 1 kg is equivalent to 2.2 pounds and 1 ton = 2000 pounds.

14. The likelihood of a dam failure causing 10 fatalities has been assessed at 0.1 events per year. Stated another way, every 10 years we can expect one dam failure killing at least 10 people. Does this assessment seem reasonable? If you are unable to make a judgment from experience, you might do a little library research.

15. The maximum steam pressure that any cylindrical vessel can withstand depends on the radius, thickness, and intrinsic strength of the container. It can be shown that these variables are related by "maximum pressure divided by intrinsic strength equals thickness divided by radius." Boiling-water reactor vessels are about 20 feet in diameter and are made from six-inch thick stainless steel having an intrinsic strength of 100,000 pounds per square inch. Show that the maximum pressure that the vessel can contain is 5000 pounds per square inch. Compare this with typical operating pressures in a BWR reactor (see Section 10.2).

16. When a coin is flipped the probability that the side with a head on it will fall face up is ½. This is because there are two ways that the coin can end up and one event (i.e., head) that is a success. If you flip three coins at once the laws of probability predict that the probability that all three will produce a head is

$$\frac{1}{2} \times \frac{1}{2} \times \frac{1}{2} = \frac{1}{8} \ .$$

There are eight ways that the coins can end up and one event (i.e., three heads) that is a success. You can prove this by figuring out the possible combinations of heads and tails when three coins are flipped. Do this by completing the table at the right.

If some nuclear accident depends on the simultaneous failure of the closing of three valves and the probability of any single valve failing is one chance in a hundred (1/100), what is the probability that the accident will happen?

Coin 1	Coin 2	Coin 3
T	T	T
H	T	T
T	H	T
—	—	—
—	—	—
—	—	—
—	—	—
H	H	H

SOLAR ENERGY

11

Picturesque windmills have pumped water and produced electricity in rural America for many generations. As the search for reliable energy sources continues, modern wind turbines may find increasingly important roles. Photograph by Blair Pittman. (Courtesy of the EPA.)

11.1 MOTIVATION

Solar energy is by far our most available energy source. And although our lives absolutely depend on it for food production and we call on it for a multitude of things ranging from suntanning to clothes drying, these uses barely tap its potential contribution to our energy supplies. Solar energy could make a major impact on our energy economy by (1) providing space heating, space cooling, and hot water for buildings, (2) producing clean fuels, and (3) generating electricity. The potential is stimulating. About 18 percent of the 74 million-billion (74×10^{15}) Btu of energy spent in the United States in 1976 was used to provide heat. If only 10 percent of this energy had been derived from the sun, an energy equivalent to 230 million barrels of oil would have been realized. In money terms, the savings would have been over one billion dollars. Since some 34 billion-trillion Btu of solar energy falls on American rooftops each year, the possibility of using solar energy for heating is not unreasonable. Recognizing this, the federal government is strongly supporting the development of solar heating (and cooling) systems for buildings. "The Solar Heating and Cooling Demonstration Act of 1974" passed in September 1974 calls for two major demonstrations of the practical large-scale use of solar energy: (1) solar heating technology within three years, and (2) combined solar heating and cooling technology within five years. Although complying with this Act will require a major technological and social effort, we can look forward to its rewards.

11.2 GENERAL CONSIDERATIONS

Solar energy is often said to be free. It is, in the sense that you do not have to pay a monthly bill for the use of a certain number of joules of energy. On the contrary, a user may have to pay a monthly bill to liquidate the initial cost of the solar energy collecting system. From this viewpoint, solar energy is not free and a prospective user must decide if in the long run it is more economical to install a solar energy system or to use some alternate. For some applications, the decision is fairly clear cut—for others, it is marginal, at best. It is wrong to think that adapting a house or a building to solar space heating is an easy task. On the other hand, household hot water heating is a solar energy application that is generally practical for any house.

A house or a building is a practical thermodynamic system. Heat added to the house tends to increase the temperature; heat leaked from the house tends to decrease the temperature. If the heat added equals the heat lost, the temperature does not change. Maintaining a constant temperature in a house requires that heat losses be balanced by heat input from the heating system. Heat losses occur from conduction through the walls, ceilings, floors, and windows, convection currents up chimneys and out through cracks and openings, and radiated heat. Historically, energy has been relatively cheap and detailed attention was not always given to preventing heat losses in homes and buildings. Economic considerations are rapidly changing this viewpoint for all forms of space heating. It is easy to see why heat loss considerations are especially important for solar heated buildings.

Typically, a contemporary home at a latitude about that of New York City requires about 100,000,000 Btu of heat during the heating season (roughly

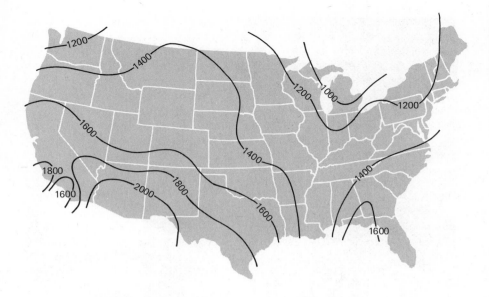

Figure 11.1 Average annual solar energy in Btu/ft² per day.

November through March). The average solar energy input in winter on an inclined south-facing surface is about 1000 Btu per square foot per day (Fig. 11.1). A reasonable solar collector may convert 60 percent of the solar input into thermal energy and about 70 percent of this energy produced may be utilized. Thus the actual solar power available as heat to the house is about

$$\frac{1000 \text{ Btu}}{\text{ft}^2 \cdot \text{day}} \cdot 0.6 \cdot 0.7 = \frac{420 \text{ Btu}}{\text{ft}^2 \cdot \text{day}}.$$

Assuming a 5-month heating season, the daily heat load is

$$\frac{100,000,000 \text{ Btu}}{150 \text{ days}} = \frac{670,000 \text{ Btu}}{\text{day}},$$

and the collecting area required is

$$\frac{670,000 \text{ Btu/day}}{420 \text{ Btu/ft}^2 \cdot \text{day}} = 1600 \text{ ft}^2.$$

This is equivalent to the area of a rectangle 16 feet wide and 100 feet long. Few houses have a south-facing roof with an area this large. Reducing the heat load by one half reduces the collecting area by one half—a much more reasonable size. We will discuss more about how this is done when we consider energy conservation in the concluding chapter.

Given a building in which heat losses have been minimized, it is often possible to obtain a substantial amount of the heat requirement from the sun without mounting solar collectors on the roof. It requires that care be taken to retain the solar energy that enters through properly oriented windows and other spe-

11.3 HEATING AND COOLING WITH SOLAR ENERGY

Figure 11.2 The black, vertical cylindrical structures are designed to store thermal energy. Sunlight entering the building is absorbed by the black surfaces. The absorbed energy raises the temperatures of the medium confined by the cylinders. At night, the thermal energy absorbed during the daytime is radiated to the interior of the house. (Photograph courtesy of Kalwall Corporation, Manchester, New Hampshire.)

cially designed openings. Reliance is made on the ability (heat capacity) of a volume of water, rocks, or concrete to retain the energy captured during the day. A scheme like this is termed passive. Figure 11.2 shows a passive system that collects solar energy impinging on cylindrically shaped columns painted black to optimize absorption of incoming solar radiation. At night, the window areas are covered to prevent thermal losses to the outside of the house. Thermal energy stored during the daytime is radiated to the interior of the house at night.

An active solar heating system, like a conventional heating system, provides heat on call. It does this using solar collectors that feed thermal energy into a storage area. Generally, solar collectors exploit the "greenhouse" effect for collecting solar energy. A typical design is shown in Fig. 11.3. After penetrating through Earth's atmosphere, solar radiation is 7 percent ultraviolet, 47 percent visible, and 46 percent infrared. This radiation is transmitted readily through a glass cover plate and is absorbed by a black metal sheet. This black metal sheet radiates energy just as any object does. Because its temperature is around 600K the radiation is primarily long-wavelength infrared which cannot penetrate through the glass cover plate. Consequently, the interior of the collector warms. Air or a liquid circulating through tubing bonded to the absorbing surface allows heat to flow to a thermal energy storage area. Insulation between the bottom surface of the black energy absorbing sheet minimizes thermal energy losses by conduction. The glass cover plate limits heat transfer by convection. The liquid-cooled system is shown diagrammatically in Fig. 11.4. The collector is oriented toward the south to optimize collection of the sun's energy. The collection area for a well-designed house is roughly one third of the floor space area to be heated. About three gallons of water are required for each square foot of collector area. The space taken up by the tank is a little over 1 percent of the floor area of the house. Heat for the building is drawn from the storage tank by conventional hot water or forced air heating systems. An auxiliary heating system is provided should the sun's energy be insufficient.

In the air-cooled system, heat is removed from the collectors by air flow. Energy is removed from the air stream by passing it over an appropriate solid medium, such as ordinary rocks. Since the heat capacity of rocks is much less than water, (Table 6.1) about 40 percent more rocks (by weight) than

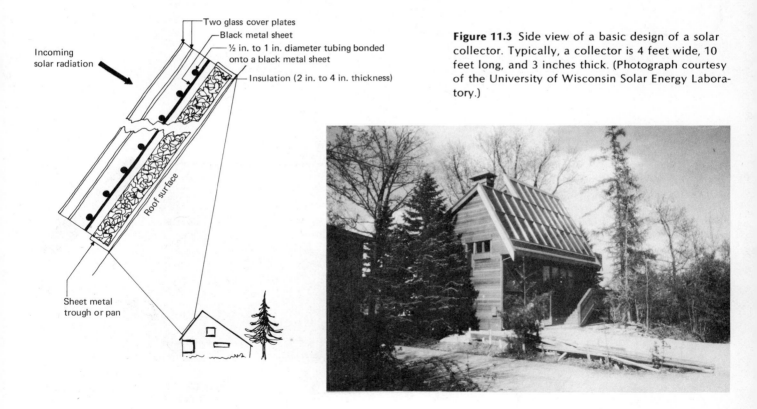

Figure 11.3 Side view of a basic design of a solar collector. Typically, a collector is 4 feet wide, 10 feet long, and 3 inches thick. (Photograph courtesy of the University of Wisconsin Solar Energy Laboratory.)

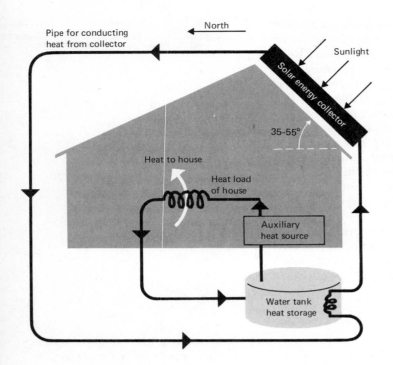

Figure 11.4 Schematic drawing of a solar heated house with liquid-cooled collectors and water for thermal energy storage. Heat is removed from the collectors by an appropriate fluid and transferred to the water tank. Heat is drawn from the tank as needed. An auxiliary heat source is used when the tank contains insufficient thermal energy.

Figure 11.5 Schematic drawing of a solar heated house with air-cooled collectors and rocks for thermal energy storage. Heat is removed from the collectors by circulating air over them. Normally the energy is fed directly to the house (path A). If more energy is available than needed, it is fed into the rock bin storage unit (path B).

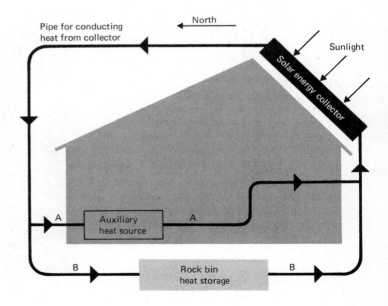

water are required for a given system. The air-cooled system is shown diagrammatically in Fig 11.5. If heat is needed during the day, air is circulated from the house to the collectors and back to the house. If heat is not needed, it is circulated into the rock storage system. Heat can be extracted from the rock system on demand.

Regardless of the time of year, every home requires hot water for a multitude of purposes. Using a conveniently located water heater energized by a flame or electric heating element, water at about 25 °C from the local water system is warmed to about 60 °C. Every kilogram of water warmed requires 1000 calories (1 kcal) for each degree Celsius warmed (see Sec. 6.2). Thus, 35 kcal of energy are required to warm each kilogram from 25° to 60 °C. If the water can be preheated more economically by some auxiliary source, then a monetary savings can be effected. If the preheater fails for some reason, one always has the built-in system to rely on. Solar energy can perform this preheating function and often is able to provide all of the energy. For this reason residential water heating is the most feasible current application of solar energy. To illustrate, let us apply our accumulated knowledge and techniques to the required decision-making processes in the evaluation of a solar energy system.

Solar energy systems can be obtained commercially or constructed personally if one is sufficiently skillful. Whether purchased or home-built, one needs to know the collector efficiency, that is, the fraction of the incident solar radiation converted into thermal energy. Then one needs to know the percentage of days with sunshine and how much solar power impinges on the collector for a given orientation of the collector. Ideally, the collector should be oriented continually toward the sun. While this is possible, it is mechanically complex and is normally economically unattractive. The next best choice is to orient the collector at some fixed angle that provides a reasonable trade-off between the difference in sun angle from winter to summer. A south-facing orientation equal to 15° plus the geographic latitude of the installation is recommended (Fig. 11.3). While this is generally feasible in new construction, retrofitting a house for solar energy may not allow this choice. Then one must locate the collector in the best available position. A number of sources exist for the energy available for different geographic locations and collector orientations.* Finally, one needs to know the initial temperature of the water. This will, of course, depend on location and season. At a latitude of 42° (New York City) water temperature is about 25 °C in summer and about 20 °C in winter.

With this background information about the energy source, one must determine the required collector area. Answers to the following questions are essential.

1. How much hot water is desired?
2. What is the desired hot water temperature?
3. What fraction of the input energy is to be obtained from the sun?

---

* "Applied Solar Energy," Aden B. Meinel and Marjorie P. Meinel, Addison-Wesley, Reading, Mass. (1976).

A typical hot water temperature is 140°F (60°C) and a typical household uses about 20 gallons (75 kg) of hot water per person per day. The fraction of energy obtained from the sun is determined in large part by how much a user will pay. One usually assumes 100 percent and then evaluates the situation on financial grounds.

Let us calculate the collector size required for the following assumptions:

number of occupants = 4
initial water temperature = 25°C
hot water temperature = 60°C
usage rate = 20 gallons/person · day = (75 kg/person · day)
percentage of sunshine days = 80%
collector efficiency = 60%
energy availability = 3400 kcal/m² day (winter)

Using these data we have

$$
\begin{aligned}
\text{thermal energy available} \atop \text{for heating water} &= \text{solar energy} \atop \text{incident on collector} \\
&\times \text{collector efficiency} \\
&\times \text{fraction of days} \atop \text{that are sunshine} \\
&= 3400 \ \frac{\text{kcal}}{\text{m}^2\text{day}} \cdot (0.6)\,(0.8) \\
&= 1600 \ \frac{\text{kcal}}{\text{m}^2\text{day}} \ ;
\end{aligned}
$$

$$
\begin{aligned}
\text{thermal energy required} \atop \text{to heat water} &= \text{water required} \atop \text{per person} \\
&\times \text{specific heat} \\
&\times \text{temperature} \atop \text{difference} \\
&\times \text{number of persons} \\
&= \frac{75 \ \text{kg}}{\text{day/person}} \cdot \frac{1 \ \text{kcal}}{\text{kg}°\text{C}} \cdot 35°\text{C} \cdot 4 \ \text{persons} \\
&= 10{,}500 \ \text{kcal/day}.
\end{aligned}
$$

Hence, the collecting area required is

$$\frac{10,500 \ \frac{kcal/day}{1,500 \ \frac{kcal}{m^2 day}}} = 7 \ m^2 .$$

Having calculated the required collector area for winter when the incident solar power is least, the system will provide more than enough energy during the warmer months. On a yearly basis the energy provided by the sun for hot water heating is

$$10,500 \ \frac{kcal}{day} \cdot 365 \ days \cdot \frac{1}{860} \ \frac{kW\text{-}hrs}{kcal} = 4460 \ kW\text{-}hrs.$$

If the conventional hot water heater were electric and electricity cost 4¢/kW-hr, then the solar energy would be worth

$$4460 \ kW\text{-}hrs. \ \frac{\$0.04}{kW\text{-}hrs} = \$178.$$

If the collector system cost \$1000, then it would take about 6 years to recover the cost at the assumed cost of electricity. However, energy costs will surely escalate and the collector would pay off in less than 6 years.

Although these simple calculations are approximate and depend on a variety of simplifying assumptions, they do show that residential hot water heating is feasible with solar energy. Whether or not the public opts to pay the required initial investment remains to be seen. Similar but decidedly more involved calculations can be made for utilizing solar energy for space heating. In a properly designed new home, solar space heating is generally feasible. Properly designed means making the house as energy efficient as possible. Retrofitting a home for solar space heating is usually difficult because the house is generally not oriented correctly to take advantage of the sun and considerable work is required to make the house energy efficient.

The idea of heating a building with solar energy is obvious. Interestingly, a building can also be cooled with solar energy. Basically the solar energy is used to run a refrigerator. Refrigerators remove heat from some chamber and deposit this heat plus the energy required to do the extraction somewhere away from the chamber. In a building, the chamber to be cooled is its interior. There are different types of cooling systems. Some, like a common household refrigerator, use an electric motor to do the work. Others do not require a motor or engine. All are based on the same general principle of a circulating gas (such as ammonia in a refrigerator) that is easily converted to a liquid under pressure (Fig. 11.6). The refrigerator motor does the work of compressing gaseous ammonia into the liquid state. Heat released in the process is shunted to the exterior. The liquid ammonia then flows to the evaporator in which it returns to its natural gaseous state. In doing this, it extracts heat from the substances it is intended to cool. The cycle is repeated over and over. In a household refriger-

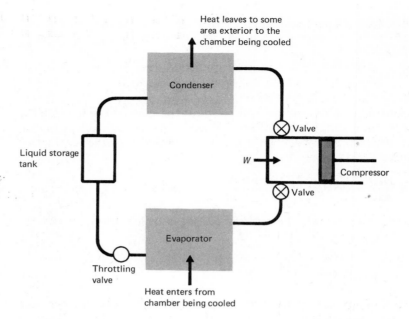

Heat leaves to some area exterior to the chamber being cooled

Condenser

Valve

W

Compressor

Valve

Liquid storage tank

Evaporator

Throttling valve

Heat enters from chamber being cooled

Figure 11.6 The cooling cycle in an ordinary refrigerator. When the liquid evaporates to a gas, it draws heat from the chamber being cooled. After being pumped from the evaporator the gas is fed to the condenser where it loses heat in condensing back to a liquid. The liquid is fed back to the evaporator and the cycle is repeated.

ator, the evaporator is located in the ice compartment and the condenser at the rear of the refrigerator. Although the evaporator is usually hidden from view, you can usually see the condenser.

While the technicalities of solar heating and cooling are well understood, it will be difficult to integrate them into our society because of the reluctance of the public to accept new technology. Implementation could be accelerated if tax incentives were provided or if long-term, low-interest rate loans were made available. Although the widespread utilization of solar energy will take time, it will come. If 20 percent of the energy needs of the United States were supplied by solar energy in 2020, the arduous development path would be worthwhile.

11.4 GENERATING ELECTRICITY WITH SOLAR ENERGY

Generating Electricity from Solar-derived Thermal Energy

The solar power impinging on a square area only 12 inches on a side is about 60 watts. This is equivalent to the electric power required to operate a typical light bulb. It is impossible to convert all this solar energy into electric energy. To achieve 60 watts of electric power from a conversion of solar power at 10 percent efficiency would require a solar power input of 600 watts. A square receiving area of about 38 inches (or about three feet) on a side would be required and this begins to get large. Still if the solar power falling on one percent of the area of the United States were converted to electric power at an efficiency of only one percent, the power produced would exceed the present electric power output of all generators in the United States. If all 50 states in the

United States were of equal area, the equivalent of one half of one state would be sacrificed as a solar energy receiver. Obviously this is a price which has to be considered carefully.

A competitive solar power system would have to produce about 1000 megawatts of electric power. Hence any system, regardless of how the conversion of solar energy is effected, necessarily requires a collecting area of a few miles on a side. An obvious way of achieving this conversion is to concentrate the solar energy by means of lenses or mirrors and convert it to thermal energy. The heat could then be channeled into a conventional steam turbine system. This ploy is faced with the fact that the solar energy is not available on a 24-hour basis whereas the demands for electric energy are. There are several possible ways to circumvent this difficulty. It is conceivable that the electric energy could be stored using batteries. Or some of the electric energy could be used to pump water to an elevated reservoir. Then, as energy was needed, the water could be used in an artificial hydroelectric system (Chapter 12). A possibility that is being seriously considered is to use the electric energy output to produce a fuel that could then be transported and used as needed. One such scheme dissociates water into hydrogen and oxygen. The hydrogen is then used as an efficient nonpolluting fuel. The most advanced system at this time converts solar energy to internal energy in a thermal reservoir for use in a conventional steam turbine-electric system.

A Solar-thermal-electric Power System

A proposed solar-thermal-electric energy conversion scheme is shown in Fig. 11.7. It differs from a conventional fossil fuel system only in the way the heat is produced. On the average, about 3500 megawatts of solar radiation fall on an area of about 14 square kilometers in southwestern United States. Converted to electric power at an efficiency of about 25 percent results in an output characteristic of a modern coal-fired or nuclear-fueled electric power plant.

A typical collector will be cylindrically shaped and will consist of a central metallic pipe surrounded by a glass pipe of larger diameter (Fig. 11.8). The metal pipe is coated with a highly absorbing material and is separated from the outer glass pipe by a vacuum to prevent heat from being conducted away from the inner pipe. Visible radiation focused onto the inner pipe easily passes through the glass and is absorbed. Thermal radiation from the metal is mostly infrared and will not penetrate the outer glass. The thermal energy acquired by the collector is transferred to a thermal storage area by a heat transfer substance such as liquid sodium flowing through the metallic collector. The energy can be stored as latent heat or as thermal energy in an appropriate reservoir. The latent heat principle is as follows. When a solid changes to a liquid, heat has to be added. This is called the heat of melting (or heat of fusion). It amounts to 80 calories per gram for changing ice to water. When a liquid changes to a solid, heat must be extracted. This is called the heat of solidification. Again, it amounts to 80 calories per gram for changing water to

Figure 11.7 The major components of a solar electric plant. Except for the manner of producing thermal energy, it is very much like a conventional coal-burning or nuclear-fueled electric power plant.

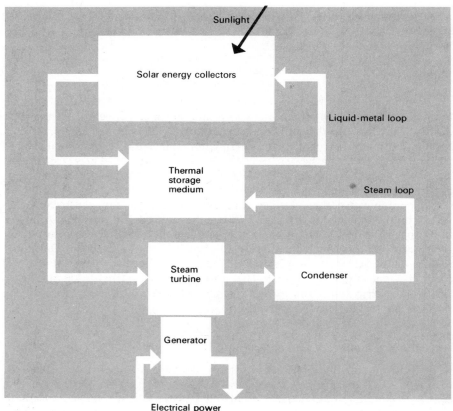

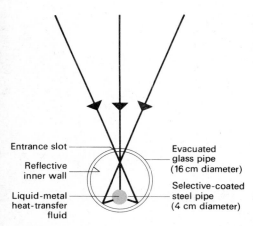

Figure 11.8 Schematic depiction of the way solar radiation would be focused on a pipe containing a heat-transfer medium. The evacuated glass pipe prevents heat from being conducted away from the heat-transfer pipe.

ice. The idea in the solar energy system would be to use heat derived from solar energy to melt a solid. Certain salts have been proposed. Then the energy could be extracted by changing the liquid back to a solid. The thermal storage area would also function as an energy reservoir for those times when the solar radiation was absent. For this reason, the areas selected should not be costly and should be favored with a maximum amount of sunlight. There are vast areas in the southwestern United States that satisfy these requirements. Like any other electric power system, these solar "farms" would not be without environmental impact. They would be subject to the thermal waste energy problem like any steam turbine system, they could upset the local desert ecology, and they could conceivably produce local weather changes. However, in an era of fossil fuel power plant systems with their environmental problems, solar energy systems such as these must be considered as viable alternatives.

The first American system employing this concept is being built in southern California near the city of Barstow. More than 1500 mirrors, each having a surface area of about 40 square meters, will focus solar radiation on a

Figure 11.9 An artist's concept of a 10-megawatt solar-driven electric power plant. Solar radiation impinging on the collectors is reflected and focused on the top of the centrally located tower. The thermal energy generated at the top of the tower is transferred to the buildings below where it is converted to electric energy by conventional means. (Drawing courtesy of Department of Energy.)

configuration of cylindrical tubes mounted atop an 86-meter high tower (Fig. 11.9). A computer-controlled mechanical system continually orients the mirrors toward the sun. An electrical output of about 10,000,000 watts will be produced. An oil/rock thermal storage system will permit the system to produce about 7,000,000 watts for four hours without sunlight. The facility is expected to be operational by early 1981. Successful operation of this facility would pave the way for commercial solar power plants with electric power outputs of 50,000,000 watts and larger.

Generating Electricity with Solar Cells

A photovoltaic cell, commonly called a solar cell, converts electromagnetic (solar) energy directly to electric energy. There is no steam turbine to contend with and there are no moving parts of any kind. Miniature photovoltaic cells are used routinely as light sensors in the light-metering mechanism of cameras. Arrays of solar cells provide electric power for many satellites in the space program. While the microscopic physics of photovoltaic cells is difficult, the salient features can be understood with the physics background at hand.

Metals such as copper and aluminum are referred to as conductors because electrons flow easily through them. Glass, wood, plastic, and other electrically similar materials do not readily conduct electrons. They are termed non-conductors (or insulators). There is a class of materials termed semiconductors having conductive properties intermediate to insulators and conductors. The elements germanium and silicon and compounds such as gallium arsenide, copper sulfide, and cadmium telluride are noteworthy semiconducting materials. Modern electronic innovations such as computers, hand calculators, and hi-fi sets owe their success to microscopic electrical elements made of semiconductors. Semiconductors are categorized as either n-type (negative) or p-type (positive). In an n-type semiconductor, the current mechanism involves negative charges (electrons); in a p-type semiconductor, the current mechanism effectively involves positive charges. This motion of positive charges is called hole conduction. A structurally continuous sandwich of p- and n-type materials is called a p-n junction. A photovoltaic cell is made of a p-n junction (Fig. 11.10). When solar energy is absorbed on one surface of the cell, electron-hole pairs are liberated. Electrons are attracted by an internal electric field into the n-region; holes are attracted to the p-region. A charge separation, and therefore a potential difference, evolves from the resulting separation of positive and negative charge. An electrical device connected across the cell provides a path for charge flow. As long as the cell is illuminated it behaves like a battery producing a potential difference of about 0.4 to 0.6 volts. While the potential difference developed by a single photovoltaic cell of a given type does not depend on the size of the cell, the electric power produced will. A tiny solar cell used in the automatic exposure mechanism of a camera may have a sensitive area of about 4 square millimeters. Manufacturing considerations preclude building a single cell of arbitrary size. However, smaller

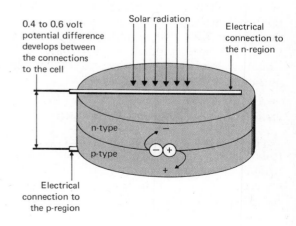

Figure 11.10 Schematic drawing of a solar cell. The cell is about as thick as a dime (about 1 millimeter). Depending on the electric power output, the area varies from a fraction of a square centimeter to a few square centimeters. Absorption of a photon of light liberates an electron (−) • hole (+) pair. An internal electric field across the pn junction forces the negative charge into the n-region and the positive charge into the p-region. The accumulation of negative charges in the n-region and positive charges in the p-region gives rise to a potential difference between the n- and p-regions. The solar cell behaves like a small battery.

Figure 11.11 (a) Series connection of batteries of solar cells. The net electric potential is the sum of the individual components. (b) Parallel connection of batteries of solar cells. The electric potential is unchanged but the parallel combination of four components yields four times as much energy as a single component.

(a)

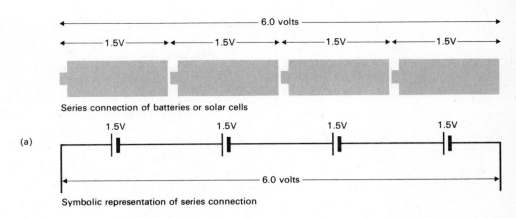

Series connection of batteries or solar cells

Symbolic representation of series connection

(b)

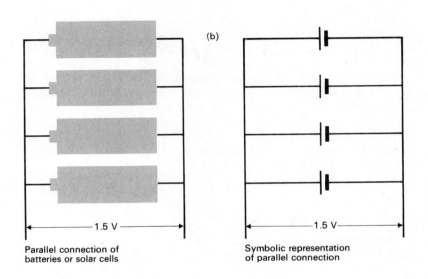

Parallel connection of batteries or solar cells

Symbolic representation of parallel connection

(c) A commercial solar cell array. Each quarter section of a disc is an elementary solar cell producing 0.4 volts. Thirty six of these cells connected in series produce 14.4 volts. The array can deliver 2.16 watts of electric power. This is sufficient to power a small transistor radio. (Photograph courtesy Sensor Technology, Chatsworth, California.)

(c)

units can be connected into arrays (see Fig. 11.11) to increase either or both the potential difference and power. The Skylab manned space satellite was powered with a solar cell array producing 20,000 watts of electric power. This is sufficient to provide elecricity for about seven homes.

Photovoltaic cells in cameras are usually made from cadmium sulfide (CdS). They are relatively cheap but have a light-to-electric energy conversion efficiency of only 4–5 percent. Solar cells for the space program are made from silicon (Si) and have energy conversion efficiencies of 10–15 percent. Being an elemental constituent of sand, silicon is extremely abundant. However, several stages of purification are required to make electronic-grade material. Coupled with a tedious fabrication process, this tends to inflate the production costs of silicon solar cells. Presently the cost of producing electric power with solar cells is about 10 times that of conventional methods. Hopes are high for increasing the efficiency to around 20 percent and decreasing the manufacturing costs. Even if this is achieved, the solar cell energy systems still face the energy-storage problem of all solar systems. Electric energy derived from solar cells finds some application in remote areas requiring only daylight operation. An irrigation facility in Nebraska is powered with a 25,000-watt solar cell array. A 250,000-watt system is ticketed for a community college in Blytheville, Arkansas.

A large-scale (1000-megawatts) solar cell facility would have the same land-use problem of any other solar energy collector of similar power output. A scheme that avoids this particular problem envisions 12-square-mile arrays of solar cells in orbit around the earth (Fig. 11.12). The system takes advantage of the larger concentration of solar energy outside the earth's atmosphere. It would orbit at an altitude of about 22,000 miles from the earth's surface and would be synchronized with the earth's rotation so that it would always appear directly above some position on the earth. Being far out in space, the solar input is available on a nearly continuous basis. The electric energy generated would be transmitted to earth by electromagnetic radiation in the form of microwaves. The receiver would be an open-mesh antenna six miles in diameter. Such a system conceivably could generate and supply enough electric power to satisfy the needs of a city the size of New York City. While the fabrication costs of the solar cell arrays, the expense of the launching, and the engineering expenses preclude the use of these space systems in the near future, investigative efforts are being conducted.

Figure 11.12 A satellite orbiting Earth could conceivably collect solar energy, convert it to electric energy, and transmit microwave energy to Earth. In order to serve a community on Earth, the satellite must maintain its position over the area it is serving.

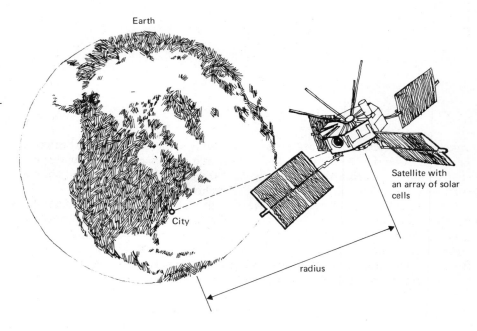

Earth

City

Satellite with an array of solar cells

radius

After the rays of the sun have made their trek through the earth's atmosphere, about 1 percent of the energy reaching the earth is converted to atmospheric motion. Witnesses of hurricanes and tornadoes know that this energy is not distributed uniformly; however, there are regions where the winds are reasonably continuous. The most noteworthy areas are the New England and Middle Atlantic Coasts, along the Great Lakes, the Gulf Coast, and the Aleutian Islands, and through the Great Plains, Rocky Mountains, and Cascade Mountains. Wind power in these regions is estimated to be 100 million megawatts. If only a fraction of this could be converted to electric power, it would easily match the 1977 United States capacity of approximately four-hundred-thousand megawatts.

The idea of using wind-driven electric generators is quite old. In the 1920s and early 1930s, wind-powered electric generators were familiar landmarks on farm buildings. Wind-driven water pumps are still widely used in the Great Plains region. Abandoned windmill towers, now often supporting TV antennas, are common sights in rural America. But the Rural Electrification Administration established by President Franklin D. Roosevelt virtually ended the rural use of wind power. Reasonably large-scale wind-powered electric generators have also been constructed. A 1.25-megawatt wind-powered generator was operated for 16 months during World War II at "Grandpa's Knob" in central Vermont. Power generation stopped after a bearing failure in 1943. Returned to operation in 1945, one of the machines eight-ton blades fractured and separated. The project was never revived largely because of the widespread acceptance of electric power from conventional sources. A serious interest in wind power has returned in this era of diminishing fossil fuels and air pollution problems.

The role of a wind turbine is to convert the kinetic energy of wind into rotational mechanical energy. Clearly, one expects the energy imparted to the rotating turbine to increase as the wind speed increases. Similarly, larger diameter blades intercept more of the wind and therefore impart more energy to the shaft of the turbine. Therefore, we reason that the power developed by a wind turbine will increase as the propeller diameter and wind speed increase. More detailed considerations show that the maximum possible power that can be extracted from an air stream is

$$P = 0.5\,D^2 v^3 \text{ watts,}$$

where D is the diameter of the propeller in meters and v is the speed of the wind in meters per second. With a fairly strong 8-meter-per-second (18-mile-per-hour) wind and a propeller 40 meters (131 feet) in diameter, the maximum power available is 409,600 watts (409.6 kW). A practical wind-turbine generator system converts about one fourth of this to electric power. An average household requires about 3 kW of electric power. Therefore, a single generator of this magnitude could service a community of about 30 homes. In remote areas, wind turbines could be very practical. But in urban areas it is clear that a large number of fairly large wind turbines would be required.

As a prelude to development of commercial wind-powered generators with one to two megawatt electrical outputs, the National Aeronautics and

Figure 11.13 The 0.1-megawatt wind-powered electric generator engineered by the NASA Lewis Research Center, Cleveland, Ohio. The experimental wind turbine has a truss tower 100 feet tall and rotor blades that span 125 feet, tip to tip. The horizontal cylinder at the top contains the power transmission and controls systems. The wind turbine was tested at the NASA Plum Brook Station near Sandusky, Ohio in late 1975. Similar units that will supply as many as 50 homes with electricity are being field tested. (Photograph courtesy of NASA Lewis Research Center.)

Space Administration (NASA) has constructed a 100-kilowatt (0.1 megawatt) prototype (Fig. 11.13). The rotor blade is 125 feet in diameter and the tower is 100 feet high. It commences generating electricity in a wind of 3.6 meters per second (8 miles per hour), and achieves the 100-kilowatt output for a wind speed of 8.0 meters per second (18 miles per hour). After successful operation of this system in late 1975, the development of four larger systems followed. Each of the first two will generate 200 kilowatts of electricity. The third and fourth machines will produce 1,500 and 2,000 kilowatts, respectively. The first field test of this series of generators will take place in the rural community of Clayton, New Mexico.

Wind turbines of the NASA design, having a horizontal axis of rotation, must face the wind for optimum operation. The symmetrical Darrieus rotor (Fig. 11.14) avoids this problem. Although wind turbines of this type are not as developed as the propeller type, they do show promise and are being pursued.

Field testing the wind turbines will provide valuable practical experience and will, in large part, determine the future of large wind turbines. What to do when the wind dies down remains a challenging problem because of the difficulty of storing electric energy. Large-scale utilization of batteries and pumped-storage systems have already been mentioned. Storing energy as compressed air is another. But one of the most interesting and potentially most useful methods is the storage of hydrogen.

Figure 11.14 An experimental Darrieus-type wind turbine generator. Standing seven stories high, the system with its 17-meter-diameter rotor can produce up to 60 kilowatts of electric power in a 28-mile-per-hour wind. (Photograph courtesy of Sandia Laboratories, Albuquerque, New Mexico.)

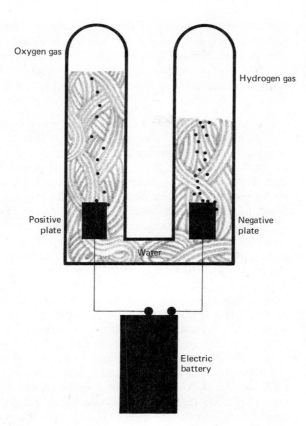

Figure 11.15 Water can be separated into hydrogen and oxygen if electric charge flows through it. Note that the volume of hydrogen is twice that of oxygen because each water molecule contains two atoms of hydrogen and one of oxygen.

From the standpoint of energy, gaseous hydrogen is not a lot different from the very popular natural gas used for household heating and cooking. Its energy content is about 325 Btu per cubic foot as compared with 1030 Btu per cubic foot for natural gas. When burned with oxygen the end product is ordinary water. When burned with ordinary air it also produces oxides of nitrogen. With few modifications hydrogen can be used in place of gasoline in an internal combustion engine. There is some concern about the explosive hazards associated with hydrogen but large quantities are safely and routinely shipped, burned, and used for other purposes. Because water is a compound (H_2O) formed from hydrogen and oxygen, it is possible to reverse the process and separate water into its atomic constituents. One of the most convenient ways of doing this is to pass an electric current through water (Fig. 11.15). Thus the electrical output from a wind-driven generator could be used to dissociate water into hydrogen and oxygen. The hydrogen could be stored as a compressed gas or cooled liquid and shipped to the energy market or piped to market using existing natural gas pipelines. The market could be conventional electric power plants or other industries. Professor William E. Heronemus of the University of Massachusetts has proposed a wind-powered system that could supply most of the New England electricity market by 1990. Wind-powered generators of megawatt capacity would be mounted on 500-foot-high floating stations and anchored offshore. The electric output would dissociate water into hydrogen and oxygen. The hydrogen would be pumped into underground storage tanks for use onshore as fuel for conventional power plants.

The use of wind energy may seem primitive. It is, nevertheless, a significant, nonpolluting source and must be exploited in the future.

11.6 ENERGY FROM THE OCEAN

Our discussion in Chapter 6 led us to the conclusion that the oceans contain an enormous amount of thermal energy but that it could not be extracted by a heat engine whose operating temperature matches the surface temperature of the ocean. However, from absorption of solar energy the surface is warmer than the ocean depths. Near the lower Atlantic Coast and in the Gulf of Mexico the temperature decreases from about 25 °C at the surface to about 5 °C at a depth of 1000 feet. A condition exists for extracting heat from the upper part of the ocean, converting some of the energy to useful work, and rejecting the remainder to the cooler deep region. As remote as this system appears, engineering studies indicate that there are no technical reasons precluding the construction of a 1000-megawatt electric power plant deriving input thermal energy from the ocean. In principle, it would function like a steam-turbine–electric-generator system. But because heat will be extracted at about 20 °C (293K) and rejected at about 10 °C (283K), water cannot be used as a working fluid because it could not be vaporized. A fluid which vaporizes at 20 °C and produces a high vapor pressure is required. Liquid ammonia is a possibility. The maximum thermal efficiency of such a system would only be 3.4 percent because of the small temperature difference between the hot and cold sources. A conceptual design is shown in Fig. 11.16. The structure would be about 1000 feet tall and would be anchored in the ocean. Warm water would flow into the top part of the system, then past the heat exchangers that would transfer heat to a boiler that would vaporize the working fluid. The vapor would drive a turbine generator in the customary fashion. After passing through the turbine the gas would be condensed to a liquid and returned to the boiler.

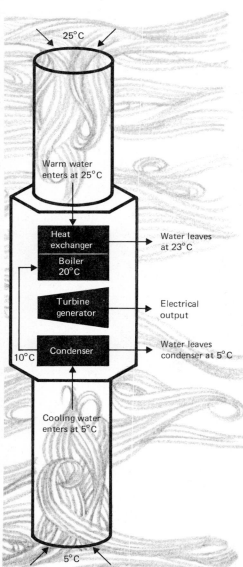

Figure 11.16 Conceptual design of a solar seapower system, operating between ocean temperature levels of 25° and 5°C. Water drawn from the ocean surface layer enters a heat exchanger where energy is transferred to a boiler that vaporizes an appropriate liquid. After passing through the turbine, the vapor is liquefied by a condenser cooled with water pumped from 1000 to 2000 feet below the surface.

The attractive features of this system are (1) it is essentially pollution free, and (2) it makes use of a virtually limitless resource, namely the sun, which replenishes the internal energy of the ocean surface with its heating rays. Constructing such a system in the ocean presents some unique engineering problems. Present United States plans call for construction of a 1-megawatt system in 1979, a 5-megawatt system in 1980–81, and a 100-megawatt demonstration plant in 1983–84.

11.7 BIOMASS ENERGY

A century and a half ago, American homes and factories were fueled primarily by wood. Even today, this solar-derived biomass energy source provides more household heating than the solar collectors which have captured the imaginations of many. If taken seriously, biomass products may provide a variety of solid, liquid, and gaseous fuels in quantities that compete with conventional types. While agricultural and lumber wastes constitute a significant ready-made biomass energy source, plants such as wood, sugarcane, and algae can be harvested for their energy content.

Burned for its heat content wood yields about 8000 Btu per pound— roughly that of the highly attractive Western coal. Additionally, wood has essentially no sulfur and burns with little ash residue. In Chapter 2 we discussed how hydrocarbon-based fossil fuels evolve from decomposition of once-living plants and animals in an oxygen deficient environment. A similar anaerobic (not requiring oxygen) process can be devised for biomass products to produce methane (CH_4), the dominant component of natural gas. The annual production of nearly 300-million tons of agricultural wastes (particularly corn stalks, husks, and cobs) and nearly 30-million tons of manure is a huge ready-made source for anaerobic production of methane. Because natural gas is so widely used in crop drying, the conversion of wastes to methane is appealing. While the wood products industry derives about 40 percent of the total energy used from bark and mill wastes, some 24-million tons of unused mill wastes and 83-million tons of discarded tree trimmings are generated annually. These wastes like nearly all biomass products could be burned for their energy content or converted to ethanol, which is an alternative to gasoline.

TOPICAL REVIEW

1. In what sense is solar energy "free"?
2. What is meant by passive use of solar energy?
3. Discuss the salient features of an active solar heating system?
4. Describe the construction and principle of a solar energy collector.
5. Distinguish between water-cooled and air-cooled solar energy collectors.
6. What physical principle is exploited in a cooling system like a refrigerator or house air conditioner?
7. What are some potential large-scale uses of solar energy?
8. How might solar energy be used to generate electricity?

9. What are some advantages and disadvantages of generating electricity from solar energy?

10. Where does the energy for winds come from?

11. Is it realistic to ever expect significant production of electricity from wind energy?

12. Present a solution to the question, "What do you do in a wind-powered system when the winds die down?"

13. What is the energy mechanism that produces temperature variations near the surface of an ocean?

14. What is the attraction of ocean thermal energy for producing electricity?

15. Describe a possible method for extracting energy from an ocean.

16. What is biomass energy?

17. What are some attractive features of biomass energy?

REFERENCES

Solar Energy for Heating and Fuel Production

1. "Applied Solar Energy: An Introduction," Aden B. Meinel and Marjorie P. Meinel, Addison-Wesley, Reading, Mass. (1976).

2. "Solar Heating Design," William A. Beckman, Sanford A. Klein, and John A. Duffie, John Wiley and Sons, New York (1977).

3. "The Solar Home Book: Heating, Cooling, and Designing with the Sun," Bruce Anderson, Cheshire Books, Harrisville, New Hampshire (1976).

4. "Solar Research and Technology," L. O. Herwig in *Physics and the Energy Problem—1974,* M. D. Fiske and W. W. Havens, Jr. (eds.), American Institute of Physics, New York (1974).

Solar Energy for Electric Power Production

1. "Solar Thermal Electricity: Power Tower Dominates Research," *Science* **197,** 353 (22 July 1977).

2. "Photovoltaics: The Semiconductor Revolution Comes to Solar," *Science* **197,** 445 (29 July 1977).

3. "Solar Energy Research: Making Solar After the Nuclear Model?" *Science* **197,** 291 (15 July 1977).

4. "The Coming Age of Solar Energy," D. D. Halacy, Jr., Harper Row, New York (1973).

Energy from the Winds

1. "Wind Energy: Large and Small Systems Competing," *Science* **197,** 971 (2 September 1977).

2. "Wind Energy Conversion Systems," Report No. NSF/RA/W-73-006, December 1973. Available for $6.50 from National Technical Information Service, Springfield, Virginia 22151.

3. "Windmills: The Resurrection of an Ancient Energy Technology," Nicholas Wade, *Science* **184,** 1055 (7 June 1974).

4. "Windmills," Julian McCaull, *Environment* **15,** No. 1, 6 (January/ February 1973).

Energy from the Sea

1. "Ocean Thermal Energy: The Biggest Gamble in Solar Power," *Science* **198,** 178 (14 October 1977).

2. "Solar Sea Power," Clarence Zener, *Physics Today* **26,** No. 1, 48 (January 1973).

3. "The Sea Plant, A Source of Power, Water and Food without Pollution," J. H. Anderson, *Journal of Solar Energy* **14,** No. 3 (February 1973).

4. "Ocean Temperature Gradients: Solar Power from the Sea," W. D. Metz, *Science* **180,** 1266 (1973).

Biomass Energy

1. "Photosynthetic Solar Energy: Rediscovering Biomass Fuels," *Science* **197,** 745 (19 August 1977).

11.2 General Considerations

1. Suppose that someone builds a structure that prevents solar radiation from impinging on a neighbor's solar energy collectors. Do you think that there should be legal restrictions against such structures?

2. If solar heating can be shown to work in a cold climate, what will determine whether homeowners will or will not opt for it?

11.3 Heating and Cooling with Solar Energy

3. What are some ways that solar energy is routinely used in and around a home?

4. What problems will arise in trying to introduce solar home heating to the public?

5. Water and rocks are commonly used to store energy in solar-heated buildings. What advantages does water have over rocks?

6. Why is a solar collector oriented in a south-facing direction?

7. The sun is lower in the sky in winter than it is in the summer. Show how a properly designed window awning allows solar radiation to enter in the winter but not in the summer.

8. Why can a tree be a good summer solar radiation shield for a house but still allow significant solar radiation to impinge on the house in winter?

9. When the outdoor temperature is 25 °F, the temperature between a storm window and a single-pane glass window can easily be 60 °F if irradiated by sunlight. What is responsible for this rise in temperature?

10. If you had a material that melted at 50 °F, how could you use it to cool a building?

11. To melt a solid requires the addition of heat. For ice it is 80 calories per gram. Is it possible to ever recover the energy that goes into melting a solid?

11.4 Generating Electricity with Solar Energy

12. What are some environmental questions that might be raised when solar energy is considered as a large-scale energy source for producing electricity?

13. Compare the priorities of using land for solar farms as against highways and roads.

11.5 Energy from the Winds

14. Wind turbines for powering large electric generators will have to be located a few hundred feet off the ground. What environmental problems will result from having to do this?

15. What energy transformations result when energy is extracted from wind to produce electricity?

16. Why is energy required to separate a water molecule (H_2O) into hydrogen and oxygen?

11.6 Energy from the Ocean

17. Would you expect electric power plants utilizing heat from the ocean to be built offshore of the North Atlantic and North Pacific coasts?

18. What environmental problems might result from electric power plants submerged in the ocean?

19. It is estimated that about 30 percent of the electric power output from an ocean thermal energy electric power plant will be needed to power the pumps that circulate water in the system. Why does such a system require such a large pumping capacity? What sort of engineering problems does this requirement present?

20. Because the temperature difference of the source and sink for an ocean thermal energy electric power plant is very small the energy efficiency at best is very small. If such a system is to produce electricity on a competitive scale why would you expect the rate at which water is taken in to be very large?

11.2 General Considerations

1. Seventy-five million-billion (74×10^{15}) Btu of energy were used in the United States in 1976.

   a) Determine the amount of energy used for space heating if 18 percent of the total energy used went for this purpose.

   b) Knowing that the energy content of a barrel of oil is 5.8 million Btu, show that 230 million barrels of oil could have been saved if 10 percent of the energy for space heating was derived from the sun.

2. The mean distance from the earth to the sun is 93 million miles. When solar radiation has propagated this distance it has spread out over the surface of an imaginary sphere of radius 93,000,000 miles. Measurements just outside the earth's atmosphere reveal that the radiation on each square foot of this sphere is 130 watts. Show that the radiation emitted by the sun is 400 billion-billion megawatts $(400 \times 10^{18}$ MW). The formula for the surface area of a sphere of radius r is $A = 12.6r^2$. There are 5280 feet in a mile.

11.3 Heating and Cooling with Solar Energy

3. A dry-roasted peanut has an energy content of about 1 food calorie (1000 calories). If the energy from burning a dry-roasted peanut is applied to 1 kilogram of water, how much will the temperature of the water rise? (Actually, a dry-roasted peanut will burn. Try it.)

4. About 1000 gallons of fuel oil are needed each year to heat a home in Central United States.

   a) Assuming a five-month heating season, how many Btus are required each day?

   b) If the solar energy input is 1200 Btu/ft² · day and a collecting system can utilize 50 percent of the input solar energy, determine the required collecting area.

5. Using Fig. 11.1, determine the average annual solar energy input for the area where you live. Using information from Table 3.2 express your result in kcal/m² · day.

6. If the tap water temperature in your area is 30°C, how many kilocalories of heat are required to increase the temperature of 20 kilograms of water to 58°C?

7. Elementary school students often demonstrate the magnitude of solar energy by constructing a solar cooker for hot dogs. It consists of a bowl-shaped structure of aluminum foil. The bowl shape allows the radiation to be focused on the hot dog.

   a) Suppose that ten percent of the radiation falling on the reflector gets absorbed by the hot dog. How big a surface area would be required to achieve a power of 100 W on the hot dog?

b) Suppose that a 1000-W household appliance cooks a hot dog in five minutes. How long would it take the solar cooker to achieve the same result?

8. The exercise below involves considerations for a solar hot water heating system. Using the example in Section 11.3 as a guide, fill in the following designated blanks:

      number of occupants = 3

      initial hot water temperature = 30 °C

      hot water temperature = 57 °C

      usage rate = 20 gallons per person per day

      percentage of sunshine days = 75%

      collector efficiency = 65%

      average energy available = 1200 Btu/ft^2 · day

      thermal energy available for heating water = _____

      thermal energy required to heat the water = _____

      collecting area required   = _____

      installed cost of collectors at \$20/ft^2 = _____

9. Glauber's salt (sodium sulfate decahydrate) that melts at 90 °F is being studied as a heat-storing medium for homes and industries. Energy is required to melt the salt but this energy can be released when the molten salt changes back to a solid. It takes 0.3 Btu to melt 1 cm^3 of Glauber's salt. When the outdoor temperature is 0 °F, a well-insulated home needs about 1 million Btu of heat for a two-day period. How many cubic centimeters of Glauber's salt would be required to store 1 million Btu of thermal energy? Compare this volume with a household refrigerator measuring 100 cm × 200 cm × 50 cm.

10. Suppose that you wanted to build a device that would convert solar radiation to electric power. If the radiation amounts to 0.1 W/cm^2 and you could convert the solar energy to electric energy with an efficiency of one percent, how big a square would be required to produce 1000 W of electric power?

11. a) If you were to place a 12-inch-square piece of paper on the ground, on the average it would receive about 65 watts of solar radiation. Approximating the United States as a rectangle 3000 miles long and 1000 miles wide, how much radiaton falls on one percent of the land area of our country?

b) If this radiant power is converted to electric power with an efficiency of one percent, show that it would exceed the 1977 electric generating capacity of about 400,000 megawatts (4×10^{11} watts).

11.4 Generating Electricity with Solar Energy

12. In early 1978, electricity could be derived from solar cells at a cost of $11 per watt of electric power. What would the cost of the solar cells be for a home requiring 2000 watts of electric power?

13. Suppose that the batteries in a small electric car had the energy equivalent of four gallons of gasoline (about 500,000 Btu) and that a recharging system using photovoltaic cells (Section 11.4) is used to recharge the batteries. If the batteries are to be recharged in a day's time and the average solar energy available is 1600 Btu/ft² · day, how big a collecting area would be required if the recharging system has an efficiency of 10 percent? Does the size required seem practical to you?

14. The data in Fig. 11.1 are expressed in Btu/ft² · day. A Btu is a unit of energy and day is a unit of time. Thus Btu/day is a unit of power.

    a) Knowing that 1 Btu = 1055 joules and 1 day = 86,400 seconds, show that 1 Btu/day = 0.0122 watts.

    b) A small car is about 5 feet wide and 10 feet long. In an area where the average solar energy input is 2000 Btu/ft² · day, about how many watts of solar power impinge on a small car?

    c) If this solar power is used to run an electric motor, what is the largest possible power output that can be obtained from this motor?

    d) If this car could function with a 10 horsepower motor, could you ever get enough solar power to operate it? (1 horsepower = 746 watts.)

15. The Skylab manned space satellite was powered with an array of solar cells producing 20,000 watts of electric power. If the energy conversion efficiency of the solar cells is 15 percent and the solar power available is 1400 watts per square meter, how big a collecting area is required? Compare this area with a classroom 7 meters wide and 12 meters long.

11.5 Energy from the Winds

Assume the power in watts available to a wind-powered electric generator is $P = 0.5 \ D^2 V^3$ where D is the diameter of the propeller in meters and V is the wind speed in meters per second. The following problems involve manipulation of this expression.

16. How much power is available to a wind-powered electric generator having a diameter of 40 meters if the wind speed is 8 meters/second?

17. Suppose that a rural home could manage with a wind-powered electric generator producing 250 watts. If the system had an efficiency of 25 percent and the wind speed was 5 meters/second, how large a diameter propeller would be required?

18. If the efficiency for generating electricity does not depend on wind speed, how will the power output of a given generator change for a doubling of the wind speed?

19. If the efficiency for generating electricity does not depend on propeller diameter, how will the power output change for a doubling of the propeller diameter?

20. The first experimental wind turbine engineered by NASA used blades 38 meters long. The maximum wind speed where it was tested is about 8 meters per second.

   a) Determine the available wind power for these conditions.

   b) The electrical output for these conditions is 100,000 watts. What is the efficiency of the system for converting wind power to electric power?

21. Most of the expressions that we have used for quantitative analyses involve terms that do not involve a numerical exponent. For example, the potential energy of water atop a dam is $E = w \cdot H$. On the other hand, the power associated with wind flowing into the blades of a wind turbine varies as the cube of the speed; i.e., P is proportional to V^3. This is very important because it means that if the speed is doubled the power will increase eight times. It is fairly easy to see why the power varies as the cube of the speed. Imagine a cylinder of air of cross-sectional area A and length L moving by the turbine blades with speed V.

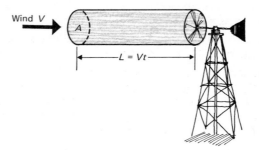

The mass of this amount of air is $M = \rho V = \rho A L$ where ρ is the density of the air (in kilograms per cubic meter, for example). The kinetic energy of this mass is just $E = \frac{1}{2} MV^2 = \frac{1}{2} \rho A L V^2$. If this cylinder flows by in a time T, the power is

$$P = \frac{E}{T} = \frac{1}{2} \rho A \left(\frac{L}{T}\right) \cdot V^2.$$

However, L/T is just the speed. So $P = 1/2 \rho A V^3$ showing that power is proportional to the cube of the speed.

   a) Does all the wind power blowing into the wind turbine get transferred to the turbine blades?

   b) How does the concept of efficiency apply to this situation?

   c) Certainly the power developed should increase as the length of the turbine blades increases. Why should you expect the power to depend on the square of the blade length, i.e., P is proportional to D^2?

11.6 Energy from the Ocean

22. In Chapter 6 we showed that the maximum thermal efficiency for a heat engine extracting heat at a temperature designated T_{hot} and rejecting heat at a temperature designated T_{cold} is

$$\epsilon_{max} = \left(\frac{T_{hot} - T_{cold}}{T_{hot}} \right) \times 100 \text{ percent.}$$

Show that the maximum thermal efficiency for a solar sea power system operating between temperatures of 20°C (293K) and 10°C (283K) is 3.4 percent.

23. If the turbine of an electric power plant deriving input thermal energy from the ocean has an energy-conversion efficiency of 2 percent, at what rate will it have to extract thermal energy in order to have a power output of 1000 megawatts?

24. Hydroelectric units at the Grand Coulee Dam discharge water at a rate of 1,870,000 pounds per second. Because of the intrinsic, low energy-conversion efficiency, an ocean thermal energy electric power plant generating 100 megawatts (remember, a large electric power plant generates 1000 megawatts) will have to process water at a rate of 100,000 gallons per second. Compare this rate with the discharge rate at the Grand Coulee Dam. (One gallon of water weighs 8.3 pounds.)

OTHER ENERGY ALTERNATIVES

12

Shown here is an electric power plant that derives its input energy from steam emanating from Earth's interior. Located at The Geysers in Northern California, this facility generates 24 megawatts of electric power. (Photograph courtesy of Pacific Gas and Electric Company.)

12.1 MOTIVATION

In this chapter we are going to discuss some alternative energy systems. Some are old, some are visionary. The production of electricity from flowing water from a dammed river has been viable for several decades and even today makes a significant impact on our energy budget. Although hydropower will continue to grow slightly, it will make less of an effect on our energy appropriations in the future. Energy from the tides has been a dream for half a century, but it is likely that none of us will live to see a single kilowatt of electric power generated from it in this country. Some systems are fascinating. Can you imagine your reading lamp being energized with electric power from a power plant whose input energy comes from garbage and refuse! Although some of these facilities may seem remote, none is beyond the realm of possibility. If priorities change or unsuspecting snags develop in the advanced electric power technologies, these systems may well come to the forefront of the energy picture.

12.2 HYDROELECTRICITY

The monumental dams on the Tennessee, Colorado, and Columbia rivers have been tourist attractions for several decades (Fig. 12.1). Their prime function, though, is to retain river water so that it can be shuttled off to a water turbine that drives an electric generator. The principle of the process is straightforward. It simply involves a way of supplying power to a generator other than by a steam turbine. The energy language is much the same. Mechanical energy in a steam turbine is produced by allowing steam to flow from an area of high thermal potential energy to an area of low thermal potential energy. Energy is transferred to the turbine in the process. From our considerations in Sections 3.6 and 3.7, we know that water at the top of a dam is in a higher state of

Figure 12.1 The Grand Coulee Dam on the Columbia River in Washington is 4173 feet long and 550 feet high. The electric power output from the generators at its base amounts to over 2000 megawatts. Roosevelt Lake, impounded by the dam and used as a recreational facility, extends for 151 miles upriver. (Photograph courtesy of United States Department of Agriculture, Soil Conservation Service.)

gravitational potential energy than water at the bottom of the dam. As the water flows from higher to lower energy states, energy is extracted by a water turbine to drive the electric generators. Elementary considerations tell us that the amount of energy available will depend not only on the difference in height of the two water levels but also on the weight of the water that flows over the dam. As an illustration of this idea, imagine the difference in energy transferred by one-pound and one-ounce boxes falling on your toe from a desk top. Expressed formally,

energy available = weight × difference in height,
$$E = w \cdot H.$$

The power available is the rate of change of this energy transformation. Expressed formally,

power = energy ÷ time,
$$P = \frac{E}{T} = \left(\frac{w}{T}\right) \cdot H.$$

Note that w/T is just the rate of flow of the water in newtons (or pounds) per second. To illustrate, the new hydroelectric units at the Grand Coulee Dam will discharge water at a rate of 1,870,000 pounds per second from a height of 285 feet. The power available amounting to 720 megawatts will be converted to electric power at an efficiency of 85 percent. Thus the electric power available from a single hydroelectric unit is comparable to a typical 1000-megawatt coal-burning or nuclear unit. If plans are fulfilled, the Grand Coulee Dam will someday generate 10,000 megawatts of electric power.

The historic motivation for hydroelectric systems seems obvious. The massive water flows were there so why not use them for generating electricity and for recreational areas? The dams could also help in flood control. Although initial construction costs are large, low operating and low maintenance costs make the systems economic. But there are prices to be borne. Fish migration is restricted by the presence of the dams. This is of special concern on the Columbia River because of the salmon traffic. The dams alter silt-flow patterns; this could cause the dams to fill up with silt rather than water. Also, the areas flooded back of the dams may have more desirable uses. When compared with air pollution from coal-burning plants and radioactivity problems with nuclear plants, the impact of the dams on the environment still seems small.

Hydroelectric development proceeded to the point of producing nearly 20 percent of the United States electric energy in 1960. Although the number of hydroelectric systems continued to grow, the role of hydroelectricity has been declining since 1960. In 1970 and 1976, hydroelectric sources accounted for 16.7 and 13.9 percent, respectively, of the total electric energy output. By 1985 it is expected to contribute 7.5 percent to the total. This decline is a consequence of having developed the best hydroelectric sites and resistance from environment-conscious groups to proposed sites.

Figure 12.2 Aerial view of Philadelphia Electric's Muddy Run pumped-storage project. The powerhouse is shown at the Susquehanna River's edge, in foreground. The horseshoe-shaped construction above the powerhouse is the intake canal for the system. This system utilizes eight 110-megawatt reversible pump–turbine combination units that function alternately as motor-driven pumps and turbine generators. (Photograph courtesy of Philadelphia Electric Company.)

One must always realize that a conventional power plant does not store electric energy. It produces electric power on demand. If the users turn off all their energy consuming devices, the power plant stops generating electricity. Under normal operating conditions, the demand varies with the time of day and with the time of year. Because a power plant operates most efficiently if its output is constant, a power company will often utilize auxiliary units, such as gas-turbine electric units burning natural gas, to handle the demand during peak periods. One way of storing energy for peak periods and making use of the power plant during slack times is to use the electrical output for motor driven pumps that force water to an elevated reservoir (Fig. 12.2). When the demand for electricity peaks, the water is drained through a hydroelectric system to produce electricity for the peak periods. Although only about two thirds of the energy is recovered, this system can still be economical. These so-called pumped-storage systems are already utilized to some extent in the United States. Projects exist at Taum Sauk in Missouri, Northfield Mountain in Massachusetts, Muddy Run in Pennsylvania, Yards Creek in New Jersey,

Cabin Creek in Colorado, Salina in Oklahoma, Kittatinny Mountain by the Delaware River, and Ludington, Michigan. These projects are not without environmental impact and objections to them are usually raised.*

Geothermal energy, as the name implies, is the internal thermal energy of the earth. It is not a new source and is not going to contribute significantly to satisfying rising energy demands. However, the drive to seek less polluting sources for conversion to electric energy has attracted attention to it.

Saying that geothermal energy is not going to be competitive does not mean that the supplies are limited. Rather, it means that most of the sources are too inaccessible to be practical. Figure 12.3 shows a model of a cross-sectional view of the earth's interior. The temperature of the earth is fairly constant (2500 °F–3500 °F) out to the inner edge of the surface crust, which varies in thickness from 2 to 30 miles. Within the crust, the temperature decreases fairly uniformly at a rate of 2 °C for each 100 meters of distance up to the earth's surface. It is believed that energy released in the decay of radioactive nuclei, in particular ^{238}U, ^{232}Th, and ^{40}K, gives rise to the thermal energy in the crust. The thermal energy stored under the continental United States to a depth of six miles is equivalent to the energy obtainable from the combustion of trillions of tons of coal. But like solar energy, this energy is so dispersed that it is economically difficult to capitalize on it. There are, however, areas where the hot molten rock (magma) from the core is forced up through structural defects and either surfaces in the form of volcanoes or is trapped close to the surface. These masses trapped close to the surface produce geysers and fumaroles (Fig. 12.4), concentrated sources of hot water or steam that can be tapped and used to heat the boiler of a conventional steam turbine. Therefore, a natural heat source is available that does not have the particulate and sulfur dioxide pollution aspects of a fossil-fuel power plant. Because this also involves a steam turbine operating at about 14 percent efficiency because of the temperature differences involved, thermal pollution is a potential problem. It is also true that all water is not directly suited as a thermal source because of high mineral and salt content. Removal of these contaminants is a possibility, but a more likely solution would be to use a heat exchanger so that the water from the geothermal source does not actually come into contact with the turbine parts. Although these minerals are contaminants as far as operation of the turbine is concerned, they could possibly be recovered and sold to appropriate markets.

Hot water and steam sources are normally found in regions of extensive geologic activity. In the United States, such regions are located in the furthermost western states and many are on federally owned land. In these areas there is an estimated 1,350,000 acres underlaid with enormous heat sources with a capacity for 15,000 to 30,000 megawatts of electric power. The only area commercially developed thus far in the United States is in the Big Geyser region of Northern California (Fig. 12.5). Pacific Gas and Electric Company (P.G. and

---

* See, for example, "Storm King Is Dead Again," *Environment*, vol. 16, no. 7 (September 1974): p. 27.

12.3 GEOTHERMAL ENERGY

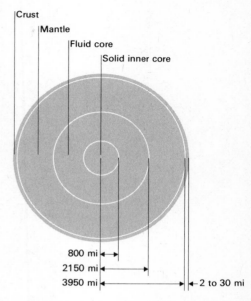

Figure 12.3 A structural model of the interior of Earth. Note how thin the crust is compared to Earth's radius.

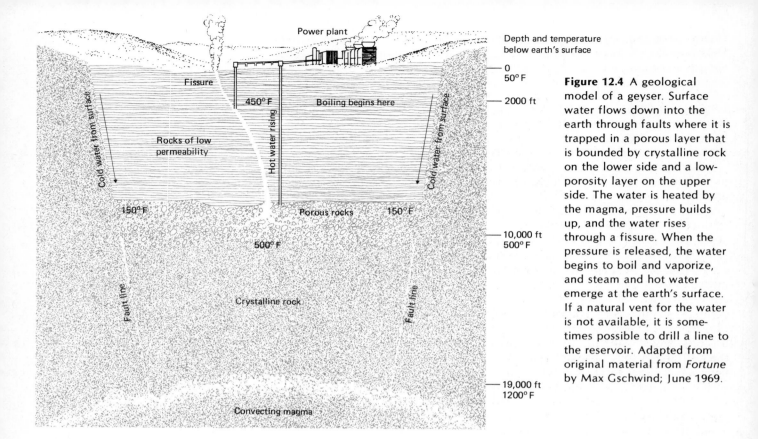

Figure 12.4 A geological model of a geyser. Surface water flows down into the earth through faults where it is trapped in a porous layer that is bounded by crystalline rock on the lower side and a low-porosity layer on the upper side. The water is heated by the magma, pressure builds up, and the water rises through a fissure. When the pressure is released, the water begins to boil and vaporize, and steam and hot water emerge at the earth's surface. If a natural vent for the water is not available, it is sometimes possible to drill a line to the reservoir. Adapted from original material from *Fortune* by Max Gschwind; June 1969.

Figure 12.5 This geothermal electric power plant in Sonoma County, California, produces 54 megawatts of electric power. The total capacity of all geothermal plants in this area amounted to 502 megawatts in 1975. This capacity is expected to increase to 908 megawatts in 1979. (Photograph courtesy of Pacific Gas and Electric Company.)

E.) has operated an 84-megawatt plant since 1960. The cost of electricity is competitive with other more conventional means and P.G. and E. plans a total of 908 megawatts by 1979. The Imperial Valley of California has been suggested as an area of possible development with an estimated capacity in the thousands of megawatts.

Another method to utilize underground energy is to open large cavities in deep, very hot rock formations in the earth. Fracturing can be done hydraulically or with explosives. Water could then be pumped into these cavities, vaporized, and the steam collected to drive a turbine. If this proves feasible, geothermal energy could contribute significantly to our overall energy budget.

Probably, more breakfast cereal is bought because of its colorful packages with catchy sayings, enclosed trinkets, and deceptive advertisements than because of its nutritional value. In fact, we have become accustomed to clever packaging for all merchandise. Countless hours are spent determining the carton that optimizes sales. But the interest in packaging declines abruptly once the product is purchased even though the disposal of the used containers and trash, in general, is one of the most perplexing of the many societal problems we face. Solid waste is generated at an average rate of five pounds per person per day in the United States. A city of a million people can produce 2500 tons of solid waste each day. It would take 500 trucks each carrying five tons to haul these wastes. To compound matters, trash production is growing at a rate of 4 percent per year. Still, in a technological era that witnessed a landing on the moon, this trash is likely to be dumped into a hole in the ground and covered with dirt. A growing shortage of land disposal sites is stimulating interest in alternate disposal schemes. One such scheme is to burn the trash for input energy to an electric power plant. Some recovery of marketable products would be done in the process.

Although this idea is relatively new in the United States, it is quite old in Europe. Over 50 years ago, Paris began burning garbage to generate electric energy and today enough is generated to supply the needs of a city of 50,000. Since 1968, a 7.3-megawatt trash-burning power plant has been operating in Milan, Italy. Additional plants are planned. In West Germany, steam heat is produced from the burning of refuse produced by nearly eleven million people.

Combustible paper products account for about 53 percent of the total weight of solid wastes. Another 7 percent is in other combustible carbon-based products. When burned, each pound of solid waste produces approximately 5300 Btu of heat. Although this is low when compared with an energy content of 13,000 Btu per pound for coal, it is still significant. Furthermore, the sulfur content of solid waste is about 0.1 percent, putting it in a class with the best low-sulfur coal. These two features have brought attention to solid wastes as an energy source of great potential.

The energy potential of solid wastes is impressive. If the 136 million tons of solid waste produced in 1970 had been burned, then it would have produced 1400 trillion Btu of heat. If converted to electric energy at an efficiency of 32

12.4 ENERGY FROM BURNING TRASH

percent, the burning would have produced 130 billion kW-hrs, which amounts to about 11 percent of the electric energy produced by conventional means in 1970.

Pioneering efforts for deriving energy from solid waste were made in St. Louis. Beginning in 1972, a prototype system operated by the Union Electric Company burned a mixture of trash and coal to generate electricity (see Fig. 12.6). The plant ultimately processed about 650 tons a day of the city's output of trash. In February 1977, the Union Electric Company announced cancellation of its solid waste utilization system that included plans for a capacity of

Figure 12.6 Trash is a source of energy for generating electricity or producing steam in a number of American cities (see Table 12.1). (Photograph courtesy Union Electric Company.)

7,500 tons a day. The reasons cited were rising costs and a recently passed Missouri law that made it difficult to finance the construction of electric power plants. Several other cities are proceeding with St. Louis-type systems, which produce a shredded-trash fuel for input to steam boilers (Table 12.1). Other cities are converting trash energy directly to steam (Table 12.1).

Table 12.1 Compilation of operational and planned trash-energy systems in the United States.

Location	Capacity (tons/day)	Year	Output
St. Louis, MO	650	1972	fuel
E. Bridgewater, MA	600	1973	"
Ames, IO	200	1975	"
Bridgeport, CO	1600	1977	"
Chicago, IL	1000	1976	"
Milwaukee, WI	1000	1976	"
New Britain, CO	1800	final design stage	"
Monroe County, NY	2000	final design stage	"
Braintree, MA	240	1971	steam
Nashville, TN	720	1974	"
Saugus, MA	1200	1975	"
Akron, OH	1000	final design stage	"
Hempstead, NY	2000	final design stage	electricity

It is also possible to convert solid wastes into synthetic gases or liquid fuels that can be transported and used in a variety of ways. Baltimore has built a plant that produces a low-energy content gas from solid wastes. It will process 1000 tons per day. The quality of gas produced is grossly inferior to natural gas. It is much dirtier and its energy content is 90 percent less. For these reasons shipping it long distances is uneconomical. A facility producing a heavy oillike fluid called Garboil from solid wastes is located in San Diego. Although Garboil cannot be refined to make gasoline, it can be used as a substitute in oil-burning electric power plants. Garboil also has the attractive feature of being easily stored and transported.

Whether burning solid wastes solely for the energy content is the best solution for a mounting problem remains to be seen. Perhaps recycling the wastes is more appropriate. However, in an era of concerns for energy resources and the environment, the burning of solid wastes for the energy content is a viable alternative.

12.5 COAL GASIFICATION AND LIQUEFICATION

"There are two things wrong with coal today. We can't mine it and we can't burn it," said S. David Freeman, Director of a Ford Foundation Energy Policy Project. There is enough coal in the United States to satisfy our energy needs for at least 500 years at present energy consumption rates—*if* it could be mined and burned. Although natural gas can be burned in an environmentally acceptable way, it is in extremely short supply. The conversion of coal to a synthetic gas takes advantage of the best features of these two fuels. As discussed in Chapter 4, the control of sulfur oxides is the primary environmental problem from burning coal. Techniques exist for removing the sulfur oxides after burning, but they are expensive and unreliable. With coal gasification it is possible to remove most of the sulfur content of the coal during the gasification process.

Coal gasification is not a new concept. Coal gas was widely used for lighting and cooking until natural gas came into prominence after World War II. However, the conversion of energy from coal gas to electricity on a commercial scale has yet to be demonstrated although the principle is relatively simple.

If coal is burned for energy-producing purposes, then one tries to supply enough oxygen to optimize the burning conditions. Ideally, only carbon dioxide (CO_2) and heat are produced. If there is insufficient oxygen, combustion is incomplete and gaseous products other than carbon dioxide are produced. Coal gasification capitalizes on incomplete combustion conditions. A certain amount of oxygen (or air) is used to instigate burning for the purpose of producing heat. The basic heat-producing reactions from carbon and hydrogen in the coal are

$$C + O_2 \longrightarrow CO_2$$

and

$$2H_2 + O_2 \longrightarrow 2H_2O.$$

When a certain temperature is reached, steam is injected into the system promoting reactions with carbon and carbon dioxide.

$$C + H_2O \longrightarrow CO + H_2$$

and

$$C + CO_2 \longrightarrow 2CO.$$

Both carbon monoxide (CO) and hydrogen (H_2) are gaseous and can be burned to produce energy. Methane (CH_4), the molecular form of natural gas, is formed to some extent.

$$C + 2H_2 \longrightarrow CH_4$$

and

$$coal + heat \longrightarrow CH_4 + hydrocarbons.$$

The resulting gas is basically a mixture of carbon monoxide and hydrogen, with lesser amounts of methane. In the simplest situation, the energy content is about 120–160 Btu per cubic foot. At further expense, distillation can increase the energy content to 500–600 Btu per cubic foot. These energy contents are low compared with an energy content of 1030 Btu per cubic foot in natural gas. However, the attractive feature is that during the gasification the sulfur content of the coal is converted to hydrogen sulfide (H_2S) that can be removed by proven methods. The resulting low-sulfur content gas can then be burned to produce heat in conventional electric power plants. Locating both the gasification system and the power plant near the coal-producing sites so as to minimize transportation of the raw material would be an extremely attractive concept. Unfortunately, this concept does not avoid the problems associated with both strip and shaft mining. There is, however, interest in gasifying coal while it is still in the ground. The principle is basically the same as used in gasification plants. Tests of this procedure have produced a combustible gas. It has, however, been of variable energy content and recovery has been poor. Nevertheless, the possibility of extracting the energy content of coal without ravaging the landscape is enough of a virtue to prompt a continued exploration of this technology.

Current research is proceeding with the development of large-scale, high-energy content coal gasification systems for producing a gas comparable in quality to natural gas and with coal liquefaction technology. The coal liquefaction process involves the reaction of hydrogen with coal to produce a liquid hydrocarbon. Industrial processes for doing this have been known for some time. A large part of the gasoline for Germany's World War II effort was produced from coal liquefaction. Although it is attractive to manufacture a product that can replace gasoline, which is refined from a dwindling natural resource, the process at present is expensive both in terms of money and energy usage. Demonstration plants for coal liquefaction and high-energy content gasification are expected by 1979. The successful development of all of these systems would greatly alleviate the plight of dwindling natural gas and oil reserves.

12.6 MAGNETOHYDRODYNAMICS (MHD)

The efficiency of any single energy conversion process is always less than 100 percent. If several sequential conversions are required to produce a desired form of energy, the overall efficiency is less than the smallest efficiency in the chain. This means that if every individual conversion stage were nearly perfect, except one, then the overall efficiency would still be low. Thus any method of producing electric energy that uses a steam turbine will be intrinsically inefficient because of the low thermal efficiency of the turbine. This point was emphasized in Chapter 5 during the discussion of the energy model of a fossil-fueled electric power plant. One wonders why a scheme isn't devised that avoids the steam turbine. Despite considerable thought and physical effort, only a single scheme has emerged as being commercially practical in this century. This method bears the intimidating name magnetohydrodynamics, abbreviated MHD.

Any quantity of charge in motion constitutes an electric current. A flow of gas or a stream of liquid could produce an electric current if the constituents were charged. A gas turbine like that used to power a jet airplane produces a stream of high-speed gas but little electric current because there is practically no net charge associated with the atoms and molecules of the gas. If the moving gas could be made to contain free electric charges, then it could produce an electric current. This condition is achievable by "seeding" the gas with atoms or molecules that readily lose their electrons in the hot gas environment. Elemental potassium and cesium or appropriate compounds that include one or the other of these elements are used for this purpose. Producing an electric

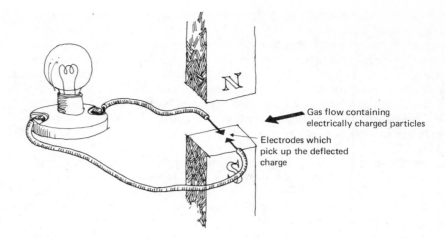

Gas flow containing
electrically charged particles

Electrodes which
pick up the deflected
charge

Figure 12.7 Principle of the magneto-hydrodynamic electric generating process. The electrically charged particles in the moving gas are deflected by the magnetic field into electrodes connected to an external electrical circuit.

current in a gas is one thing, diverting it into the electrical wires of a house is another. The transfer in an MHD system is accomplished by directing the gas through a magnetic field (Fig. 12.7). By virtue of the force experienced by the moving charges in a magnetic field, the charges are deflected laterally into metallic electrodes that are connected to the wires of an electric utility system. This is the same physical principle as in the magnetic confinement of a plasma (see Section 13.6).

By using the hot exhaust gases to vaporize water for a conventional steam turbine, an MHD system should have an overall energy conversion efficiency of between 45 and 60 percent. This is a substantial improvement over that attainable with a steam turbine system. Although a variety of fuels can be used to heat the gas to high temperature (2600°C), coal is the most attractive because it is our most abundant energy resource. As in any coal-burning process, this scheme would produce particulates and sulfur oxides. The particulate problem is minimized because the required high temperature permits more complete burning of the coal. Economics demand recovery of the seed used to enhance the charge in the gas. During this recovery operation, the sulfur oxides can also be collected. Environmentally, MHD systems are attractive in terms of both air and thermal pollution.

The technical aspects of an MHD system are far from trivial. Consider the following two examples. The electrodes that pick up the charge deflected from the gas stream must withstand temperatures as high as 2000°C. If they are cooled to prevent melting, the cooling decreases the amount of net charge in the gas. Huge electric currents are required in the massive coils of wire that produce the charge-deflecting magnetic field. It has been difficult to make an MHD system whose output power is greater than the input electric power required for the magnets. Recent advances in magnet technology using wires that have zero electrical resistance at temperatures near that of liquid helium (4.2K) may alleviate this problem.

Figure 12.8 An MHD electric power generator developed by the AVCO Everett Research Laboratory, Inc. The technician is standing in front of one pole of the deflecting magnet. The high-speed hot gases flow toward the magnet through the large duct extending the full length of the photograph. This generator is capable of producing a 250,000-watt output of electric power. (Photograph courtesy of the Avco Everett Research Laboratory, Inc.)

In 1965, an American-built MHD unit produced 30 megawatts of electric power for a short period of time. Despite encouraging results, interest waned until federal legislation in 1974 called for a demonstration of commercial feasibility. Plans are underway now to build an MHD test facility in Butte, Montana. Expectations are to construct a 50- to 100-megawatt generator by the mid 1980s and a full-size commercial facility by the 1990s. Commercial production of electric power by MHD principles has already been demonstrated in Russia where a 25-megawatt system provides electricity for the Moscow area. A 500-megawatt generator is scheduled for operation in 1982. The Soviet Union and the United States have shared substantial cooperation in MHD research.

12.7 ENERGY FROM BURNING HYDROGEN

The overwhelming emphasis in energy research and development is toward new and more efficient ways of generating electricity. This is justified in large part because electric energy usage is the fastest growing segment in the energy budget. However, only 29 percent of the energy used in 1976 went toward electric energy production in steam-powered electric power plants. Clearly, there are and will continue to be many uses of energy other than for producing electricity. As anxieties for fossil fuels grow, the concern for replacing them also grows. Many feel that hydrogen is the logical replacement and that we should be seriously proceeding toward a "hydrogen economy," with the same vigor that we are developing an "electric economy." There is considerable justification.

Hydrogen, when burned for its energy content, is an alternative to natural gas. But unlike natural gas that exists in a condition which can be burned as soon as it is removed from the earth, hydrogen must be manufactured and this requires energy. This energy is not completely recovered. So the cost per Btu for hydrogen in an era of plentiful natural gas will always be more expensive. But if natural gas supplies diminish as anticipated, hydrogen could be used and could be transported in the same pipeline manner.

About half the cost of electricity is due to transmission and distribution. Savings can be made if giant electric power complexes are built near the consumption areas but then public objections arise as do concerns for pollution. Hydrogen on the other hand is easy and economical to transport in underground pipelines. For those energy applications that could be handled by hydrogen as well as electricity, it makes some sense to produce the hydrogen from the electric output of a plant well removed from the consumption area and then pipe the hydrogen in. This process would also allow a power plant to operate at constant power, which is the most efficient way to operate. This would be particularly applicable to solar-powered electric systems because they will probably have to be in remote areas. To a somewhat lesser extent it applies to nuclear plants, especially nuclear breeders, because for safety reasons it is especially desirable to locate them in remote areas.

Hydrogen has uses other than for heating. To mention a few, it is an important ingredient in the production of fertilizers and foodstuffs. The production of synthetic fuels from organic material and trash will require large quantities of hydrogen. Although there are some hazards associated with the handling of hydrogen, they are not insurmountable. Because of the many uses of hydrogen and the clean way that it burns, it undoubtedly will play an increasingly important role in future energy scenes.

12.8 TIDAL ENERGY

Few things have captured the imagination of poets and writers more than the incessant motion of the oceans and the tantalizing ebb and flow of the tides. The thought of harnessing the enormous energy content of both the oceans and tides has pervaded the minds of humans for centuries. To some extent we have not been denied the use of this energy. Water-powered mills operating from tidal motion were used in New England in the eighteenth century. A tidal-powered sewage pump functioned in Hamburg, Germany, until 1880,

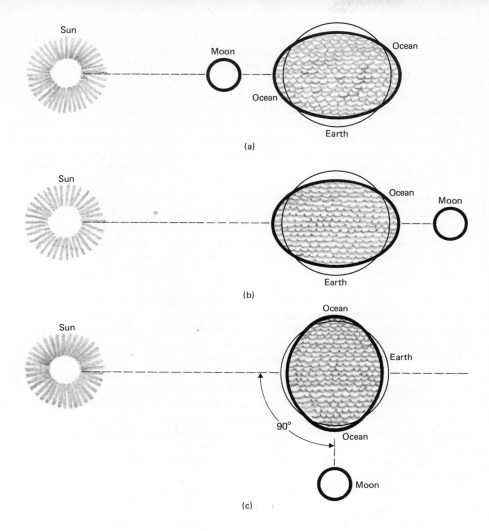

Figure 12.9 (a) When the moon is between the earth and the sun, the oceans bulge out on both sides of the earth producing a spring-tide condition. (b) Spring tides are also produced when the earth comes between the sun and the moon. (c) Neap tides are produced when the angle between the lines connecting the earth and the sun and the earth and the moon is 90°.

and a water pump installed in 1580 under the London Bridge operated successfully for two and a half centuries. These systems were eventually replaced by the more economical and convenient electric motors. The search for environmentally acceptable methods for producing electric power has stimulated renewed interest in tidal energy. Although no source exists that renders less environmental damage, tidal energy is difficult to harness and marginally economical.

The tides have their origin in the gravitational force exerted on the earth by the moon and the sun. Gravitational forces hold the earth in its orbit about the sun, and the moon in its orbit about the earth. The earth and moon make complete revolutions every 365 and 28 days, respectively. Thus there are times when the moon comes between the earth and sun. When this happens, the gravitational force exerted upon the earth by the moon and sun is maximum. The fluid ocean bulges out under this force and produces a high-tide condition (Fig. 12.9a). Interestingly a high tide also occurs on the opposite side of the earth. This is because the gravitational force between two masses decreases as the separation increases. Because the water on the side of the earth opposite the moon is farther away than the earth's solid core, the core is pulled toward the moon and away from the water causing it to bulge on the side of the earth opposite the moon. The tide falls as the earth rotates on its axis. Because the earth makes a half rotation on its axis in 12 hours, a high-tide condition will return to the initial position when the earth rotates to a position away from the moon. Thus, two high (and two low) tides occur each day. Maximum high tides also occur when the moon is on the opposite side of the earth from the sun (Fig.12.9b). These tides that occur for the aligned positions of the earth, moon, and sun are called spring tides. Tides also occur when the moon is not on a line with the earth and sun, but the tides are less pronounced. The minimum effect occurs when a line connecting the earth and moon is at right angles (90°) with a line connecting the earth and sun (Fig. 12.9c). These tides are called neap tides. Both spring and neap tides occur every 14 days because the earth, moon, and sun line up twice in a 28-day lunar month.

The idea in a tidal electric power plant is to allow water to flow into a basin during high-tide. As the water flows in, it is directed into the blades of a water turbine that turns an electric generator producing electric power. At the peak of the tide, the water gates are closed and the water is trapped in the basin. When the tide recedes, the gates are opened and water again flows through the water turbine. Power can be produced on the return of the water to the ocean by reversing the blades on the turbine.

To evaluate the power developed from a hydroelectric system we used the relation

$$P = \left(\frac{w}{T}\right) \cdot H,$$

where w/T is the water flow rate (newtons/sec) and H is the height from which it falls. Basically the same relation applies to tidal systems except that the equation has to be divided by two because the height changes as the water flows out of the tidal basin. This height is called the tidal range when discussing tidal energy systems. For tidal systems there is much less flexibility in the parameters. The time (T) is fixed by the period of the tides. If the water is used as it flows both into and out of the basin, the period is six hours. The tidal range (H) is fixed by the prevailing tidal conditions. In some contemplated tidal energy sites the average tidal range varies between 18 and 35 feet. Because both the large time and small range are working to limit the power, one must look for natural areas where the tidal basins are very large so that the total weight of the water (w) can be made large. Suppose that the basin at a given site is nine square miles (if a square, it would be three miles on a side) and that the average tidal range is 18 feet. The tidal power based on a six-hour period from this site would be 159 megawatts. Thus the power is considerable.

The first commercial tidal power plant was built in France where the Rance River empties into the Atlantic Ocean on the Brittany Coast (Fig. 12.10). Completed in 1966, it generates 240 megawatts of electric power. It produces electricity both on the entrance of water into and exit of water out of an 8.5 square mile basin having an average tidal range of 27.6 feet. Though its output is not comparable to a 1000-megawatt contemporary nuclear or coal-burning power plant, it is providing valuable operating experience for this unique type of system.

Although United States tidal power plants have been in various stages of planning for nearly 50 years, a single kilowatt of power is yet to be produced. The most celebrated planned project would utilize Passamaquoddy Bay (which links Maine and New Brunswick, Canada) as a 100-square-mile tidal basin. The impounded water would flow through 100 ten-megawatt turbogenerators into Cobscook Bay in Maine (Fig. 12.11). That seven miles of dams and 160 water gates are required indicate the complexity of the project. If built at current prices, the project would be likely to cost over a billion dollars. Such a huge cost per kilowatt of power would put it in a class with peaking units. Although the project is attractive from the environmental standpoint and makes use of a source of power which goes unused everyday, it is unlikely that it or any United States tidal power system will be built in the near future because of the capital cost.

Figure 12.10 A tidal-electric power plant located at the mouth of the Rance River in France. The water basin is shown in the upper portion of the photograph. Six water-turbine–electric-generator systems are housed in the left-most part of the retainment structure. Boats may enter the tidal basin through the opening on the right-most part of the retainment structure. The complex is capable of generating 240 megawatts of electric power. (Photograph courtesy of French Embassy.)

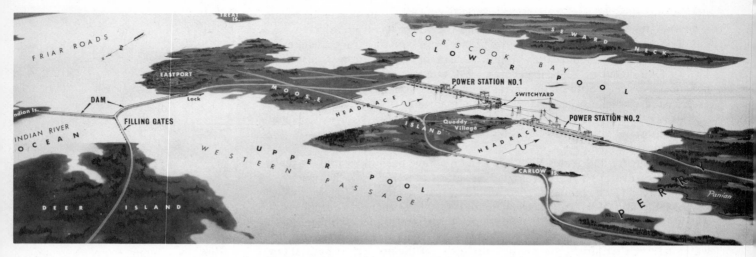

Figure 12.11 An oblique map of the proposed Passamaquoddy tidal power system. Water flows into the upper pool during high tides. At the peak of the tide, the filling gates are closed and the water is trapped in the upper pool. When the tide recedes, the water is channeled into the power stations by the headraces and empties into Cobscook Bay. (Map courtesy of Department of the Army Corps of Engineers.)

TOPICAL REVIEW

1. What is the meaning of hydro in hydroelectric and magnetohydrodynamics?

2. Describe the importance of hydroelectricity in the present and future energy scene.

3. What determines how much power is available from water flowing over a dam?

4. List some environmental concerns emanating from hydroelectric power plants.

5. What is a pumped-storage system?

6. What is geothermal energy?

7. Describe how electricity is produced from geothermal energy.

8. What are some problems with the use of geothermal energy?

9. For what reasons has trash emerged as a potential energy source?

10. To what extent is "energy from trash" realistic?

11. Can trash be used for any energy purpose other than producing electric energy?

12. Describe the coal gasification process.

13. Why is there considerable interest in gasifying coal for use in electric power plants?

14. What does MHD mean?

15. How would an MHD electric power plant differ from a nuclear power plant?

16. When can we expect to see MHD power plants in the United States? in the world?

17. What is meant by the "hydrogen economy"?

18. Name some advantages and disadvantages from burning hydrogen.

19. Name some industrial uses of hydrogen.

20. Describe how tides are produced.

21. How might tidal energy be used for producing electricity?

22. Realistically, when can we expect to have our homes lighted with tidal-produced electricity?

REFERENCES

Hydroelectric Power

1. *Energy: Electric Power and Man*, Timothy J. Healy, Boyd and Fraser, San Francisco, Calif. (1974).

Geothermal Energy

1. "Geothermal Power," Joseph Barnea, *Scientific American* **226**, No. 1, 70 (1972).

2. "Geothermal Energy," William W. Eaton. Available for 40 cents from DOE Technical Information Center, P.O. Box 62, Oak Ridge, TN 37830.

3. *Geothermal Energy* edited by Paul Kruger and Carel Otte, Stanford University Press, Stanford, Calif. (1973).

Electric Energy from Burning Trash

1. "Fuel from City Trash," Dennis Wilcox, *Environment* **15**, No. 7, 36 (September 1973).

2. "Power from Trash," William C. Kasper, *Environment* **16**, No. 2, 34 (March 1974).

3. "Tapping Resources in Municipal Solid Waste," S. L. Blum, *Science* **191**, 669 (20 February 1976).

Coal Gasification and Liquefaction

1. "The Gasification of Coal," Harry Perry, *Scientific American* **230**, No. 3, 19 (March 1974).

2. "Clean Power from Dirty Fuels," Arthur M. Squires, *Scientific American* **227**, No. 4, 26 (October 1972).
3. "Coal Research (II): Gasification Faces an Uncertain Future," *Science* **193**, 750 (27 August 1976).
4. "Coal Research (III): Liquefaction Has Far to Go," *Science* **193**, 873 (3 September 1976).
5. "Oil and Gas from Coal," Neal P. Cochran, *Scientific American* **234**, 24 (May 1976).

Magnetohydrodynamics (MHD)

1. "New Ways to More Power with Less Pollution," Lawrence Lessing, *Fortune* **82**, No. 5, 78 (November 1970).
2. *Energy: From Nature to Man*, William C. Reynolds, McGraw-Hill, New York (1974).
3. "MHD Power Generation," A. Kantrowitz and R. J. Rosa in *Physics and the Energy Problem—1974*, M. D. Fiske and W. W. Havens, Jr. (eds.), American Institute of Physics, New York (1974).

The Hydrogen Economy

1. "The Hydrogen Economy," Derek P. Gregory, *Scientific American* **228**, No. 1, 13 (January 1973).
2. "Liquid Hydrogen as a Fuel for the Future," Lawrence W. Jones, *Science* **174**, No. 4007, 367 (22 October 1971).
3. "Energy and the Future," Allen L. Hammond, William D. Metz, Thomas H. Maugh II, American Association for the Advancement of Science, Washington, D.C. (1974).

Tidal Energy

1. "Time and Tide," F. L. Lawton, *Oceanus* **17**, 30 (Summer 1974).
2. *Tidal Power*, edited by T. J. Gray and O. K. Gashus, Plenum Press, New York (1972).

SUGGESTIONS FOR FURTHER UNDERSTANDING OF CHAPTER 12

12.2 Hydroelectricity

1. For reasons of transportation and water supply, many cities are located near rivers. Why haven't most cities derived their electricity needs from hydroelectric power plants on these rivers?
2. Even though a river flowing through a city may only be able to support a small hydroelectric project, is there any reason why a city might still want to have it available?

3. The Bonneville hydroelectric plant near the end of the Columbia River utilizes a small reservoir of water and a difference in water level of 59 feet. The Grand Coulee hydroelectric plant located several hundred miles up the Columbia River utilizes a huge reservoir of water and a difference in water level of some 285 feet. Why are both facilities able to produce large power outputs with such different water facilities?

4. What are some complications that might arise when river flow is restricted by the presence of dams?

5. Does the electric generator used on a bicycle for operating a horn or headlight produce electricity all the time that the bike is in motion? Is the generator able to store electric energy for future use?

6. Why are pumped-storage hydroelectric systems not practical in all geographic locations?

7. Water in a pumped-storage hydroelectric system must flow to a lower geographic level. In a region where a natural high level is unavailable for a reservoir, how could a pumped-storage system be built below ground level?

12.3 Geothermal Energy

8. Why isn't the geothermal energy in the Big Geyser region of northern California used for household heating rather than for input energy for an electric power plant?

9. Why is the availability of water an important consideration for a geothermal electric power plant?

10. A geothermal electric power system converts thermal energy to electric energy. Why would a geothermal system be no more efficient than a contemporary coal-burning system?

11. The central core of the earth is believed to be molten and at a temperature of about 3000°F. What physical property of the earth between the core and the crust prevents the surface of the earth from being much warmer than it is?

12.4 Energy from Burning Trash

12. Based on your own experience, does the generation of five pounds of trash per person per day seem unrealistic? You might like to make your own estimate.

13. Do you believe that as our affluence increases the amount of solid wastes also increases? If you accept this, then you will likely accept the fact that about ⅓ pound of municipal waste is discarded in the United States for each dollar of gross national product.

14. Burning was once a popular way of disposing of trash. Why do you think this method is now forbidden? Would the problems of open burning also be present to some extent in power plants burning trash for energy?

15. Historically, cities have looked to suburban areas for their land-fill waste disposal sites. How has the population settlement pattern complicated the waste disposal problems of cities?

16. Although relatively few American cities (see Table 12.1) have commitments to trash-energy systems, the technological success of trash-energy systems has been demonstrated. What, then, are the barriers preventing many more cities from constructing trash-energy systems?

12.5 Coal Gasification and Liquefaction

17. The increasing reliance on foreign oil from which gasoline is refined is a perplexing American problem. What other alternatives do we have for producing gasoline or gasoline substitutes?

18. Gaseous carbon monoxide (CO) is a burnable gas that can be derived from coal. If carbon monoxide is used as a replacement for natural gas, what special precautions would be in order?

19. Coal can be used to produce substitutes for either natural gas or gasoline. From a knowledge of the relative complexities of the molecules in the gas and gasoline, why would you expect liquefaction to be the more difficult of the two processes?

20. What is the end product in the chemical reaction describing the burning of carbon monoxide (CO)?

21. Before the advent of natural and synthetic gases for heating and cooking, a relatively low-grade fuel called coal oil was common. What would you guess is the origin of this product?

22. If air is used in the initial burning process for coal gasification, what gaseous pollutant will be produced that is also associated with the internal combustion engine?

12.6 Magnetohydrodynamics (MHD)

23. Does magnetohydrodynamics (MHD) represent a new source of energy?

24. Would there be any environmental problems associated with the MHD method of generating electricity?

25. The exhaust gases from a MHD generator are at a temperature of about 3000K. Why does this allow for the possibility of a thermodynamic efficiency substantially higher than that in a coal-burning or nuclear-fueled electric power plant?

26. The United States has developed a technology for producing large magnetic fields that is unrivaled in the world. Why do you think that countries, Russia, for example, with an interest in MHD systems are interested in American magnet technology?

12.7 Energy from Burning Hydrogen

27. Although hydrogen can be burned in an internal combustion engine, what problems will be encountered in trying to use it on a large scale?

28. Why is it possible to recycle hydrogen after it is burned but not possible to recycle natural gas after it is burned?

12.8 Tidal Energy

29. How does the concentration of energy compare in tidal and nuclear power plants?

30. Suppose that the orbit of the moon was such that a line connecting the earth and moon was always at right angles (90°) with a line connecting the earth and sun. Would there be any meaning to the concept of spring and neap tides?

12.2 Hydroelectricity

NUMERICAL PROBLEMS

1. A "cube" of water 1 foot on a side weighs 62.4 pounds. If water flows over a dam at a rate of 200 ft³/second, what is the flow rate in pounds/second?

2. A city of 15,000 people wants to build a small hydroelectric power plant for emergency use of electricity. If the dam is to be 8 meters high and the flow rate 200 ft³/second, how much hydropower is available for conversion to electric power. (Remember, 1 ft³ of water weighs 62.4 pounds and 1 pound = 4.45 newtons.)

3. It is desired to build a pumped-storage system that has 100,000 kW-hrs of gravitational potential energy. The vertical height between the reservoir and turbine is 80 meters.

   a) Using Eq. 12.1, calculate the weight in newtons of the amount of water needed. One cubic meter of water weighs 9800 newtons.

   b) If the water depth is 5 meters, how big an area will be required for the reservoir?

4. The formula for determining power produced by falling water is $P = w \cdot H/T$. If weight, height, and time are expressed in pounds, feet, and seconds, respectively, then the power is in foot-pounds per second. This can be converted to watts by multiplying the answer by 1.35. Show that 720 megawatts of power are produced by a water flow of 1,870,000 pounds per second falling through a height of 285 feet.

12.3 Geothermal Energy

5. The temperature of the core of the earth is not accurately known, but the general consensus is that it is less than 10,000°F. The variation of

temperature with depth into the earth's crust is fairly well known; 1°F increase for every 100 feet down into the crust. If this variation were to continue, what would you expect for the temperature of the earth's core? Assume the distance to the earth's surface to the edge of the core is 2000 miles. This illustrates how an extrapolation can go astray.

6. The weight of six miles of the earth's crust below the United States is about 100 billion-billion pounds (10^{20} pounds). If the crust were cooled 1°F, then each pound would release about one Btu of heat. Therefore, cooling six miles of the crust 1°F would release 100 billion-billion (10^{20}) Btu of energy. Assuming the heat obtained from one ton of coal is 25 million Btu, show that the energy derived from cooling six miles of the earth's crust below the United States by 1°F is equivalent to the heat which can be produced from about 4 trillion (4×10^{12}) tons of coal. This should convince you of the enormous energy content of the earth's crust.

12.4 Energy from Burning Trash

7. If a family of four generates 4 pounds of trash per occupant, what percentage of their daily winter heat requirement of 500,000 Btu could they possibly derive from burning the trash?

8. Solid waste is generated at an average rate of five pounds per person per day. Knowing this, show that a city of one million people will produce 2500 tons of solid wastes each day.

9. One scheme for converting municipal wastes to a burnable gas would produce per ton of waste 24,000 cubic feet of gas having an energy content of 300 Btu per cubic foot. Compare the gas energy content per pound of waste with an energy content of 13,000 Btu per pound of coal.

10. You will need the following data for this problem:

    1 pound of trash = 5300 Btu of energy,
    1 kilowatt-hour = 3413 Btu.

    a) Show that the energy content of 136 million tons of trash is 1400 trillion Btu.

    b) Show that 1400 trillion Btu are equivalent to 410 billion kilowatt-hours.

    c) Assuming a conversion efficiency of 32 percent show that 1400 trillion Btu of energy will produce 130 billion kilowatt-hours of electric energy.

11. Several of the trash-energy systems listed in Table 12.1 have trash-handling capacities of about 1000 tons per day. About how many people will such a plant accommodate each day?

12.5 Coal Gasification and Liquefaction

12. A certain pilot coal liquefaction plant produces 3 barrels of a low-sulfur, utility fuel for each ton of coal processed. If the heat content of the liquid

is 5,800,000 Btu per barrel, how many Btu of oil heat is obtained for each pound of coal processed? Compare this result with the original coal heat content of about 13,000 Btu per pound.

13. A coal gasifier being built in Pike County, Kentucky will process 1½ tons of coal per hour to produce 1 million cubic feet of gas having an energy content of 160 Btu per cubic foot. Assuming that the heat energy content of the coal is 13,000 Btu per pound, show that the heat energy content of 1½ tons of coal is 39 million Btu and the energy content of the gas produced is 160 million Btu. Where does the extra energy for the gas come from?

12.8 Tidal Energy

14. The energy available from a tidal basin is related to the mass (M) of the water contained and the tidal range (H) by $E = \frac{1}{2} MgH$ where g is the acceleration due to gravity (9.8 meters/second^2). The amount of energy available per kilogram of water contained is then $E/M = \frac{1}{2} gH$.

a) Determine the energy per kilogram for a tidal range of 10 meters.

b) Compare this with coal having an energy content of 13,000 Btu per pound. From this comparison, comment on the general impracticality of tidal energy systems. (1 kilogram is equivalent to 2.2 pounds and 1 Btu = 1055 joules.)

15. a) A "cube" of water 1 foot on a side weighs 62.4 pounds. Knowing this, show that a basin 3 miles wide, 3 miles long, and 18 feet thick weighs 282 billion pounds. (There are 5280 feet in a mile.)

b) If this water is drained out of the basin in 6 hours, show that the average rate of drainage is 13 million pounds per second.

c) Using the expression

$$P = \frac{1.35}{2} \left(\frac{w}{T} \right) \cdot H$$

for determining tidal power, show that the power developed is 159 megawatts.

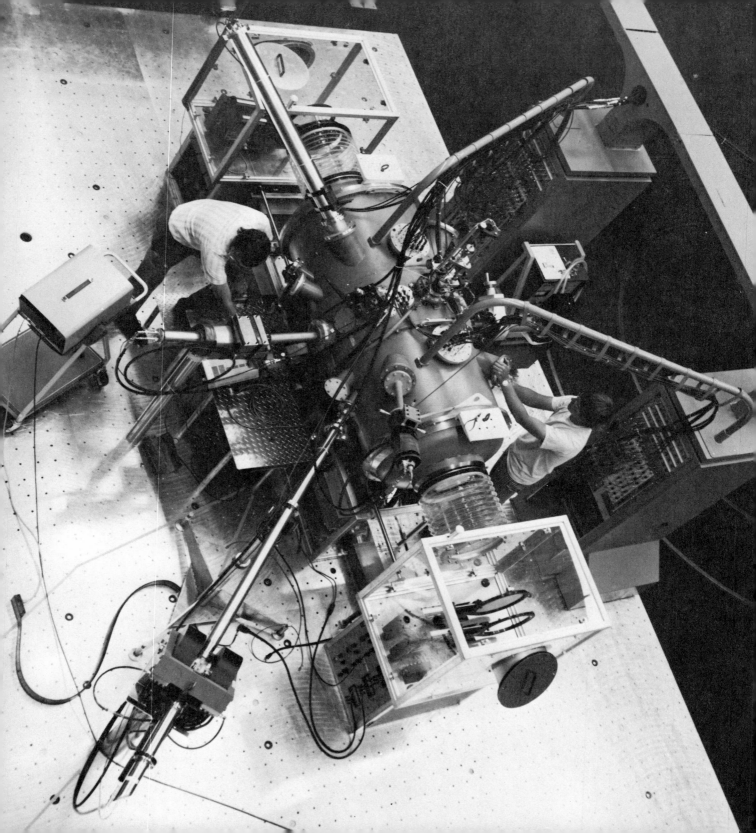

NUCLEAR BREEDER AND NUCLEAR FUSION REACTORS

13

Research on laser-induced nuclear fusion is performed with the apparatus shown here. Hopefully, these tedious, expensive endeavors will reward mankind with an abundant, socially acceptable energy supply. (Photograph courtesy Lawrence Livermore Laboratory.)

13.1 MOTIVATION

While we have discussed several methods for generating electricity on a commercial scale, none has been touted as a complete answer to long-range energy problems. Hydroelectricity is limited by the availability of useful dam sites. Coal is abundant, but there are health and environmental problems from mining, there are air pollution problems from burning it, there is concern for the "greenhouse" effect brought on by carbon dioxide from burning it, and there is the realization that there are better uses for it than burning. Cheap uranium-235 fuel is limited. If less abundant, more costly ores are exploited, electricity will be more expensive and the same rape of the landscape results as with strip mining of coal.

The nuclear breeder and nuclear fusion reactors are widely heralded possibilities for large-scale, long-term energy production. Although controversy shrouds the use of nuclear breeder reactors, they do make better use of natural uranium resources, and several foreign nations have commitments to nuclear breeder reactor technology. It remains to be seen whether the United States follows the foreign commitment. Conceivably, nuclear fusion energy technology can avoid many of the concerns put forth for the nuclear breeder reactor and may be the long-term energy solution. But even though hopes are high for nuclear fusion energy, as yet its technology is unproven. In order to respond intelligently to questions surrounding these exciting, but controversial, technologies it is important to understand their basics. Obtaining this understanding is the goal of this chapter.

13.2 THE NUCLEAR BREEDER REACTOR

To review, ^{233}U, ^{235}U, and ^{239}Pu are the only useful nuclear reactor fuels. Of these, only ^{235}U occurs naturally and it comprises only 0.7 percent of natural uranium which is predominantly ^{238}U. The isotopes ^{233}U and ^{239}Pu must be produced through nuclear reactions and transmutations. A breeder reactor starts with a loading of a fissionable fuel like ^{235}U. In addition to deriving energy by the normal nuclear fission process as in a light-water-cooled nuclear reactor, it produces fissionable nuclei like ^{233}U and ^{239}Pu from nuclear transformations of ^{232}Th (thorium) and ^{238}U. It is possible to produce more fuel than is actually "burned" in the energy-producing process. But don't be deceived into thinking that this is a perpetual motion device. The new fuel that is produced requires ^{232}Th or ^{238}U that is used up in the transmutations and must be replaced as the process proceeds.

The fuel production principle is as follows. When a nucleus such as ^{235}U or ^{239}Pu is fissioned with a neutron, there is no way to predict the composition of the reaction products. Some reaction products are more likely to be formed than others. Some reactions produce no neutrons and others produce several. Depending on the nucleus, the average number of neutrons produced per fission reaction is between two and three. One of these neutrons is required to sustain the fission chain reaction process. The remainder can be used for some other purpose. The purpose in a breeder reactor is to generate a fissionable material by arranging for a neutron to be captured by an appropriate nucleus. The nucleus that is formed in this capture process is radioactive. After two successive disintegrations, the desired stable (or very long-lived) end product is a fissionable nucleus. The nucleus that first captures the neutron is called fittingly a fertile nucleus and the process leading to a fissionable nucleus is called breeding. The isotope ^{238}U is a suitable fertile nucleus. When ^{238}U captures a neutron, the slightly more massive isotope ^{239}U is formed. This unstable isotope disintegrates by emitting a negative beta particle. Because the proton number increases by one, the species formed is an isotope of neptunium ($^{239}_{93}$Np$_{146}$). This isotope is also radioactive and also disintegrates by emitting a negative beta particle. The end product is plutonium-239 ($^{239}_{94}$Pu$_{145}$) which is a suitable nucleus for a reactor fuel. The sequence of events leading to the formation of ^{239}Pu is shown in Fig. 13.1. A similar breeding process can be devised using thorium-232 as the fertile isotope to produce fissionable uranium-233.

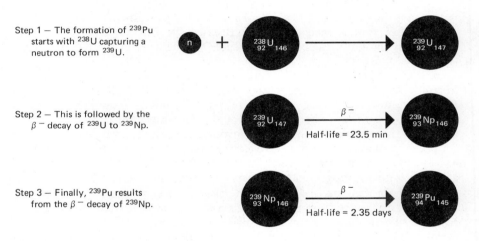

Step 1 — The formation of ^{239}Pu starts with ^{238}U capturing a neutron to form ^{239}U.

Step 2 — This is followed by the β^- decay of ^{239}U to ^{239}Np.

Step 3 — Finally, ^{239}Pu results from the β^- decay of ^{239}Np.

Figure 13.1 Steps leading to the production of plutonium-239 from the breeding of uranium-238.

It often happens that novel schemes for producing energy have such low production rates that they are impractical. Although the mechanism by which ^{239}Pu is made from neutron bombardment of ^{238}U is attractively simple, it is important to know just how fast it can be produced. One way of characterizing the production rate is to determine how long it takes to produce twice as many plutonium atoms as there were fissionable atoms present when the reactor started operating. This time is called the doubling time.

Depending on reactor design, the doubling time for a very efficient nuclear breeder reactor would be about six years. This means that a very efficient breeder reactor would produce enough fuel to refuel itself and another similar reactor after six years of operation. The doubling time for first generation breeder reactors is expected to be 20 to 25 years. Clearly, nuclear breeder reactors could substantially extend the lifetime of uranium fuel.

13.3 THE LIQUID METAL FAST BREEDER REACTOR

On several occasions we have stressed that the chance of nuclei fissioning when bombarded with neutrons depends sensitively on the energy of the neutrons. The same chance aspect prevails in the attempt to have ^{238}U capture neutrons to make ^{239}U. Nature has ordained that neutron capture (*not* fission) by ^{238}U is significant only for energetic (or fast) neutrons.* Fast neutrons are required if substantial breeding is to occur. (Recall that slow neutrons are required in conventional water-cooled reactors.) Because both breeding and fission reactions are desired, the fission reactions must be initiated by fast neutrons even though it is intrinsically more difficult. Water cannot be used as a coolant because of its ability to slow (or moderate) the speed of neutrons. The alternatives to water for a coolant are limited. Molten sodium and molten salt are possibilities. Liquid sodium is used as the coolant in most breeder reactors. These reactors are designated LMFBR which means liquid metal (for cooling) fast (for energetic neutrons) breeder reactor.

Sodium has several desirable features. Even though it is a metal, its melting point is only 210°F. Thus it will melt at the temperature of boiling water. It does not boil until it reaches 1640°F. Because the maximum temperature of sodium circulating in the core of a reactor will be around 1150°F, there is a substantial temperature margin before boiling occurs. While sodium is in the liquid state, the pressure that it exerts on the walls of a container is insensitive to temperature even when the temperature reaches 1150°F. Remember that in a pressurized water reactor the pressure rises to about 2250 pounds per square inch when the water temperature is 600°F. The only pressure on the liquid sodium is that required to circulate it around the reactor core. Thus concerns for pressure-induced pipe and valve leaks and failures are minimal.

Liquid sodium is an excellent heat conductor. While removing heat from a reactor it does not significantly capture or moderate neutrons needed to sustain the chain reaction and to instigate the breeding reactions. Fission fragments that do escape from the fuel rods and leak into the sodium coolant are readily tied up chemically making it difficult for them to escape to the environment. While it seems that sodium is a reactor designer's dream, there are disadvantages. Liquid sodium will oxidize and burn if exposed to the air and will react violently producing hydrogen if exposed to water. Continual bombardment of the sodium coolant produces the radioactive sodium isotopes ^{22}Na and ^{24}Na, which can accumulate to a hazardous extent. So, despite the fact that few problems are encountered in the pressure considerations involved with containing liquid sodium, extensive precautions still must be taken to ensure that the sodium does not escape.

Heat is extracted from the reactor core by liquid sodium, but steam is required for the turbine which drives the electric generator. This is accomplished by heat transfer from the sodium-cooling circuit to a water-cooling circuit. As a precaution against radioactivity leaking into the water circuit, an intermediate liquid sodium cooling circuit is used (Fig. 13.2).

---

* A neutron with energy greater than 10,000 electron volts (10 keV) is called a fast neutron.

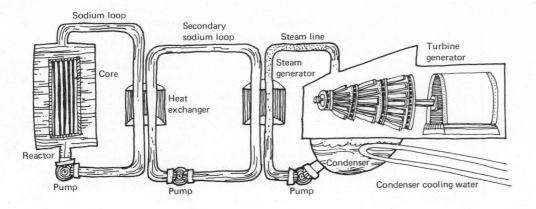

Sodium loop

Secondary
sodium loop

Steam line

Turbine
generator

Steam
generator

Core

Heat
exchanger

Reactor

Pump

Pump

Pump

Condenser

Condenser cooling water

The feedback method (Sec. 10.6) used to control the light-water nuclear reactor is also used in the fast breeder reactor. If the output power increases above some desired level, a signal is fed back to the input and action is taken to decrease the power. For example, control rods can be inserted to limit the availability of neutrons for fission reactions. Routine control of a fast breeder reactor and a water-cooled reactor are somewhat different. One reason is that the time between the birth of a neutron and when it is captured or produces fission is substantially less in a fast breeder reactor. Another reason is that there are fewer delayed neutrons in the fast breeder reactor if ^{239}Pu is used for the fuel. Routine control is manageable unless a malfunction occurs that produces a feedback that tends to increase the output. A void (or bubble) in the liquid sodium is such a condition. To a small but significant extent, sodium in the coolant captures neutrons. If a void occurs, there are more neutrons available for fission because fewer neutrons are captured. The extra neutrons tend to increase the output of the reactor. Increased output produces more heating and possibly more voids that may increase the power still further. The process takes place in so small a time that conventional controls cannot operate. A control is needed that operates just as fast as the void mechanism, but which produces an effect that diminishes the output power. Such a control utilizes a principle named the Doppler effect.*

Imagine a situation where you are riding in a car listening to the sound from the horn in a parked car. As you pass the car, the pitch (frequency) of the sound changes suddenly. This phenomenon is caused by the Doppler effect. It is due to the fact that the relative speed between you and the crests of the sound waves changes abruptly when you pass the stationary car.† Molecules in a solid are not free to migrate but they do vibrate. The frequency and speed of

Figure 13.2 The heat-transfer circuits in a fast breeder reactor. Thermal energy generated in the reactor core by nuclear fission reactions is transferred to a heat exchanger. This closed loop retains any radioactive material acquired in the heat transfer. A second closed liquid sodium loop transfers heat to a steam generator that vaporizes water for the steam turbine.

---

* Named for Christian Johann Doppler, Austrian physicist (1803–1853).

† Relative means the speed of the wave crests as "seen" by you in the car. If two cars are traveling side by side on an interstate highway at speeds of 50 miles per hour each, then the speed of one car relative to the other is zero. On the other hand, if two cars approach each other doing 50 miles per hour then their relative speed is 100 miles per hour.

their vibrations depend on temperature. The greater the temperature, the greater the frequency and speed. Thus the relative speed between an impinging neutron and a nucleus attempting to capture it depends on temperature. Because the chance that a nucleus in the fuel element will capture a neutron depends sensitively on the relative speed, the chance also depends on temperature. As the temperature increases the chance of capture increases. More captures mean that fewer neutrons are available for fission. When the output of a reactor suddenly increases and the temperature of the core goes up, this effect instantaneously comes into play and tends to reduce the output. Valuable time is gained so that conventional controls can be brought into play.

The fuel for a breeder reactor, unlike that for a water-cooled reactor, is highly enriched. Whereas the fuel in light-water-cooled reactors is 3%–5% fissionable nuclei, the percentage is 15%–30% for a nuclear breeder reactor. In addition, the active core is very compact. It may occupy a volume of only a few cubic meters (or cubic yards, if you want). These features lead to a concern called secondary criticality; a concern which is insignificant in light-water reactors. Criticality in a light-water reactor is attained after the fission neutrons have been sufficiently reduced in energy. If the coolant which also serves as the moderator is accidentally removed, then the mechanism for achieving criticality is removed (negative feedback) and the chain reaction tends to stop. However, in a nuclear breeder reactor the sodium functions only as the coolant and if it is suddenly removed, the reactor is still critical until other sensors detect that something is wrong. If a meltdown of the core occurs, there is the remote possibility of the fuel assuming a different geometric arrangement and developing a critical mass that can produce an uncontrolled chain reaction. This is the secondary criticality. Consequently, the design considerations of fast breeder reactors are more stringent. Because water and sodium are highly reactive, an emergency cooling system obviously cannot employ water. Current designs take advantage of the exceptional characteristics of sodium for conducting heat. In one design the reactor is sealed in a pool of sodium. In another the heat is transferred out of the reactor into a vat of sodium.

13.4 STATUS OF NUCLEAR BREEDER REACTOR PROGRAMS

Russia, France, and England have successfully operated nuclear breeder reactor electric power plants having electric power outputs ranging from 235 to 350 megawatts (Table 13.1). Second-generation systems are being built by all three countries. Japan and West Germany have systems in the construction stage.

Although an operational nuclear breeder reactor electric power plant was built in the United States in 1963, there has not been a commitment to an intensive breeder reactor program (Table 13.2). This initial reactor named for Enrico Fermi was permanently terminated in 1966 after a blockage in a sodium cooling line produced a partial meltdown of its core. No other commercial facility followed until plans were announced in February 1972 for a major nuclear breeder reactor electric power plant to be built on the Clinch River in eastern Tennessee (Fig. 13.3). Intended for completion between 1978 and 1980, it was to have an electric power output between 300 and 500 megawatts

Table 13.1 Status of foreign nuclear breeder reactor electric power plants.

Location	Name of system	Electric power output (megawatts)	Operational date
Russia	BN-350	350	June 1972
France	Phenix	233	December 1973
England	PFR	250	August 1976
Russia	BN-600	600	1978
West Germany	SNR-300	282	1981
Japan	Monju	300	1984
France	Superphenix	1200	

Table 13.2 American nuclear breeder reactors and their power characteristics.

Location	Name of system	Thermal output (megawatts)	Electric output (megawatts)	Operational date
Idaho Falls, Idaho	EBR-II*	62.5	16.5	1963
Monroe County, Michigan	Enrico Fermi	200	67	1963 (terminated 1966)
Clinch River, Tennessee	Demo No. 1	750–1250	300–350	?

* Experimental Breeder Reactor

Figure 13.3 Conceptual drawing of the proposed nuclear breeder reactor electric power plant to be built on the Clinch River in eastern Tennessee. Although similar facilities are a reality in several European countries, enthusiasm for the American project has wavered with presidential administrations. (Photograph courtesy of Westinghouse Electric Corporation.)

and was to be financed by both federal and private sources. Interest in the project has varied with presidential administrations and while it may be completed eventually, the date is highly uncertain.

Although the liquid metal fast breeder technology is the most advanced, there are other alternatives using the breeder concept that have merit and are receiving nominal support. Two examples are the thermal breeder using fuel bred from ^{232}Th and the gas-cooled fast breeder reactor (GCFBR).

The thermal breeder derives its name from its use of slow (or thermal) neutrons. Fissionable ^{233}U would be bred from ^{232}Th using slow neutrons (see Problem 13.2). Recognizing that a standard pressurized water nuclear reactor could be converted to a nuclear breeder reactor operating on the conversion of ^{232}Th to ^{233}U, plans were begun in 1965 for converting a reactor at Shippingport, Pennsylvania.* Now completed, its breeding and energy-producing characteristics are being studied. Although this project has not been as widely publicized as the fast breeder reactor, its importance should not be minimized because not only is it a way of making better use of limited naturally occurring fissionable nuclei but it avoids the ^{238}U–^{239}Pu breeding cycle.

The GCFBR utilizes gaseous helium as the heat transfer medium. This scheme is already used commercially in a type of ^{235}U-burning reactor. Helium has the advantage of being chemically inert. It is also very difficult to form radioactive products through neutron bombardment of helium. The main disadvantage is that the helium has to be compressed to pressures about 100 times atmospheric pressure, a process entailing extremely reliable pressure vessels and auxiliary equipment. However, experience with helium in conventional reactors indicates that these problems are solvable.

13.5 SPECIAL PROBLEMS WITH HANDLING PLUTONIUM

It is not the function of a light-water reactor to produce ^{239}Pu through breeding. Nevertheless, it does produce some because of the ^{238}U content of the fuel. This plutonium is valuable, and when the fuel rods are reprocessed it is extracted. Extracting plutonium is not simple because it is extremely dangerous. Not only does it produce deleterious chemical effects on the body, but it is radioactive and decays by emitting alpha particles. Internal to the human body, alpha particles are among the most damaging nuclear particles (Table 10.3). If plutonium is taken into the body in soluble form, it concentrates in bones and the liver and tends to remain. If taken into the lungs as small particles, for example, it can produce intense local damage and possibly induce cancer. For these reasons, plutonium processing is done by remote control methods. The amount of plutonium available for recovery in 1973 was about 500 kilograms (about 1100 pounds). This is expected to rise to 21,000 kilograms (46,000 pounds) in 1980. Considering that the maximum allowable plutonium burden on the body is set at 0.6 of a billionth of a kilogram, the potential hazard is obvious even without nuclear breeder reactors. Inherent

---

* The first commercial nuclear electric generating station became operational in Shippingport, Pennsylvania in 1957. It utilized a pressurized water reactor that operated successfully for 17 years.

dangers exist both at the reprocessing plant and in the transportation of the spent fuel rods and the processed plutonium. Prudent policy can minimize the chance of accidents, but accidents may still occur because of the human element involved.

If the nuclear breeder reactor proves successful and if we proceed to an energy economy based upon it, then the problem with plutonium will become magnified tremendously. For example, the plutonium radioactivity in a nuclear breeder reactor will amount to about one million curies and the stores of plutonium are anticipated to be around 720,000 kilograms in the year 2000.

While the dangers associated with the handling of plutonium are serious enough, even more serious is the possibility that plutonium might be stolen and used by terrorist groups. A convincing argument can be made for the impossibility of making a bomb from the ^{235}U fuel used for light-water reactors. The concentration is only a few percent and, to make a bomb, nearly 100 percent pure ^{235}U is needed. No one, save the federal government, has the enrichment facilities and processing technology to make ^{235}U bomb material. However, the plutonium in a fuel rod is highly enriched and only 5 kilograms of ^{239}Pu are needed to make a bomb like the one that destroyed Nagasaki in World War II. Granted, considerable danger would be involved in processing the plutonium and considerable technical knowledge would be required in order to construct a bomb, but the possibility cannot be ruled out. Even if a bomb were not made from stolen plutonium, the thieves could demand high ransom for its return. Whether these problems can be solved remains to be seen. But certainly other energy options, though not so highly promising as the nuclear breeder reactor, must be pursued vigorously if in so doing we can avoid the potential problems associated with a plutonium economy.

13.6 THE NUCLEAR FUSION PROCESS

Energy is released in any nuclear reaction in which the total mass of the reacting nuclei exceeds the total mass of the products after the reaction. This follows from the principle of conservation of energy and the fact that mass is equivalent to energy ($E = mc^2$). Splitting a heavy nucleus into fragments is but one of many energy-releasing nuclear reactions. It is possible to sort of reverse the fission process and combine the very lightest of nuclei together to form other nuclei whose combined mass is less than the total mass of the reacting nuclei. This reaction is called nuclear fusion. Fusion normally involves the isotopes of hydrogen and helium. The interaction of two protons to form a deuteron, a positive beta particle, and a neutrino takes place routinely in the sun and is illustrative of the nuclear fusion process. Pictorially,

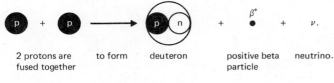

2 protons are fused together   to form   deuteron   positive beta particle   neutrino.

In symbols,

$$_1^1H_0 + {}_1^1H_0 \longrightarrow {}_1^2H_1 + \beta^+ + \nu.$$

The sum of the masses of a deuteron, beta particle, and neutrino is 0.001amu less than the combined masses of two protons. This mass is converted into 0.93 MeV of energy as a result of the fusion reaction. Although the energy released per reaction is small compared with the energy released in a nuclear fission reaction, it is still several thousand times more than the energy released in a chemical reaction.

While it is simple to write down the equation for the proton–proton fusion reaction and to determine the energy released, it is no easy task to actually achieve the fusion process. Because each interacting proton possesses a positive charge, a mutual repulsive electric force tends to separate them. Energy must be supplied to overcome this electric repulsion, and it must be sufficient to get the particles to within the interaction distance (about 0.0000000000001 centimeters) of the nuclear force. Even when this is achieved, the fusion process is very improbable because of the variety of possible nuclear reactions. In the laboratory, the necessary energy can be supplied by a proton accelerator such as a cyclotron. In the sun (and any practical nuclear fusion energy source), the energy is supplied thermally requiring a temperature of millions of degrees. This necessary temperature and the high density of protons for large numbers of collisions are available only in the interior core of the sun. Because of the low reaction probability and large ignition temperature, the proton–proton reaction as a practical man-made energy source is extremely remote.

Several other fusion reactions have the same energy-producing character as the proton–proton reaction and have greater reaction probability. The deuteron–deuteron and deuteron–triton reactions are of particular interest (Fig. 13.4). The deuteron–deuteron reactions are extremely attractive because

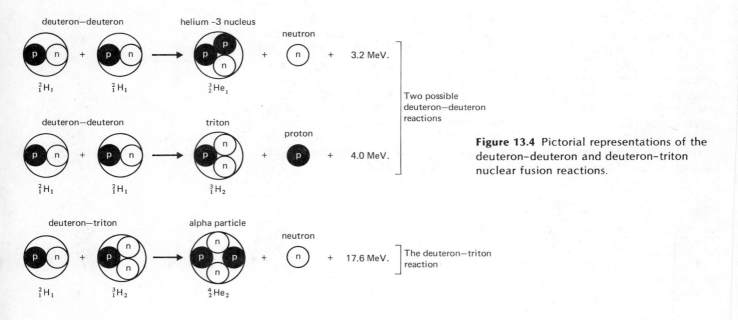

Figure 13.4 Pictorial representations of the deuteron-deuteron and deuteron-triton nuclear fusion reactions.

deuterium is not radioactive and occurs naturally. Although only one of every 7000 naturally occurring hydrogen isotopes is deuterium, there is essentially an unlimited supply in the oceans because of the enormous amount of hydrogen in the water. While tritium does not occur naturally and is radioactive, the deuteron–triton reaction is more appealing than the deuteron–deuteron reaction because the energy release is substantially larger and the ignition temperature is lower (about 40,000,000 °C compared with about 400,000,000 °C). As impressive as these basic considerations may be, producing an ignition temperature of millions of degrees for a sufficient length of time and containing the "nuclear fire" seem technologically inconceivable. Nevertheless, the practical application of fusion energy appears to be within our grasp and it offers a nearly unlimited source of energy without the bulk of the radioactivity problems associated with nuclear fission (Fig. 13.5).

Figure 13.5 A controlled nuclear fusion research device called the Princeton Large Torus. Successful operation of systems like this one is an important step toward bringing fusion-produced electricity to fruition by the turn of this century. (Photograph courtesy of Plasma Physics Laboratory, Princeton University.)

13.7 PRACTICAL CONSIDERATIONS FOR NUCLEAR FUSION REACTORS

An ordinary flame burns at a temperature of about 1000°C. The thermal energy associated with this temperature is about 0.1 eV. This is sufficient to excite some atoms to higher energy states, but not enough to remove an electron from an atom. Conversely, the thermal energy associated with a nuclear fusion fire is much more than enough to completely remove all electrons from the atoms of the gas. In this condition the gas is termed a plasma. Once kindled, the natural tendency for a fusion fire is to expand—thereby tending to extinguish itself. So the reactions must be contained if energy is to be derived. The time of containment of the fusion reactions is called the confinement time.

A burning process cannot be instigated unless a sufficiently high ignition temperature is provided. While this ensures that the fire starts, it does not guarantee that it will sustain. Once the fire is self-sustaining, one hopes to derive more energy than was required to kindle the fire. For example, a lighted match provides both an ignition temperature and a certain amount of energy for lighting a gas stove. Once lit, more energy is derived from the burning gas than was provided by the match. A nuclear fusion fire, once ignited with an appropriate ignition temperature, becomes self-sustaining when energy production exceeds energy loss. Those mechanisms such as conduction and convection that remove energy from any burning process are also important in nuclear fusion. There are others unique to nuclear fusion; for example, radiation from the rapidly moving charged particles in the plasma. The goal of nuclear fusion research is to minimize these energy losses so that a net gain in energy may be achieved.

The ignition temperature, the confinement time, and the density of the ions in the plasma are the crucial characteristics that determine whether or not the fusion process in question will be a useful energy source. Energy balance considerations reveal that the ion density multiplied by the confinement time must be at least 100 trillion (10^{14}) sec/cm³ for systems utilizing the deuterium–tritium reaction. This is called the Lawson criterion.* If the plasma used has an ion density of 1000 trillion (10^{15}) per cm³, then it must be contained for at least 0.1 sec. All research on fusion energy sources is centered on maximizing this product of ion density and confinement time and creating the required ignition temperature. The containment of a 40,000,000°C plasma is a challenging problem because the plasma must not contact the walls of a confining structure. It is not that there is concern that the hot plasma will melt the structure. The energy content of the plasma is just too low. But if the plasma does strike the walls, it loses energy, an action which extinguishes the self-sustaining fusion process. The force that a pipe would exert on the plasma to balance the force exerted by the plasma on the pipe is simulated with a magnetic force. These systems are referred to as magnetic bottles (Fig. 13.6). All sorts of exotic configurations have been devised to try to achieve this containment. Most are based on the physical principle that a charge experiences a force when it moves in a magnetic field. The idea is to have moving charged particles interact with an imposed magnetic field (or fields) in such a way that a force is exerted on the ions so as to contain the plasma. This permits the

---

* Named for British physicist J. D. Lawson.

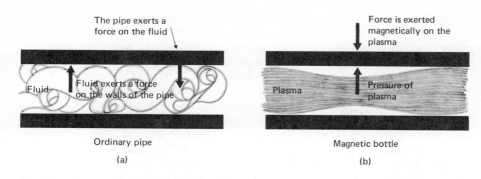

The pipe exerts a force on the fluid

Fluid

Fluid exerts a force on the walls of the pipe

Ordinary pipe

(a)

Force is exerted magnetically on the plasma

Plasma

Pressure of plasma

Magnetic bottle

(b)

Figure 13.6 (a) In an ordinary pipe, the walls exert a containing force on a gas under pressure. This force is the reaction to the force exerted on the walls by the gas. (b) In a magnetic bottle, a magnetic force contains the charged ions of the plasma.

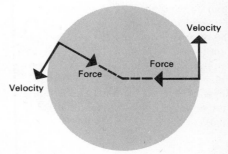

Velocity

Force

Force

Velocity

Figure 13.7 An object will travel in a circular path when the net force is perpendicular to its velocity.

plasma to be kept completely away from any confining structure that would extract energy from it.

This confinement can be understood more quantitatively in the following manner. When the net force on an object is always perpendicular to its instantaneous direction of travel, the object will travel in a circular path (Fig. 13.7). Examples of this behavior include the circular orbit of a satellite about the earth (Sec. 4.2) and the movement of an object attached to a string when it is swung in a circle. If a charged particle moves in a direction perpendicular to a magnetic field, the particle experiences a force perpendicular to both the magnetic field and its direction of travel (Fig. 13.8). Hence it moves in a circular path. The larger the magnetic field, the greater the force and the smaller the radius of the orbit. The charged particle moves in a circle around the magnetic field lines. If the direction of travel is not perpendicular to the magnetic field, then the force is still perpendicular to the direction of travel and the magnetic field, and the particle will now move in a helical path (Fig. 13.9). It is this principle that is used to regulate both the density and temperature of ions in a plasma.

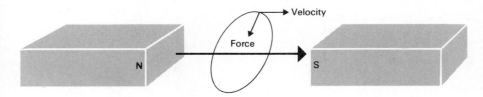

Velocity

Force

N

S

Figure 13.8 The magnetic field is directed from the N to the S pole. The charge moves at right angles to the direction of the magnetic field and revolves in a circle. This is the principle employed in cyclotrons used to accelerate nuclei such as protons and alpha particles to very high speeds.

While the principle is straightforward, instabilities develop for which there is no explanation in elementary terms. These instabilities lead to leakage of the plasma ions from the magnetic field and prevent achievement of the necessary confinement time.

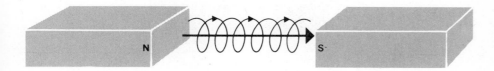

N

S

Figure 13.9 If the direction of travel of the charged particle is not perpendicular to the magnetic field direction, then the particle moves in a helical path. To visualize this motion, consider the following analogy. If a person rotates a mass on the end of a string, then an observer nearby sees the mass moving in a circle. If the person starts walking with the rotating mass, then the observer sees the mass moving in a spiral path.

13.8 TYPES OF NUCLEAR FUSION REACTORS

Figure 13.10 The Tokamak confinement geometry. The plasma constitutes the secondary winding of a transformer (Sec. 5.7). When a current is sent through the primary winding, a large current is produced in the plasma. This current heats the plasma producing the required ignition temperature. Current in a coil surrounding the plasma produces a magnetic field that contains the plasma. Several Tokamak systems are being studied in the United States (see Table 13.3).

Three types of magnetic confinement research programs receive the bulk of the federal support for fusion development. The principles of the devices being developed in these programs are referred to as Tokamak, magnetic mirror, and theta pinch. Schematic diagrams of the machines and their basic operating principles are shown in Figs. 13.10, 13.11, and 13.12. Although all three are very promising concepts, Tokamak systems receive about two thirds of the federal support for fusion research. Tokamak is an acronym from Russian words for toroidal magnetic chamber. The Russians announced in 1969 that a Tokamak had achieved a confinement time of 0.02 seconds using an ion density of 70 trillion per cm³. This is still well below the Lawson criterion for a useful energy source, but represents a major breakthrough in the state of the art. It was realized in the United States that a research device at Princeton University could be converted to the Tokamak principle. This was accomplished and successful operation began in July 1970. The Russian results were confirmed. Four laboratories conduct the bulk of the research on Tokamak systems in the United States (Table 13.3). A test reactor burning deuterium and tritium is expected to be operational in 1982. If successful and if a prototype

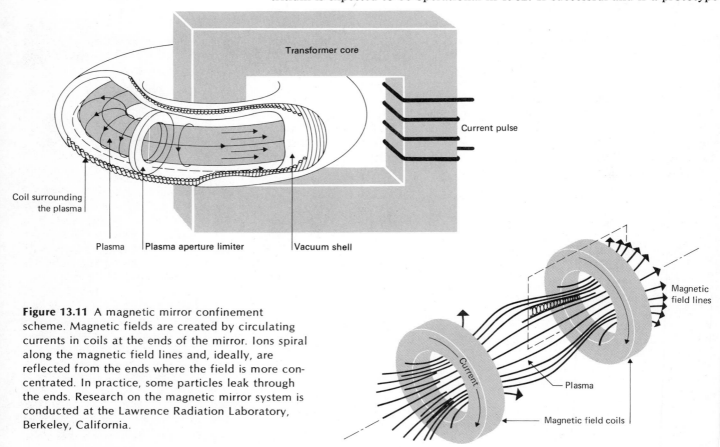

Transformer core

Current pulse

Coil surrounding the plasma

Plasma   Plasma aperture limiter   Vacuum shell

Figure 13.11 A magnetic mirror confinement scheme. Magnetic fields are created by circulating currents in coils at the ends of the mirror. Ions spiral along the magnetic field lines and, ideally, are reflected from the ends where the field is more concentrated. In practice, some particles leak through the ends. Research on the magnetic mirror system is conducted at the Lawrence Radiation Laboratory, Berkeley, California.

Magnetic field lines

Current

Plasma

Magnetic field coils

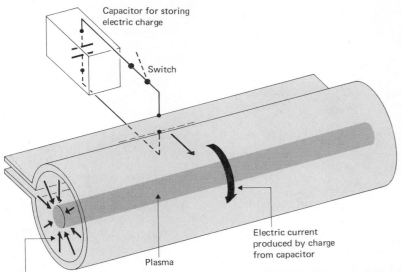

Capacitor for storing
electric charge

Switch

Electric current
produced by charge
from capacitor

Plasma

Magnetic pressure
on plasma

Figure 13.12 The theta pinch confinement geometry. A tube containing the plasma is surrounded with a metallic cylinder that has been split along its wall. A huge electric charge creates an electric current of over 100,000 amperes when switched into the cylinder from a charge-storing device called a capacitor. The resulting current creates a magnetic field that compresses (pinches) the plasma toward the axis of the cylinder. The compression heats the plasma and creates the fusion conditions. Research on the theta pinch confinement scheme is conducted at the Los Alamos Scientific Laboratory, Los Alamos, New Mexico.

Figure 13.13 A portion of the energy-storing capacitor bank for the Scyllac theta pinch fusion device. The relative sizes of the bank and the workers gives some indication of the magnitude of the project. (Photograph courtesy of Los Alamos Scientific Laboratory.)

Table 13.3 Names and types of nuclear fusion systems being researched in the United States.

Location	Name of system	Type
Princeton, NJ	Princeton Large Torus (PLT)	Tokamak
Oak Ridge, TN	ORMAK	Tokamak
Cambridge, MA	Alcator	Tokamak
La Jolla, CA	Doublet II A	Tokamak
Berkeley, CA	2 X-II B	Magnetic mirror
Los Alamos, NM	Scyllac	Theta pinch

power reactor is operational by 1985 and a system generating net electrical power is demonstrated by the early 1990s, then a demonstration nuclear fusion power plant producing several hundreds of megawatts of electric power may be functional in 1998.

Even if the fundamental aspects of the nuclear fusion process are solved, there are many engineering problems with no simple solutions. Structural damage and induced radioactivity due to intense neutron bombardment will be particularly difficult problems to solve. If both the fundamental and engineering problems can be solved, the first generation of practical fusion devices for producing electric energy will probably use the fusion reactions as a source of heat energy for a conventional steam turbine system (Fig. 13.14). Deuterium and tritium are likely to be used as fuel.

Figure 13.14 A possible scheme for producing electric energy from a fusion reactor. Deuteron–triton fusion reactions are instigated in the central core. Neutrons produced by the reactions are absorbed in the lithium blanket that surrounds the central core. The thermal energy generated in the lithium blanket is transferred to a heat exchanger that vaporizes water for a steam turbine. Tritium formed from neutron-induced reactions in the lithium blanket is recirculated back into the reactor for fuel for the fusion reactions. Containment of the plasma is provided by magnetic-field windings that surround the entire structure.

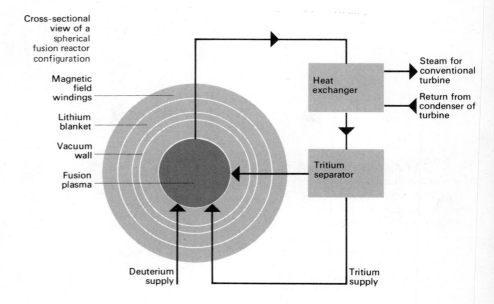

Because the neutron possesses the bulk of the 17.6 MeV of energy released in the deuteron–triton reaction, it is a matter of extracting the energy, converting it to heat, and transferring it to the turbine. This can be done using liquid lithium (or some lithium compound) to absorb the neutrons and produce tritium and helium via nuclear reactions. The tritium could then be extracted and recirculated back to the plasma for fuel.

13.9 ENVIRONMENTAL CONSIDERATIONS AND OUTLOOK

The nuclear fusion reactor solves one of the major drawbacks of the nuclear fission reactor—the production of large quantities of radioactive wastes in the form of fission fragments. The alpha particles produced from the absorption of neutrons by lithium are stable against nuclear decay. By capturing electrons, the alpha particles can be converted to helium, recovered, condensed to a liquid, and used as a low temperature coolant (4.2K) in a variety of industries. The tritium is recycled and used as fuel. However, the tritium is not recycled immediately. It must be reclaimed from the lithium coolant and this may require a hold-up time of about a day. As a result, about 100 million curies of tritium may be held up in the process. This is comparable to the amount of radioactivity in the fission products of a breeder reactor. Tritium emits only a low-energy negative beta particle so there is little problem in shielding it. But tritium, like hydrogen, is difficult to contain and will diffuse to some extent through the walls of a container. Because tritium has the chemical characteristics of hydrogen, it can combine with oxygen to form water. In this state, it is biologically hazardous. The tritium problems are not without solution, but nonetheless constitute a major radioactivity situation in fusion reactors which will utilize the deuteron–triton or deuteron–deuteron reactions.

The nuclear fusion reactor avoids the problems of storage of radioactive fission fragments, requires no critical mass of fuel that might lead to a nuclear explosion, extends the lifetime of fuel supplies to millions of years, makes no demands on a type of fuel that may have other national priorities, and will probably be more thermodynamically efficient. Like the nuclear breeder reactor, it emits no particulates or chemical compounds to the atmosphere. Looking well into the future, the nuclear fusion reactor offers the possibility of direct energy conversion into electric energy.* If this is achieved, the efficiency could rise to, perhaps, 90 percent and the thermal pollution problem using steam turbines would be eliminated. The high temperature technology gained in the development could be used in a multitude of ways.

Although the Tokamaks are an exciting prospect at this time and research is forging ahead on this type of machine, there is no guarantee that it is the nuclear fusion scheme that will be used ultimately. What is termed inertial confinement of nuclear fusion reactions is an equally exciting prospect.

---

* "The Prospects for Fusion Power," William C. Gough and Bernard J. Eastlund, *Scientific American* vol. 224, no. 2 (February 1971): p. 50.

13.10 INERTIAL CONFINEMENT OF NUCLEAR FUSION REACTIONS

When a nucleus at rest emits an alpha particle, for example, the nuclear remanence recoils in a direction opposite to the travel direction of the alpha particle.

Nucleus
at rest

Residual
nucleus

Alpha
particle

The explosion of a bullet produces a similar result. The resulting motions are a consequence of Newton's third law of motion. The force of the recoiling nucleus on the alpha particle causes it to move in one direction and an equal but oppositely directed force of the alpha particle on the residual nucleus causes it to move diametrically opposite. If you could arrange for a surface layer of particles on a sphere to explode radially outward, then the inner core would experience a reaction force from each escaping particle. The core would not move because the reaction forces all balance, but a shock wave would propagate inward through the core, and the core would absorb the energy of the disturbance. Sufficient energy will produce significant heating of the core. Inertial confinement of nuclear fusion reactions exploits this principle. A fusible spherical pellet of deuterium and tritium, for example, is irradiated uniformly with a pulse of some appropriate radiation or particles. The impact of the radiation causes a surface layer to escape outward (ablate). An inward-rushing shock wave heats the pellet sufficiently to instigate nuclear fusion. Because the pulse duration is a fraction of a billionth of a second, the inertia of the pellet is sufficient to confine the fusion reactions long enough for a net gain in energy. Light beams from lasers and beams of energetic electrons are being used for the radiation impinging on the pellet.

Laser is a word derived from the first letter of the words Light Amplification (by) Stimulated Emission (of) Radiation. The lasers of interest for the fusion process produce an extremely energetic beam of electromagnetic radiation, which is often visible. However, in two features it is very different from a beam of light from the sun or from an incandescent light bulb. First, it is composed almost entirely of waves having the same wavelength; if it were visible, it would appear as a pure color. Second, the peaks and valleys of the individual waves all line up. Radiation of this type is said to be coherent. Radiation from an incandescent light bulb is incoherent because there is a random relationship between the peaks and valleys of the individual waves. The mechanism for emission still evolves from a transition from a higher energy state to a lower energy state.

Gas lasers composed of a mixture of neon and helium and with continuous power outputs of one or two milliwatts are commonly used in light experiments in an elementary physics laboratory. The lasers used in fusion research produce extremely energetic flashes of radiation. For example, the energy in a pulse may be 1000 joules and the flash may last for only a tenth of a billionth (10^{-10}) of a second. This corresponds to an instantaneous power of ten trillion watts. Lasers 10 to 100 times more powerful are in the planning stage. Thus the power is enormous. The lasers may be of either a gas or a solid type. Carbon dioxide is often used in gas lasers. A glass with a distribution of neodymium atoms is routinely used in solid lasers. It is because of the huge power outputs from a laser that it is useful for igniting a fusion reaction. The laser would be focused on a suitable sample of a fusible material that would absorb the energy and, hopefully, instigate the fusion process.

A proposed scheme (Fig. 13.15) would focus a pulse of laser radiation on a tiny solid deuterium-tritium pellet about one millimeter in diameter. (A dime is about one millimeter thick.) The intense energy absorbed by the pellet creates a pressure disturbance which propagates inward rapidly compressing and heating the pellet. The pellet is brought to ignition temperature, and fusion reactions develop. Pressures generated by the fusion reactions cause the pellet to explode. Neutrons from the fusion reactions possess the bulk of the energy produced. This energy is extracted by nuclear reactions in a lithium blanket which surrounds the pellet. A new pellet would be inserted and the process continued. Although laser fusion is far from a practicality, it is a promising scheme and its development is being pursued vigorously.

Electron bombardment performs much the same function as irradiation with intense laser light. Electrons accelerated through potential differences of around one million volts and producing currents of a half a million amperes

Figure 13.15 Laser-fusion power cycle concept. The laser beam strikes a fusible pellet placed near the center of the spherical structure. Energy from the fusion process is transferred to the lithium blanket and then to a conventional steam generator. (Drawing courtesy of the Los Alamos Scientific Laboratory.)

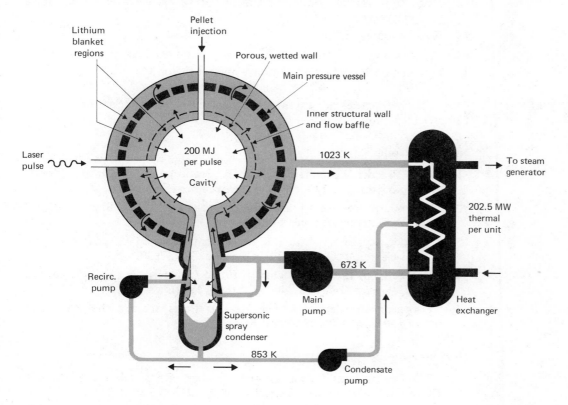

can be focused to a spot a few millimeters in diameter. Absorption of the electron energy instigates a shock wave in much the same fashion as in laser-induced fusion.

TOPICAL REVIEW

1. Describe the process of nuclear breeding.
2. How does a nuclear breeder reactor differ from a light-water reactor?
3. What is the significance of the nuclear breeder reactor?
4. What does LMFBR stand for?
5. Name some desirable and undesirable features of liquid sodium when used as a heat transfer medium.
6. What is meant by "secondary criticality"?
7. Discuss the effort being devoted toward the development of a commercial nuclear breeder reactor?
8. Describe the nuclear fusion process.
9. What is meant by ignition temperature? confinement time? plasma? plasma density?
10. How is plasma confined in a nuclear fusion reactor?
11. Describe the probable form of the first practical nuclear fusion reactor.
12. Name some advantages and disadvantages of using nuclear fusion for a practical energy source.
13. What is a laser?
14. Describe inertial confinement of nuclear fusion reactions.
15. How can the laser be used to induce nuclear fusion?

REFERENCES

The following articles are useful for expansion of the material presented in Chapter 13.

Nuclear Breeder Reactors

1. "Fast Breeder Reactors," Glenn T. Seaborg and Justin L. Bloom, *Scientific American* **223,** No. 5, 13 (November 1970).
2. "Plutonium: Reactor Proliferation Threatens a Nuclear Black Market," Deborah Shapeley, *Science* **172,** 143 (1971).
3. "Breeder Reactors," Available for 40 cents from DOE Technical Information Center, P.O. Box 62, Oak Ridge, TN 37830.
4. "Energy from Breeder Reactors," Floyd L. Culler, Jr., and William O. Harris, *Physics Today* **25,** No. 5, 28 (May 1972).
5. "European Breeders (II): The Nuclear Parts are not the Problem," *Science* **191,** 368 (30 January 1976).

Nuclear Fusion

1. "The Prospects of Fusion Power," William C. Gough and Bernard J. Eastlund, *Scientific American* **224,** No. 2, 50 (February 1971).

2. "The Tokamak Approach in Fusion Research," Bruno Coppi and Jan Rem, *Scientific American* **227**, No. 1, 65 (July 1972).

3. "Controlled Nuclear Fusion," Samuel Glasstone, Available for 40 cents from DOE—Technical Information Center, P.O. Box 62, Oak Ridge, TN 37830.

4. "Fusion Power by Laser Implosion," John L. Emmett, John Nuckolls, and Lowell Wood, *Scientific American* **230**, No. 6, 24 (June 1974).

5. "Fusion Research (I): What is the Program Buying the Country," *Science* **192**, 1320 (25 June 1976).

6. "Fusion Research (II): Detailed Reactor Studies Identify More Problems," *Science* **193**, 38 (2 July 1976).

7. "Fusion Energy in Context: Its Fitness for the Long Term," J. P. Holdren, *Science* **200**, 168, (14 April 1978).

13.2 The Nuclear Breeder Reactor

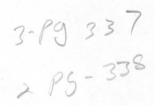

SUGGESTIONS FOR FURTHER UNDERSTANDING OF CHAPTER 13

1. In an ordinary nuclear reactor using ^{235}U for fuel, what is the minimum number of neutrons that can be released in a fission reaction necessary to maintain a self-sustaining process?

   In a breeder reactor using ^{239}Pu for fuel and ^{238}U for fertile nuclei, what is the minimum number of neutrons released in a fission reaction necessary to maintain a self-sustaining process as well as replacement of the spent fuel?

2. How does the basic energy-conversion method in a breeder reactor differ from that in a conventional light-water reactor?

3. What practical role does the half-life of ^{239}U and ^{239}Np play in the breeding process shown in Fig. 13.1?

4. According to the theory of nuclear fission, $^{231}_{92}$U$_{139}$ would be a suitable fuel for a nuclear chain reaction. It could be produced very much like ^{233}U and ^{239}Pu.

   $$^{230}_{90}\text{Th}_{140} + {}^{1}_{0}\text{n}_1 \longrightarrow {}^{231}_{90}\text{Th}_{141}$$

   $$^{231}_{90}\text{Th}_{141} \longrightarrow {}^{231}_{91}\text{Pa}_{140} + \beta^- + \bar{\nu}$$

   $$^{231}_{91}\text{Pa}_{140} \longrightarrow {}^{231}_{92}\text{U}_{139} + \beta^- + \bar{\nu}$$

   What time considerations might rule this process out as a possible source of nuclear fuel?

5. It is more than just coincidence that the useful nuclear fuels discussed—^{233}U, ^{235}U, ^{239}Pu—have an odd number of nucleons. Another nucleus in this category is ^{239}Np that is formed from the decay of ^{239}U (see Sec. 13.2). Why, then, isn't ^{239}Np used as a nuclear reactor fuel?

13.6 The Nuclear Fusion Process

6. Why could solar energy be also categorized as nuclear energy?

7. Why would the gaseous diffusion method used to separate ^{235}U from ^{238}U also be a useful technique for separating deuterium from hydrogen?

8. Storing nuclear fission fuels requires some care because of the possibility of creating a critical arrangement. Why do you not have this same worry when storing nuclear fusion fuels?

13.7 Practical Considerations for Nuclear Fusion Reactors

9. A person traveling in a circle while on a merry-go-round is analogous to a charged particle traveling in a circle in a magnetic field. What is the force that keeps the person on the merry-go-round traveling in a circle?

10. It is extremely difficult to attenuate a beam of neutrons with a material absorber. What special problem does this pose for deriving the energy released in deuterium–tritium nuclear fusion reactions?

11. All materials will become radioactive if exposed to a "rain" of neutrons of sufficient intensity. Why, then, might we expect considerable radio-activity to be generated in the containment structures for nuclear fusion systems?

12. In a nuclear fission reactor, thermal energy is produced in a continuous fashion. But in the design concepts of contemporary nuclear fusion reactors, energy is produced in bursts or pulses. What problems might this cause in a commercial electric power plant deriving energy from nuclear fusion reactions?

NUMERICAL PROBLEMS

13.2 The Nuclear Breeder Reactor

1. This problem emphasizes the differences in concentration of energy in an ordinary energy source and a nuclear breeder reactor. A coil on a 1000-watt heating coil has a volume of about 500 cubic centimeters. The active core of a 1000-megawatt nuclear breeder reactor has a volume of about three cubic meters. Compare the watts per cubic centimeter for these two systems.

2. Fill in the steps for generating fissionable ^{233}U starting from ^{232}Th.

$$^{232}_{90}Th_{142} + n \longrightarrow \underline{\hspace{2cm}}$$

$$\underline{\hspace{2cm}} \longrightarrow \underline{\hspace{2cm}} + \beta^- + \bar{\nu},$$

$$\underline{\hspace{2cm}} \longrightarrow {}^{233}_{92}U_{141} + \beta^- + \bar{\nu}$$

13.3 The Liquid Metal Fast Breeder Reactor

3. Sodium-23 ($^{23}_{11}Na_{12}$) is the only stable form of natural sodium. Sodium-22 ($^{22}_{11}Na_{11}$) and sodium-24 ($^{24}_{11}Na_{13}$) are radioactive and are produced in the liquid sodium that circulates around the core of a breeder reactor. Show how these isotopes can be made by bombarding stable $^{23}_{11}Na_{12}$ with neutrons.

13.4 Status of Nuclear Breeder Reactor Programs

4. Calculate the efficiencies of the breeder electric power plants listed in Table 13.2. How does the efficiency of the proposed Demo No. 1 unit compare with the efficiency of a conventional LWR power plant?

13.6 The Nuclear Fusion Process

5. Using the data below show that 17.6 MeV of energy are released when a deuteron and a triton are fused to form an alpha particle and a neutron.

$$triton = 3.016050 \text{ amu}$$

$$deuteron = 2.014102 \text{ amu}$$

$$alpha \text{ } particle = 4.002603 \text{ amu}$$

$$neutron = 1.008665 \text{ amu}$$

$$1 \text{ amu} = 931 \text{ MeV}$$

6. Although only one out of every 7000 hydrogen isotopes is of the deuterium variety, there are still over a million-trillion-trillion-trillion (10^{42}) deuterium atoms in the oceans. If these deuterium atoms were used in deuterium–deuterium fusion reactions and each reaction liberated a millionth of a billionth (10^{-15}) of a Btu of energy, show that the total energy released is over a 100 million (10^8) Q ($1Q =$ billion-billion (10^{18}) Btu). How does this amount of energy compare with that available from fossil fuels?

7. A mass in grams equal to the atomic weight of an elemental substance always contains the same number of atoms. This number is equal to 6.02×10^{23} and is called Avogadro's number. Thus 235 grams of ^{235}U contains Avogadro's number of ^{235}U atoms. Why could you derive more fusion energy from one gram of deuterium than fission energy from one gram of ^{235}U?

13.7 Practical Considerations for Nuclear Fusion Reactors

8. Early versions of the Tokamak fusion devices typically produced plasmas of 10 trillion (10^{13}) particles per cm^3, temperatures up to 10,000,000 K, and confinement times of about 0.03 sec. How close are these machines to satisfying the Lawson criterion for a useful energy source?

13.8 Types of Nuclear Fusion Reactors

9. One stable isotope of lithium is $^6_3Li_3$. Show how a triton (a tritium nucleus) and an alpha particle (a helium nucleus) can be produced when ^6Li is reacted with a neutron.

13.10 Inertial Confinement of Nuclear Fusion Reactions

10. The energy in a pulse of electromagnetic radiation from a certain laser is 1000 joules. If the pulse lasts for one tenth of a billionth (10^{-10}) seconds, show that the instantaneous power is 10 trillion (10×10^{12}) watts.

THE POTENTIAL AND IMPORTANCE OF ENERGY CONSERVATION

14

Electric energy, the most versatile and possibly the most precious of the many forms of energy, is propagated over transmission lines such as these. Efficient use of electric energy is an important aspect of our energy policy. Photograph by DeWys, Inc.

14.1 MOTIVATION Several messages have emerged as we look back on this study of energy for a technological society. Not the least of these is the simple fact that our country is consuming more energy than it produces and that reliance on external sources is increasing. Dependency on external sources can be reduced, perhaps eliminated, by developing alternatives. But even given the incentive and financial backing, development takes several years of time. So the gap between energy production and energy consumption will continue to grow unless the rate of consumption is reduced.

Generally, conserving connotes "doing without" and it is not easy to persuade people to do without. Perhaps a more realistic approach is to view energy conservation as "doing better." If energy is literally wasted, then doing better might mean doing without. But more often it means taking advantage of the best method for a given task. Elementary algebraic concepts (Sec. 2.2) elucidate these features. The number of gallons of fuel used by a car during a trip equals the gallons used per mile of travel multiplied by the length (miles) of the trip:

$$\text{gallons} = \left(\frac{\text{gallons}}{\text{miles}}\right) \times (\text{miles}).$$

Energy conservation strives to reduce the quantity on the left side of the equation, that is, gallons. One way is to reduce the number of miles traveled. But then you may not have achieved your goal, which is to travel a certain distance. This is the fallacy in saving energy by being frugal but not achieving a task. The goal may still be attained and a fuel savings effected if a more efficient mode, using fewer gallons per mile, were opted for. That may mean selecting an unfamiliar vehicle and adopting a different lifestyle. For example, commuting to work might be more energy efficient if one were able to use either an electric car or mass transportation rather than a private automobile with an internal combustion engine. The best energy-conservation innovations are for naught unless they can be implemented. Ultimately this may mean legislation. Short of legislation, no scheme can be sold to the public unless it makes economic sense. It is estimated that relatively simple conservation measures could produce an energy savings equivalent to 7.3 million barrels of oil each day by 1980.* In energy terms this saving amounts to nearly two thirds the projected amount of imported oil in 1980. If in 1980 oil were to cost $12 per barrel, which appears conservative, then this would create an annual savings of nearly $32 billion. This amount of money becomes even more impressive when it is compared with an annual budget of a few billion dollars for energy research and development. Most energy conservation schemes are based on physical principles that we have discussed already. The application of these principles motivates this final chapter of our study of "Energy for a Technological Society."

---

*"The Potential for Energy Conservation," Office of Emergency Preparedness, October 1972.

For purposes of identifying the areas of potential energy conservation, it is convenient to divide energy consumption into four sectors—residential and commercial, transportation, electric utilities, and industry. The utilization of the contemporary forms of energy in these sectors was shown in Fig. 1.1. About half of the energy fed into the system ends up rejected to the environment unused.* Some of the rejection of energy is due to the intrinsic inefficiency of some of the energy conversion processes used. But a substantial fraction of it is wasted in the strict sense of the word and this is why opportunities exist for significant energy saving. Let us now examine each of the energy-consuming sectors for ways of conserving energy through application of simple physical principles. The treatment is illustrative and not exhaustive. You may want to extend your knowledge of this subject by examining the references listed at the end of the chapter.

14.2 POSSIBLE AREAS WHERE ENERGY CAN BE CONSERVED

Energy projections for 1980 indicate that residential and commercial establishments will use about 17.5 million-billion (17.5×10^{15}) Btu of energy. About 63 percent of this, or 11 million-billion Btu, will be used for space heating and cooling in structures most of which have temperature-controlling systems. By simply lowering the thermostat 2° in winter and raising it 2° in summer it is estimated† that 12 percent of the energy projected for space heating and cooling could be saved. Although this may seem unreasonable it is understandable. Once the temperature of a dwelling has been brought to the desired temperature, the purpose of a heating system is to provide the heat that is lost to the cooler exterior through the walls and windows. As we have seen (Sec. 7.5), the rate at which heat is lost depends on the structure of the walls and windows and on the difference in temperature between the inside and outside of the dwelling. Historically, the accepted ideal temperature for a dwelling is 70°F. So if the average outside temperature is 55°F, the average temperature difference is 15°F. If the thermostat setting were lowered 2°F so that the temperature inside is 68°F, then the average temperature difference is 13°F. Because heat furnished by the heating system is proportional to the temperature difference, then with the lower thermostat setting it would have to supply only $13/15 = 0.87$ or 87 percent as much heat. Thus there would be a 13 percent saving in energy. Thus, with only the effort required to change a thermostat setting a trivial amount, substantial energy could be saved.

Regardless of the quality of a structure, lowering the thermostat setting effects a reduction in energy use. But a more important question is, "Are the heat losses inordinate for the structure even with a reduced thermostat setting?" An answer to this question requires an examination of how heat is conducted and channeled to the outside. Our study (Sec. 6.2) of the conduction of

14.3 ENERGY CONSERVATION IN THE RESIDENTIAL AND COMMERCIAL SECTOR

---

* There are differences of opinion on how much of the energy is rejected. However, it is always about 50 percent or more.

† "Energy Conservation," G. A. Lincoln, *Science,* vol. 180 (13 April 1973): p. 155.

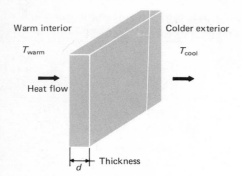

Figure 14.1 A slab of some material intercepting the flow of heat from a warm to a cool region. The slab might be a window, a wall, a ceiling, or a floor of a house. The heat flow (for example, Btu's per hour) depends on the thickness of the slab, the area of a face of the slab, the temperature difference, and the type of material composing the slab.

heat through a thickness of material (Fig. 14.1) led us to conclude that the flow of heat depends on

1. the temperature difference between the faces of the material,
2. the area *(A)* facing the heat source,
3. the physical structure of the material, and
4. the thickness *(d)* of the material.

Summarized in equation form,

$$\frac{E_{lost}}{t} = \frac{kA(T_{warm} - T_{cool})}{d},$$ (14.1)

where E_{lost} is the heat lost through conduction in a time period designated as t. The thermal conductivity (k) reflects the fact that some materials are better heat conductors than others. The larger the value of k, the better the heat conducting properties of the material. While Eq. 14.1 is in a form suitable for understanding the essence of heat conduction, it is rather awkward for practical heat loss calculations in a building. Homeowners may want to estimate the heat lost through the windows of their houses during a typical heating season. Knowing this they can get some idea of the fuel cost. To answer this type of question, let us rewrite Eq. 14.1 as

$$E_{lost} = \frac{k}{d} A(T_{warm} - T_{cool}) \cdot t,$$ (14.2)

and take the time period (t) to be the length of the heating season. Both the length of the heating season and the temperature difference will depend on the locality. Both will be larger in northern United States than in southern United States. The product of temperature difference ($T_{warm} - T_{cool}$) multiplied by the length of the heating season is called "heating degree days." Assuming that a house is maintained at 65 °F, there are about 4000 heating degree-days in central U.S.* In northern United States there are about twice as many heating degree-days. The factor that 4000 has to be multiplied by to get the heating degree-days for a particular area is called the "heating factor," symbolized H.F. (Fig. 14.2). The area (A), thickness (d), and type (k) of a wall or window will vary from house to house. So one usually determines the heat flow through an area of one square foot of a particular type of material. Knowing this, then the heat loss through, say, 100 square feet of the same material is 100 times greater. The quantity d/k that appears in Eq. 14.2 is designated the R-value (Table 14.1). It is a measure of how one square foot of a certain thickness of material resists the flow of heat through it. It is usually expressed as $ft^2 \cdot hour \cdot °F/Btu$. In an area having 4000 heating degree-days the energy required by an average heating system to account for the heat loss through one square foot of a material having an R-value of 1 is

$$E_{lost} = \frac{1 ft^2 (4000 °F \cdot days)}{1 ft^2 \cdot hr \cdot °F/Btu} \cdot \frac{24 hrs}{day} = 96,000 \text{ Btu}.$$

---

* A reference temperature of 65 °F is used because an average house needs no heat from a furnace to maintain an indoor temperature of 70 °F when the outside temperature is 65 °F.

Table 14.1 Insulation R-values for the common structural materials. A more complete compilation can be found in *Handbook of Fundamentals,* published by The American Society of Heating, Refrigerating and Air-Conditioning Engineers (ASHRAE).

	Material	Thickness (inches)	R–Value $\left(\dfrac{ft^2 \cdot hr \cdot {}^\circ F}{Btu}\right)$
Air film and spaces:	Air space, bounded by ordinary materials	¾ or more	.91
	Air space, bounded by aluminum foil	¾ or more	2.17
	Exterior surface resistance	—	.17
	Interior surface resistance	—	.68
Masonry:	Sand and gravel concrete block	8	1.11
		12	1.28
	Lightweight concrete block	8	2.00
		12	2.13
	Face brick	4	.44
	Concrete cast in place	8	.64
Building materials—General:	Wood sheathing or subfloor	3/4	1.00
	Fiber board insulating sheathing	3/4	2.10
	Plywood	5/8	.79
		1/2	.63
		3/8	.47
	Bevel-lapped siding	1/2 × 8	.81
		3/4 × 10	1.05
	Vertical tongue and groove board	3/4	1.00
	Drop siding	3/4	.94
	Asbestos board	1/4	.13
	3/8″ gypsum lath and 3/8″ plaster	3/4	.42
	Gypsum board (sheet rock)	3/8	.32
	Interior plywood panel	1/4	.31
	Building paper	—	.06
	Vapor barrier	—	.00
	Wood shingles	—	.87
	Asphalt shingles	—	.44
	Linoleum	—	.08
	Carpet with fiber pad	—	2.08
	Hardwood floor	—	.71
Insulation materials (mineral wool, glass wool, wood wool):	Blanket or batts	1	3.70
		3 1/2	11.00
		6	19.00
	Loose fill	1	3.33
	Rigid insulation board (sheathing)	3/4	2.10
Windows and doors:	Single window	—	approx. 1.00
	Double window	—	approx. 2.00
	Exterior door	—	approx. 2.00

SOURCE: Data are from "Project Retrotech," a home weatherization guide published by the Department of Energy, Washington, D.C. 20461.

For estimation purposes we round 96,000 to 100,000. If the area and heating factor are different from 1, then the energy required changes proportionally. If the R-value is different from 1, then the energy required changes inversely. Thus, in general, the energy required is approximately

$$E = \frac{A \cdot \text{H.F.}}{R} \cdot 100,000 \text{ Btu.} \tag{14.3}$$

To see how this works let us assume the following:

single pane window, $R = 1.0$
area $(36'' \times 54'') = 13.5$ square feet
heating factor $= 1$.

Then

$$E = \frac{13.5 \cdot (1.0)}{1.0} \times 100,000 \text{ Btu}$$
$$= 1,350,000 \text{ Btu.}$$

A 42-gallon barrel of heating oil provides 5,800,000 Btu of energy. Hence about 10 gallons of oil would need to be burned to provide the energy which was conducted through the window. If oil costs 50¢/gallon, then the cost would be $5 for the entire heating season. This cost could be cut in half by installing a double pane window having an R-value of 2. It would take a few years to pay for the window at $2.50/year but energy costs will escalate and the pay-off period will decline. Additionally, if the energy source is electricity, the savings are about twice as much.

For every doubling of a given thickness of insulation, the heat flow through it will be halved. But regardless of how thick the insulation becomes, there is no thickness that will completely stop all heat flow through it. Therefore, the thickness of insulation used is a pragmatic decision. One has to decide if an additional amount of insulation will pay for itself through reduced fuel costs. In order to assist a homeowner or builder in this decision, guidelines for suggested amounts of insulation have been determined. These guidelines are presented as a suggested R-value for a type of construction in a particular heating zone (Fig. 14.2). Table 14.2 lists the recommended ceiling and floor R-values for the six heating zones specified in the United States. If in your area the suggested R-value for a floor is R-19, then the sum of the R-values in the construction materials of the floor should add up to 19.

It is a tedious but straightforward job to estimate conductive heat losses through walls, ceilings, and floors, but it requires that you evaluate only the surface areas and type of construction. There are other heat losses that are not as easy to calculate. Any opening, such as a crack or chimney, will allow convective heat flows (air currents) and a loss of energy to the outside. Attention to detail in sealing cracks and openings is essential. Even if windows and doors

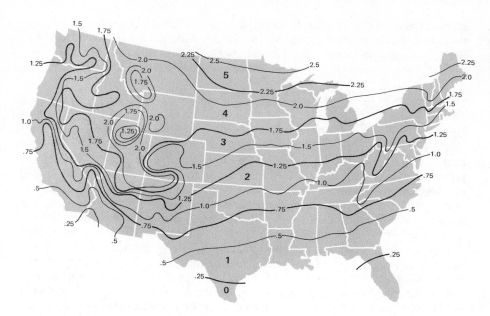

Figure 14.2 District heating factors and heating zones in the United States. A heating zone is an area bounded by two heavier black lines. Six heating zones numbered 0 through 5 have been delineated. The heating conditions are about the same for all communities located in a given heating zone.

Table 14.2 Recommended ceiling and floor R-values for the six heating zones specified in the United States. To find the heating zone where you live, see Fig. 14.2.

Heating Zone	Ceiling	Floor
0,1	R-26	R-11
2	R-26	R-13
3	R-30	R-19
4	R-33	R-22
5	R-38	R-22

fit tightly and openings are filled with an appropriate sealant, a complete air change occurs in a house every hour. In a poorly cared-for house, there may be three complete air changes each hour. Careful attention to conductive and convective heat losses will minimize the energy that needs to be supplied by the heating system. Whether one is able to afford the optimum always reduces to economics. Making a house energy efficient always involves an outlay of capital. Hopefully, the investment is returned in the long run. A homeowner who cannot obtain the initial capital must go with an energy-inefficient dwelling and pay a price premium in monthly installments. In a mobile society where a house is occupied by a given owner for short periods, the owner will usually opt for energy inefficiency because a comparatively short occupancy precludes the recovery of the capital investment. An owner who can continually feed energy into the structure can get by. But if the energy source is limited or prohibitively expensive, then there is no recourse but to shore-up the structure. Such is the case with solar energy. The amount of solar power falling on the roof of a house is marginally enough to satisfy the heating requirements. Therefore, a necessary condition for solar space heating to work is that the house be energy efficient. Solar heating should always start from this premise, otherwise a user will be extremely disappointed.

The thermal energy associated with any energy source can be written as

$$\text{energy} = \left(\frac{\text{energy}}{\text{molecule}}\right) \times (\text{number of molecules}).$$

Because the energy of a molecule is directly related to the kelvin temperature (see Sec. 7.3), we are accustomed to thinking that a sizable quantity of energy necessarily implies a high temperature. For example, a heating coil on an electric stove is "red hot" and the flame of the burner on a gas stove will quickly sear your hand if you accidentally place it in the flame. Both the coil and burner provide substantial energy and are at a very high temperature. But it follows that there can still be significant energy in a low-temperature system if the number of molecules is large. Even on a freezing day the earth and the air surrounding it contain massive amounts of energy because the volume of either is extremely large. Experiences tells us that this thermal energy will not flow spontaneously into a home which is at a higher temperature. The second law of thermodynamics guarantees this. On the other hand, there is nothing in the laws of physics that says we cannot do work (expend energy) and extract the thermal energy from the earth and deposit it in a dwelling. The basic question is, can we put more energy into the dwelling than was used to extract the energy from the ground? If the answer is yes, then there is a net energy gain and we would be wise to do it *if it can be justified economically*. If the answer is no, then one would be better off to use the input energy for the system as energy for the dwelling. The heat pump (Fig. 14.3) discussed in physics terms in Chapter 6 does just what we have outlined—it pumps heat from the cooler exterior of a building and deposits it in the interior of the building. Its principle (Sec. 6.5) is based on the standard refrigeration cycle. We characterize the heat pump's (HP) effectiveness* as

$$E_{\text{HP}} = \frac{\text{total amount of heat added to the warm region}}{\text{work required}}. \tag{14.4}$$

---

* Because a refrigerator and a heat pump are basically the same type of device, the effectiveness of each one is often defined in the same way. This was done, for example, in the first edition of this book. Strictly speaking, a refrigerator is used to remove heat from a cold region and a heat pump is used to provide heat to a warm region. Taking this into account, their effectivenesses as defined in Section 6.3 are somewhat different. The maximum effectiveness of a refrigerator is

$$\frac{T_{\text{cool}}}{T_{\text{warm}} - T_{\text{cool}}}$$

and the maximum effectiveness for a heat pump is

$$\frac{T_{\text{warm}}}{T_{\text{warm}} - T_{\text{cool}}}.$$

One also finds that what we have called effectiveness is called coefficient of performance (COP) in many texts and in the commercial sector.

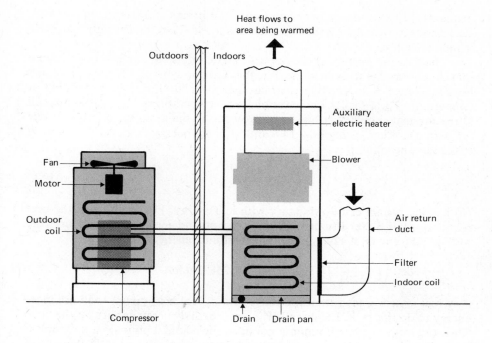

Outdoors | Indoors

Heat flows to area being warmed

Fan

Motor

Outdoor coil

Compressor

Auxiliary electric heater

Blower

Air return duct

Filter

Indoor coil

Drain   Drain pan

Figure 14.3 Layout of a heat pump installation. The system functions like a household refrigerator. An easily vaporized liquid absorbs heat from the exterior. The vapor is forced into the interior where it is condensed to a liquid. The thermal energy released in condensation remains to be circulated through the building. Energy is literally pumped from the cooler outdoors to the interior of a building being heated. An auxiliary electric heater is provided as a supplementary heat source.

Thermodynamic considerations showed us that no heat pump can be more effective than

$$E_{HP}(maximum) = \frac{T_{interior}}{T_{interior} - T_{exterior}}, \tag{14.5}$$

where $T_{interior}$ and $T_{exterior}$ are the kelvin temperatures of the interior and exterior of the building. Now for some numbers. Suppose that the outside temperature is 41 °F (278K) and the inside temperature is 68 °F (293K). Then

$$E_{HP}(maximum) = \frac{293}{293 - 278} = \frac{293}{15} = 20.$$

This would mean that one unit of input energy would cause 20 units of energy to be deposited in the building. A practical heat pump cannot perform like the ideal heat pump. Still it is often possible to achieve effectivenesses of between 2 and 4, and this is significant. For example, electric heating is the most expensive form of household space heating because the efficiency of converting the energy of coal to electricity is about 33 percent. If the electric energy were used to run a heat pump rather than to produce heat directly with a heating coil, then, in effect, the energy lost at the power plant can be recovered. Installing a heat pump requires a capital outlay and its worth must be judged on an economic basis. It is important to note that the effectiveness of a heat pump decreases as the temperature difference between the outside and inside increases. For this reason heat pumps are not practical in very cold climates. Technical improvement in heat pumps has made them increasingly attractive in mild climates.

There are important philosophical implications for the ideal heat pump that do not emerge without some thought. Thermodynamically the task of any heating system is to provide thermal energy to the inside of a building at temperature T_{interior} when the lower outdoor temperature is T_{exterior}. Let us call Q the thermal energy that has to be provided to maintain the required inside temperature. Then, using Eqs. 14.4 and 14.5, the energy that must be supplied to the ideal heat pump is

$$W = Q \left(1 - \frac{T_{\text{exterior}}}{T_{\text{interior}}} \right) . \tag{14.6}$$

No heating system can do this task with less energy. For example, if the heat requirement is 5,000 Btu/hr and the exterior and interior temperatures are 41 °F (278K) and 68 °F (293K), then the minimum energy requirement is

$$W = 5000 \left(1 - \frac{278}{293} \right) = 5000 \left(\frac{15}{293} \right) = 256 \, \text{Btu/hr}.$$

Any real system—gas furnace, electric heater, real heat pump—will use more energy to do this task. Let us call W_{actual} the actual energy used to do the task. Then the ratio of the minimum amount to the actual amount is a measure of the performance of the system. As in Chapter 6, we call this ratio the second law efficiency.

$$\text{Second law efficiency} = \frac{\text{minimum amount of energy required}}{\text{actual amount of energy used}} . \tag{14.7}$$

Using Eq. 14.6,

$$\text{Second law efficiency} = \frac{W}{W_{\text{actual}}} = \frac{Q}{W_{\text{actual}}} \left(1 - \frac{T_{\text{exterior}}}{T_{\text{interior}}} \right) . \tag{14.8}$$

Let us focus on the term Q/W_{actual}, where Q is the heat required for the task (heating the building), and W_{actual} is the amount of energy actually expended to do the task. Thus this ratio is the ordinary efficiency that we have designated as the first law efficiency.

$$\text{Second law efficiency} = (\text{first law efficiency}) \times \left(1 - \frac{T_{\text{exterior}}}{T_{\text{interior}}} \right) . \tag{14.9}$$

Let us see the implications of this. A typical household gas furnace system has an efficiency of 70 percent, that is, 70 percent of the energy derived from the burning gas appears as useful heat in the house. The second law efficiency for this system for the temperatures used earlier is

$$\text{Second law efficiency} = 70 \cdot \left(1 - \frac{278}{293} \right)$$

$$= 3.6\%.$$

What may we conclude from the enormous difference between the first and second law efficiencies? It means simply that the gas furnace is very inappropriate for the designated task of supplying energy between two areas having a temperature difference of 15K. Viewed another way, the flame temperature of 2000K for the burning gas is simply not required for the assigned task. Certainly, the gas furnace does the job but at the expense of using a much higher temperature than is required. There are other more productive uses of the high temperature than for providing household heat. For example, one might use a high-temperature gas to run a gas turbine that drives an electric generator. The rejected heat energy, which is at a much lower temperature than the gas entering the turbine, can then be used for space heating. Such a scheme is called cogeneration of energy and has important implications for energy conservation. Although heat pumps and cogeneration seem to be technologcially sensible, they are not easy to implement; but it is innovations like these that offer the greatest prospects for real contributions to energy conservation.

Although some conservation measures involve acceptance and development of new technology, others are easily effected. If these measures are taken seriously, it is estimated that by 1980 about 20 percent of the energy projected for space heating and cooling could be saved. By 1990 the percentage figure will rise to 30 percent.

14.4 ENERGY CONSERVATION IN THE TRANSPORTATION SECTOR

Of the four main areas of energy usage, transportation offers the greatest opportunity for energy conservation at the personal level. Transportation accounts for about one fourth of total energy use in the United States. About 97 percent of this energy comes from burning gasoline and lower forms of petroleum. Passenger transportation accounts for about 60 percent of the total energy used for transportation purposes. Automobiles account for nearly 90 percent of the energy used for passenger transportation. Nearly half the energy used by automobiles involves trips of less than ten miles. The automobile and truck are the favorite forms of commuting to work. Some 87 percent of the workers choose this mode. Nearly 70 percent of these workers choose to travel alone*

It is no accident that the private automobile commands this respect among the populace. The automobile is a symbol of freedom at the most personal level. At their convenience, a large fraction of the American public has both the option and the resources to drive a multitude of short mileage trips or to travel thousands of miles on the interstate highway system. While reasons abound for fostering the person–automobile relationship, some simple facts force us to reassess this love affair. Pollution is one aspect already discussed. Energy efficiency, to be examined now, is another.

Seeking an energy efficient form of transportation is akin to shopping for food where we try to get the most amount of sustenance for the least amount

---

* Because of the changing nature of these data, the numbers are only approximate. The data were taken from "Transportation Energy Conservation Data Book: Edition 2," by D. B. Shonka, A. S. Loeble, and P. D. Patterson, Oak Ridge National Laboratory, Oak Ridge, Tennessee (October 1977).

of money. When energy is of concern, we shop for a transportation vehicle that will carry the greatest number of passengers for as large a distance as possible with the least amount of fuel. Thus, a way of defining a passenger efficiency is

$$\text{passenger efficiency} = \frac{\text{(passengers transported)} \times \text{(miles traveled)}}{\text{gallons of fuel used}}. \quad (14.10)$$

Note that this involves miles traveled divided by gallons of fuel used. This ratio is the customary figure of merit for gasoline use by an automobile. Thus if an automobile obtains 12 miles for each gallon of gasoline consumed while carrying 2 passengers, its passenger efficiency is

$$\text{passenger efficiency} = (2) \cdot (12) = 24 \; \frac{\text{passenger miles}}{\text{gallon}}. \quad (14.11)$$

Now this number does not mean a lot until it is compared with other forms of transportation. This comparison (see Fig. 14.4) shows that the automobile, as used, ranks near the very bottom of the list.

As revealed by Eq. 14.10, two things have contributed to the low passenger efficiency for the private automobile. One is the conscious decision of an owner to generally include only himself in the vehicle (the average number of persons per vehicle is between one and two in urban driving). The other is the trend toward using vehicles getting fewer miles per gallon. Whether we decide to include more passengers remains to be seen. But a reversal of the trend toward using less efficient automobiles is a reality. The federal government has mandated a steady yearly increase in the fuel economy of vehicles used on American highways. As it stands now, the fuel economy of 1985 model vehicles must be 129 percent better than that of the 1974 models. These standards will not be met by simply making the internal combustion engine more efficient. That limit is determined by the laws of physics and there is not a lot of room for improvement. Some gain is possible through using a different type of engine, such as a diesel engine, but the biggest change will involve the type of cars and the way we use them. This is because the trend to massive cars and high speeds have fostered excessive gasoline consumption. Application of basic physical principles illustrates why.

When you exert a force to push a box on a flat surface, it is analogous to a car being "pushed" along a flat highway. Everyday experience reveals that the energy that must be put forth depends on the weight of the box (car), the speed that it moves, the distance that it is moved, and the nature of the surface of the box and floor. We would anticipate that the energy required increases as the weight of the box (car) and distance traveled increase. In a quantitative analysis, it turns out that for a given speed the energy is directly proportional to weight and distance.* Thus for a given automobile moving with constant speed, the energy (or, equivalently, gasoline) used per mile of travel will depend directly on the weight. If the weight is reduced, the gasoline used per mile

---

* *Problems of Our Physical Environment*, Joseph Priest, Addison-Wesley, Reading, Mass. (1973).

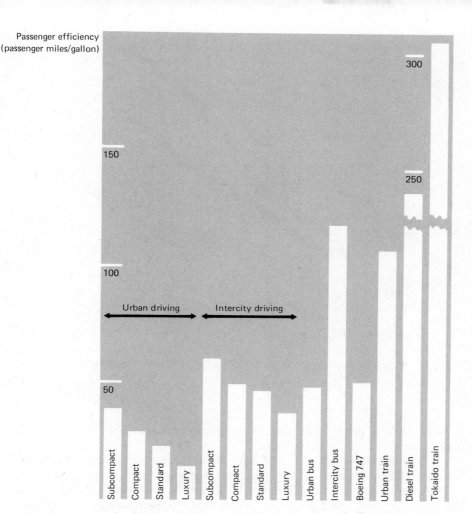

Fig 14.4 Passenger efficiencies for some transportation modes of interest. The values were measured for the year 1972. Note the significant differences in efficiency for subcompact and luxury cars in urban driving. This difference is not as pronounced for intercity driving. This is because the luxury car carries more passengers and performs better in this type of driving. The energy advantages for mass transit are evidenced by the high passenger efficiencies for bus and train travel. The data were taken from "Transportation Energy Conservation Data Book: Edition 2," by D. S. Shonka, A. S. Loeble, and P. D. Patterson, Oak Ridge National Laboratory, Oak Ridge, Tennessee.

of travel will decrease. Figure 14.5 shows data compiled by the EPA for both 1974 and 1977 model cars. The direct relation between fuel economy and vehicular weight is clear. Under normal driving the number of miles obtained per gallon of gasoline will double if the vehicular weight is cut in half. In an era

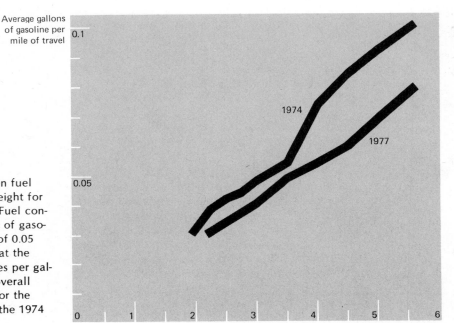

Figure 14.5 Relationship between fuel consumption and automobile weight for combined city/highway driving. Fuel consumption is expressed in gallons of gasoline per mile of travel. A value of 0.05 gallons per mile would mean that the vehicle obtained 1/0.05 = 20 miles per gallon of gasoline used. Note the overall improvement in fuel economy for the 1977 models as compared with the 1974 models.

of high gasoline prices and federally mandated fuel economy, this has instigated a trend toward lighter vehicles.

A moving automobile is held back by a force exerted on it by the air through which it moves. You can feel this force when you hold your hand out of the window of a moving car and you can perceive that the force increases as the speed increases. As a result the energy used also increases as the speed increases. The exact dependence on speed is fairly complex but a car traveling between 75 and 80 miles per hour will get about half the number of miles per gallon as one traveling 50 miles per hour. This is why a 55-mile-per-hour speed limit, if enforced, can be an extremely effective energy conservation measure. In addition, there is the consolation of knowing that reduced speeds will significantly lower the number of accidents and fatalities.

That mass transportation can be significantly more energy efficient than the private car is graphically illustrated in Fig. 14.4. Some progress has been made toward bringing about a shift to more efficient energy systems. Examples are the Bay Area Rapid Transit (BART) in the Oakland–San Francisco area and the METRO system in Washington, D.C. But mass transportation systems are extremely expensive and widespread utilization will be slow in coming. At present, using buses is probably the best way to solve mass transit

Figure 14.6 This photograph shows how the median strip of a freeway can be utilized by buses or trains to facilitate mass transit of people. (Photograph courtesy of U.S. Department of Transportation.)

problems. If buses are forced to compete for the same traffic lanes as automobiles, they are bound to be slow and uncomfortable to ride. To alleviate this, many cities are providing lanes for the exclusive use of buses (Fig. 14.6). Apart from these more drastic solutions, substantial savings could be made by simply getting more passengers in automobiles (car pooling) and walking or bicycling for many short trips.

A similar efficiency analysis can be performed for bulk freight transportation. The idea is to transport as much bulk weight as possible over the greatest distance for the least amount of energy. Expressed mathematically,

$$\text{freight efficiency} = \frac{(\text{weight in tons}) \times (\text{distance traveled})}{\text{gallons of fuel expended}}. \qquad (14.12)$$

A truck carrying a ten-ton cargo and getting six miles for each gallon of gasoline consumed would have a freight efficiency of 60. Figure 14.7 shows the freight efficiencies for several common modes of transport. As for passenger efficiency, there is a dramatic difference in efficiencies. Clearly there are reasons such as expediency and reliability for choosing a given form of transportation. However, for those situations where time is not dominant and all other

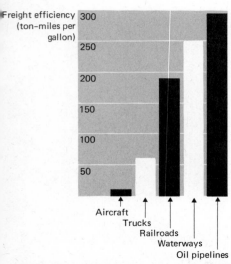

Freight efficiency (ton–miles per gallon)

- Aircraft
- Trucks
- Railroads
- Waterways
- Oil pipelines

Figure 14.7 Freight efficiencies for some selected modes of transportation. From "System Energy and Future Transportation," Richard A. Rice, *Technology Review* (January 1972).

factors are equal, considerable energy can be saved by choosing the most efficient energy method.

The idea of feedback has been used on several occasions. A change in the output of any system can conceivably affect its input in either a positive or negative fashion. Automobiles constitute the output of a very influential, complex, industrial system. Alterations in the character of automobiles can produce consequential changes in the system. For example, if the public decides for energy conservation reasons that it no longer wants the massive cars that the industry is tooled to produce, then the industry falters until it can be retooled to make smaller cars. Consequently unemployment rises in the automobile industry and in the multitude of industries tied to it. A general decline results in the economy.

Another feedback problem of concern is the effect of automobile pollution controls on fuel economy. Compared with automobiles with no controls, 1974 model vehicles experienced a 10-percent reduction in fuel economy because of pollution controls.[*] With the advent of catalytic converters it seems reasonable that engine performance would be lowered still further because of restricted gas flow in the exhaust system. However, this need not be the case for the following reason.[†] The amount of pollutants emitted in an engine depends on how the engine is adjusted (tuned) to operate. In the absence of catalytic converters, engines were adjusted to produce the least amount of pollutants. This adjustment, however, did not yield the best fuel economy. Since the catalytic converters are extremely effective for removing carbon monoxide and hydrocarbons, the engine can be adjusted for maximum performance rather than minimum pollutants. Thus the negative effect produced by restricting the gas flow can be compensated by proper adjustment of the engine. Figure 14.5 illustrates how the performance of automobiles improved between 1974 and 1977.

14.5 ENERGY CONSERVATION IN THE INDUSTRIAL SECTOR

Industry consumed about 25 percent of the total energy used in 1977. Roughly half of the energy was used for heating processes and production of steam. Since industry is a major energy consumer and much of the energy is used for relatively simple processes, the potential for significant energy savings is large. However, the practical implementation of conservation schemes is difficult. Perhaps a five-to-ten-percent reduction could be made in the anticipated 1980 demand. This would be accomplished mostly by replacing old inefficient equipment, better design of new equipment, and more conscientious maintenance of existing equipment. In the longer term, the heat rejected from electric power plants could be used or, possibly, a self-contained total energy system that would produce electric energy on site with the rejected heat going directly into heating processes and steam production.

---

[*] "Exploring Energy Choices," Ford Foundation Energy Policy Project (1974).

[†] See, for example, *Consumer Reports*, April 1974, p. 349.

Although energy use in the electric utilities and industrial sectors was about the same in 1976, it is anticipated that electric utilities will account for about 38 percent of all energy use in 1990. Thus the potential for energy conservation will accelerate for many years. The ultimate savings will come when new technologies utilizing the likes of solar, geothermal, wind, and fusion energy come on line. But as emphasized many times, these probably will not have an impact until after 2000. In the meantime we must seize on other conservation measures.

A modern coal-fired electric power plant is nearly 40 percent efficient. However, the overall efficiency for electric power production from existing plants is about 33 percent. Thus considerable savings could be made by upgrading existing plants. Although the advent of widespread use of nuclear reactors will relieve the load on fossil fuel resources, it will increase the burden on the environment for absorbing waste heat because light-water reactors are only about 32 percent efficient. Other types of nuclear reactors like the breeder and high-temperature gas-cooled reactors could increase the efficiency to around 40 percent. This generation of reactors, though, will not arrive for at least a decade. Even with these advanced reactors, significant energy will always be wasted because of the use of steam turbines. And the steam turbine is bound to be around for several decades. Thus, in the long run, it is imperative to devise schemes for making use of the rejected heat from electric power plants.

Electric power transmitted at 345,000 volts for 200 miles is 98 percent efficient. Even though the transmission efficiency is high, the energy loss is not trivial because the total energy transmitted is large. The transmission of electric energy is the most expensive of all energy-transmission systems. The reason is that the transmission lines, their supporting structures, and the land that they remove from other uses are extremely expensive. Currently, some seven million acres of land are used for overhead transmission lines. If electric energy demand quadruples by 2000, the land problem will worsen. Although the land may be available, environmental opposition to the structures is mounting. Installing the transmission lines underground would remove most of these objections but then costs would soar. Furthermore, since considerable heat is generated in conventional underground transmission lines, their capacity is lower because the earth cannot remove the heat as efficiently as the air around overhead transmission lines. Three schemes have been proposed for increasing the capacity of underground transmission. One method provides a compressed-gas insulation that more effectively removes the heat produced. The other two cool the conductors so as to reduce their electrical resistance. Because power loss is directly related to electrical resistance, energy savings could result. These latter two technologies are extremely interesting because the temperatures considered are about $-320°F$ and $-453°F$. These are the temperatures of liquid nitrogen ($-320°F$) and liquid helium ($-453°F$). The resistance of a common electrical conductor decreases uniformly as the temperature decreases. Typically the resistance will decrease by a factor of ten in going from room temperature to $-320°F$. So the power loss would decrease by a factor of ten. But realize that energy is required to cool the cable. In fact as much energy is used to cool the cable as is gained through reduction in heating loss. However, the safe capacity of the line for carrying electric current increases significantly and this is the attractive feature. There are some metals whose electrical resistance will completely vanish at extremely low temperatures. These conductors are referred to as superconductors. A niobium–tin metal alloy is a superconductor often cited for use in a power transmission line. The resistance of this alloy vanishes at $-427°F$. Again, the energy required to cool the wires offsets the gain from heat energy losses but the safe current-carrying capacity increases dramatically (Fig. 14.8). Cooled transmission lines of the type described here are not in practical use at this time. However, they are being seriously considered and research and development is progressing on them.

Figure 14.8 Comparison of the power-carrying ability of several types of power transmission cables. Note that a range of ability is given for each type. Adapted from Figure 31, *Energy and the Future,* Allen L. Hammond, William D. Metz, and Thomas H. Maugh II, American Association for the Advancement of Science, Washington, D.C. (1973).

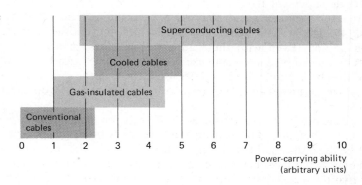

14.7 CONCLUDING REMARKS

These few illustrations of possible ways to save energy clearly show the potential and importance of conservation. Although most of the schemes involve action by the energy industry, there are many opportunities for personal energy conservation. When the urge arises to travel 80 miles per hour or to take an extended hot water shower or to excessively cool or light a house, you should always bear in mind that the energy source very likely took millions of years to produce and that the supply is rapidly depleting. When you have the option to purchase a product in a disposable bottle or aluminum can you should realize that energy was required to make the container and that energy is required to dispose of it. It is incumbent on us to recycle these products as energy and resource conservation measures. You should think seriously about walking or bicycling rather than motoring for the many short trips now taken with automobiles. Although our technological society has given us many problems resulting from energy conversion, it has also given us many benefits. It is important, though, that we control this technology rather than letting it control us. Energy conservation and more effective utilization of energy is a fantastic way to start toward the achievement of this goal.

1. How was energy consumption categorized for purposes of evaluating potential areas for conservation?

2. Roughly, how much of the total energy converted goes into useful forms?

3. What happens to the energy that goes unused?

4. What is the ultimate disposition of virtually all the energy converted?

5. Why does lowering the thermostat during heating periods tend to conserve energy?

6. What is insulation? How is it used in a building?

7. What is a heat pump? What is the attractive feature of a heat pump?

8. Explain why electric household heating is not clean.

9. Distinguish between passenger efficiency and freight efficiency.

10. How does the weight and speed of an automobile affect its gasoline consumption?

11. What economic effects can occur if a major change is made in the character of automobiles?

12. Why might the addition of catalytic converters not affect the fuel economy of an automobile?

13. Name some ways that industry might be able to conserve energy.

14. Why will there always be considerable waste of energy as long as a steam turbine is used in the conversion process?

15. What are some problems associated with the transmission of electric power?

16. What is a superconductor? How might a superconductor be utilized in the electric power industry?

The following articles are useful for extension of the material presented in Chapter 14.

1. "Energy Conservation," G. A. Lincoln, *Science* **180**, 155 (13 April 1973).

2. "Energy Conservation through Effective Utilization," Charles A. Berg, *Science* **181**, 128 (13 July 1973).

3. "Residential Energy Use Alternatives: 1976 to 2000," Eric Hirst, *Science* **194**, 1247 (17 December 1976).

4. "Transportation Energy Conservation Policies," Eric Hirst, *Science* **192**, 15 (2 April 1976).

5. *Science* **184**, No. 4134 (19 April 1974). This issue devoted entirely to energy contains several articles on energy conservation.

6. "Superconductors for Power Transmission," Donald P. Snowden, *Scientific American* **226**, No. 4, 84 (April 1972).

7. "System Energy and Future Transportation," Richard A. Rice, *Technology Review*, p. 31 (January 1972).

14.3 Energy Conservation in the Residential and Commercial Sector

1. Minimizing heat loss from a building is like trying to keep warm when you go outside in the cold. How does the success of keeping warm depend on the type and thickness of your wearing apparel?

2. Why do you intuitively expect that heat lost by conduction through a wall depends on the difference in temperature between the interior and exterior of the building?

3. From your personal reactions do you think you could tell whether the temperature in a room was 70°F or 68°F?

4. Suppose that you are considering buying one of two houses. The houses are nearly equivalent except one costs $3000 more because of improved insulation and a more efficient heating system. You can save on the cheaper one but fuel bills will be higher. What choice would you make?

5. Why would heat pumps be more effective when the temperature difference between the inside and the outside of a building is small?

6. In Section 14.3 it was shown that lowering a thermostat setting during heating periods tends to conserve energy. How does raising the thermostat setting during cooling periods tend to conserve energy?

7. How is the R-value for a heat conductor analogous to the resistance value for an electrical conductor?

8. Ohm's law, discussed in Section 5.3, relates electric current to potential difference and electrical resistance ($I = V/R$). In what ways is the heat flow equation (Eq. 14.1) similar to the equation for Ohm's law?

14.4 Energy Conservation in the Transportation Sector

9. What features of the automobile entice 86 percent of the working community to use it for commuting to work?

10. Using Eq. 14.12 as a guide, define an energy efficiency for transporting nuclear reactor containment vessels.

11. What social and economic forces exist which hinder construction of mass transit systems?

12. When you hold your hand outside the window of a moving car does the force exerted on your hand depend on the orientation of your hand? On the basis of this experience, would you expect all automobiles to experience the same retarding force when traveling at the same speed?

13. What considerations might enter into choosing a form of transportation other than energy conservation?

14.6 Energy Conservation in the Electric Utilities Sector

14. Usually, electric power use peaks around 5:00 P.M. Why does this peaking contribute to the inefficient use of energy in the electric utilities industry?

1. A newspaper article suggests that every 1° lowering of a thermostat will produce a 5 percent increase in energy savings. Assuming that the outdoor and indoor temperatures are 50°F and 70°F show that this is indeed the case as long as temperatures do not fall too far below 70°F.

2. What fraction of the total energy used for air conditioning could be saved if the thermostat setting were raised 3°F from its nominal 70°F setting? Assume that the outdoor temperature is 85°F.

3. If the thickness of insulation in the walls of a house is doubled, how much change would you expect in the heat loss through the walls?

4. The R-value for ⅜-inch plywood is 0.79 (Table 14.1). Would you expect the R-value for ¾-inch plywood to be greater than or less than 0.79? Figure out the R-value for ¾-inch plywood and check your answer with that quoted in Table 14.1.

5. The ceiling of a room in a house measures 12 feet by 20 feet. The ceiling is composed of ⅜-inch gypsum board covered with 3 inches of loose fill.

   a) Calculate the heat loss through the ceiling in an area having a heating factor of 1.25.

   b) If the loose fill is increased to 6 inches, how much will the heat loss be reduced?

   c) If the house is heated by burning fuel oil, how many gallons of oil will be saved after the additional 3 inches of loose fill are installed?

6. The construction of a well-insulated house wall is shown below. Using Table 14.1, determine the overall R-value for the wall. What sort of an energy penalty would a homeowner pay if the 3½ inches of blanket insulation were not used?

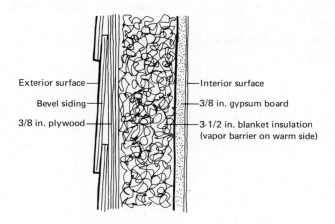

Exterior surface
Bevel siding
3/8 in. plywood

Interior surface
3/8 in. gypsum board
3-1/2 in. blanket insulation
(vapor barrier on warm side)

7. A central heating system has the task of providing energy for 100 identical apartments. How is the total energy required related to the energy used

per apartment? Without worrying about performing the assigned task, how could the total energy use be decreased? Insisting that the task be accomplished, what is the alternative if a decrease is desired for the total energy use?

8. An average house has about 1500 square feet of floor space. The walls of most contemporary houses are 8 feet high. In central United States, this house will need about 100,000,000 Btu of heat during the heating season. Figure the total area of the walls, ceiling, and floors of this house assuming that it is 50 feet long, 30 feet wide, and 8 feet high. Using Eq. 14.3, determine the average R-value for the house. Does the R-value obtained seem reasonable?

9. A certain heat pump for a house has an effectiveness of 3. How many units of heat are fed to the house for every unit of energy used to operate the heat pump?

14.4 Energy Conservation in the Transportation Sector

10. What is the passenger efficiency of a car carrying three passengers and obtaining 20 miles for each gallon of gasoline consumed?

11. A person riding a bicycle going 12 miles per hour uses about 200 Btu of energy per mile of travel.

a) Assuming that the energy equivalent of one gallon of gasoline is 100,000 Btu, determine the gasoline equivalent of 200 Btu of energy.

b) How many "miles per gallon" do you get by biking at 12 miles per hour?

c) Determine the passenger efficiency for bike riding and compare it with those shown in Fig. 14.4.

12. a) How much time does it take to travel 75 miles at an average speed of 55 miles per hour?

b) How much time is saved if the 75-mile trip is traversed at an average speed of 70 miles per hour?

Powers of Ten

APPENDIX

Hand-held calculators are enormously useful for routine numerical calculations. However, there is a limit to the number of digits incorporated in a number processed by the calculator. Generally, this number is ten. For example, the number 10,000,000,000 cannot be handled as such by most hand-held calculators because it has 11 digits. But don't despair. Most hand-held calculators will process large and small numbers if they are entered in a notation utilizing "powers of ten." The purpose of this appendix is to illustrate the importance of the "powers of ten" notation and to show how they are used to great advantage in modern hand-held calculators.

The range of numbers encountered in assessing energy concepts is staggering. For example, the radius of the nucleus of the aluminum atom is 0.00000000000036 cm, and the U.S. electric energy production in 1976 was 2,036,000,000,000 kW-hr. Multiplying, dividing, and just keeping track of numbers like these is tedious without good bookkeeping.

A number like 2,036,000,000,000 represents a measurement or an estimate and there is, therefore, some uncertainty in the number. If the number were known to an accuracy of one percent, which normally would be quite good, it would mean that the true value would, perhaps, be between 2,035,000,000,000 and 2,037,000,000,000. At best, only the first four digits, that is, 2, 0, 3, and 6, are significant. The remaining zeros denote the relative size or magnitude. For example, the number might be written as 2036 billion kW-hr. Billion denotes the magnitude and is used as a bookkeeping procedure, albeit not a particularly practical one in the sense that it does not lend itself to easy multiplication and division. To illustrate a better method, consider the numbers 0.0015 and 1500. These numbers could also be recorded as

$$1.5 \div 1000 \quad \text{and} \quad 1.5 \times 1000.$$

The 1000 figure denotes the magnitude. Now 1000 can be generated by multiplying $10 \times 10 \times 10$. A shorthand way of symbolizing this operation is 10^3. Using this notation, we can write

$$0.0015 = 1.5 \div 10^3 = \frac{1.5}{10^3}$$

$$1500 = 1.5 \times 10^3.$$

It is conventional and meaningful to write $1/10^3$ as 10^{-3} so that $1.5/10^3$ becomes 1.5×10^{-3}. The 3 in 10^3 indicates the third power of 10. The beauty of this method is that the appropriate power of ten can be obtained by counting the positions the decimal point is moved.

$1.5 \times 10^{-3} = 0.0015$ Decimal point of 1.5 is moved 3 places to left.

$1.5 \times 10^3 = 1500.$ Decimal point of 1.5 is moved 3 places to right.

To convince yourself of the utility of the method, express the following numbers as a number times ten to a power.

0.00000001 cm (approximate diameter of an atom)

30,000,000,000 m/sec (speed of light)

Express the following numbers in decimal form.

9.11×10^{-28} g (mass of the electron)

3.31×10^4 cm/sec (speed of sound in air at one atmosphere pressure and 0 °C)

You should be convinced of the usefulness of the method if for no other reason than compactness. The real utility is in multiplication and division because it reduces much of the effort to that of addition and subtraction. To illustrate, consider

$(1000) \times (1000) = 1,000,000$ (long method)

$10^3 \times 10^3 = 10^{3+3} = 10^6$ (short method).

The general rule is $10^a \cdot 10^b = 10^{a+b}$, where a and b are the powers of 10. There are no restrictions on a and b; they can be positive, negative, and even fractional. The reason for writing $1/10^3$ as 10^{-3} can now be seen.

Example:

$$\frac{1,000,000}{1000} = 1000 \text{ (long method)}$$

$$\frac{10^6}{10^3} = 10^6 \cdot 10^{-3} = 10^{6-3} = 10^3 \text{ (short method)}$$

Example:

$$(5.1 \times 10^4) \cdot (3.2 \times 10^{-1}) = 5.1\,(3.2) \times 10^{4-1}$$
$$= 16.32 \times 10^3$$
$$= 1.632 \times 10^4$$

Example:

$$(5.1 \times 10^4) \div (3.2 \times 10^{-1}) = \frac{5.1 \times 10^4}{3.2 \times 10^{-1}}$$
$$= \left(\frac{5.1}{3.2}\right) = \times 10^{4+1} = 1.59 \times 10^5$$

To use numbers expressed as powers of ten in a hand-held calculator one merely has to feed the calculator the significant figures and then the power of ten. To signal that a power of ten is being entered a key generally labelled EE is pressed. For example, 5.1×10^4 would be entered as follows:

Enter 5.1 on the keyboard

Press EE key

Enter 4 on the keyboard

The display would then read

5.1   04

which is interpreted as 5.1×10^4. If this number is to be multiplied by 3.2×10^{-1}, then one would signal the appropriate multiplication key and

Enter 3.2

Press EE

Enter -1

Execute multiplication

The answer would be revealed as 1.632   04 which is interpreted as

1.632×10^4 or   16,320.

Numerical calculations are simplified enormously with hand-held calculators. Take advantage of them. To gain confidence, perform the following operations using the power of ten notation. Check your results using both a hand-held calculator and the conventional method for multiplication and division.

$(0.000303) \cdot (43200) =$

$(250,000) \div 0.0025 =$

$(0.0101) \cdot (0.0005) \div (0.0002) =$

Devise some exercises of your own and practice the method.

Some useful numerical prefixes

Numerical unit	Prefix	Symbol	Power of ten designation
trillion	tera	T	10^{12}
billion	giga	G	10^9
million	mega	M	10^6
thousand	kilo	k	10^3
hundredth	centi	c	10^{-2}
thousandth	milli	m	10^{-3}
millionth	micro	μ	10^{-6}

Some useful physical constants	Symbol	Value
Boltzmann's constant	k	1.38×10^{-23} joules per kelvin
Speed of light	c	3×10^8 meters per second
Planck's constant	h	6.63×10^{-34} joule $\cdot$ seconds

Some useful energy and power units. Units in column 1 are convertible to equivalent units by multiplying by the numbers given in columns 4 and 5. Thus, in joules, 2000 Btu are equivalent to 2000 Btu · 1055 $\frac{\text{joules}}{\text{Btu}}$ = 2,110,000 joules.

Special energy unit	Study area of main use	Symbol	Equivalent in joules	Other useful equivalents
kilowatt-hour	electricity	kW-hr	3,600,000	3413 Btu, 860 kcal
calorie	heat	cal	4.186	
kilocalorie (food calorie)	heat	kcal	4186	1000 cal
British thermal unit	heat	Btu	1055	252 cal
electron volt	atoms, molecules	eV	1.602×10^{-19}	
kiloelectron volts	X-rays	keV	1.602×10^{-16}	1000 eV
million electron volts	nuclei, nuclear radiation	MeV	1.602×10^{-13}	1,000,000 eV
quintillion	energy reserves	Q	1.055×10^{21}	10^{18} Btu or billion-billion Btu
quadrillion	energy reserves	quad	1.055×10^{18}	10^{15} Btu or a million-billion Btu

Special power unit	Study area of main use	Symbol	Equivalent in watts
British thermal unit per hour	heat	Btu/hr	0.293
kilowatt	household electricity	kW	1000
megawatt	electric power plants	MW	1,000,000
gigawatt	total U.S. or world power production	GW	1,000,000,000
horsepower	automobile engines, motors	hp	746

Energy equivalents

1 gallon of gasoline = 126,000 Btu
1 pound bituminous coal = 13,100 Btu
1 cubic foot natural gas = 1,030 Btu
1 42-gallon barrel of oil = 5,800,000 Btu
Variations of these energy equivalent values will appear in the literature. The values quoted here are typical.

GLOSSARY

Acceleration: Time rate of change of velocity.

Actinides: Atoms, usually radioactive, produced from the bombardment of ^{235}U and ^{238}U by neutrons in a nuclear reactor.

Adiabatic process: A thermodynamic process that involves neither a gain nor a loss of heat.

Aerosol: Literally, a gaseous suspension of solid or liquid particles. Environmentally, a maximum size such as 0.0001 m is sometimes included.

Albedo: The fraction (or percentage) of incident electromagnetic radiation reflected by a surface.

Alpha particle: Originally designated as a radioactive particle emitted naturally by several elements such as radium and uranium. It is indistinguishable from the nucleus of the helium atom having two protons and two neutrons.

Alternating current (a.c.): An electric current that alternates direction at regular intervals.

Ampere (A): The ampere is the unit of electric current. One ampere is a rate of flow of charge equal to one coulomb per second ($I = Q/t$).

Arithmetic mean: For N values of something of interest, the arithmetic mean is the sum of all N values divided by the number of values.

Atom: An atom consists of a dense, positively charged nucleus surrounded by a system of electrons equal in number to the number of nuclear protons. The atom is bound together by electric forces between the electrons and the nucleus.

Atomic Energy Commission (AEC): An advisory board formed in the United States in 1946 for the domestic control of nuclear energy. Discontinued in 1974 and replaced by the Nuclear Regulatory Commission (NRC) and the Energy Research and Development Agency (ERDA). ERDA's functions were transferred to the Department of Energy (DOE) in August 1977.

Atomic mass unit (amu): A unit of mass equal to $\frac{1}{12}$ the mass of the carbon isotope with six protons and six neutrons. 1 amu = 1.6604×10^{-24} g.

Atomic number: The number of electrons (or protons) in a neutral atom.

Available energy: For given operating temperatures for a heat engine, available energy is the theoretical maximum amount of mechanical energy that can be derived.

Background radiation: Nuclear radiation in the environment that comes primarily from cosmic rays, building materials, and the earth.

Barrel: Unit of volume measurement. The barrel referred to in oil measurements contains 42 gallons.

Bay Area Rapid Transit (BART): A modern mass transit system serving the San Francisco Bay area.

Beta decay: The process in which an electron is created from the transformation of a neutron, or a positron is created from the transformation of a proton bound in a nucleus.

Beta particles: Charged particles emanating from nuclei of atoms. They are created at the time of emission. The negative beta particle is identical to an electron which orbits the nucleus. The positive beta particle (positron) is identical to an electron except that it is positively charged.

Binding energy: In nuclear physics, binding energy is the energy required to separate a nucleus into its constituent neutrons and protons.

Boiling Water Reactor (BWR): A reactor in which water in thermal contact with the reactor core is brought to a boil. The vapor from the boiling water is then used to drive a steam turbine.

Bound system: A system such as a nucleus, atom, or molecule bound (or held) together by some force(s).

Breeder reactor: A reactor which generates fissionable fuel by bombarding appropriate nuclei with neutrons formed in the energy producing operation of the reactor.

British Thermal Unit (Btu): The engineering unit of heat. The amount of heat required to raise the temperature of one pound of water by one degree Fahrenheit.

Buoyant force: An upward force exerted by a fluid on an object immersed in the fluid.

Calorie (cal): The amount of heat required to raise the temperature of one gram of water by one degree Celsius. The food calorie is equivalent to one thousand calories defined in this manner.

Capital energy: Energy that is stored in various materials within the earth. Coal, oil, and uranium are examples of capital energy.

Carbon dioxide (CO_2): A molecule containing one carbon and two oxygen atoms. It is the product of combustion of carbon in fossil fuels and is of concern because of the "greenhouse effect."

Carbon monoxide (CO): A molecule containing one carbon and one oxygen atom. It is a product of incomplete combustion of carbon in fossil fuels and is of concern because of its effect on human body functions.

Carnot cycle: A thermodynamic cycle involving four steps; two isothermal and two adiabatic changes.

Catalyst: A substance which modifies the rate of a chemical reaction without being consumed in the process.

Catalytic converter: A device using catalytic chemical reactions to convert molecules of environmental concern to harmless species.

Cavity radiator: A hollow container of arbitrary shape with a tiny hole in the container wall. When maintained at a constant temperature, it is useful for studying the fundamental aspects of thermal radiation.

Chain reaction: As applied to a nuclear reactor, a self-sustaining, multi-stage nuclear reaction instigated by neutrons and sustained by neutrons generated in the reactions.

Coal: A natural dark brown to black solid formed from fossilized plants. It is primarily carbon.

Coal beneficiation: A process that removes part of the sulfur from coal before it is burned.

Coal gasification: The conversion of coal into a gaseous fuel that is a mixture of carbon monoxide, hydrogen, and methane.

Coal liquefaction: The conversion of coal into a liquid hydrocarbon fuel. Depending on the degree of refinement, the fuel may be low grade such as coal oil or high grade such as gasoline.

Cogeneration: The generation of two forms of useful energy in a single energy conversion process. For example, a turbine may produce both mechanical energy for an electric generator and heat for a building.

Compton effect: Scattering of an X-ray or gamma ray photon by an atomic electron. The photon energy is less after the scattering.

Condensation: The change of state of a vapor to a liquid. Energy is liberated in condensation.

Condenser: A device at the exit of a steam turbine used to extract heat from water vapor condensing to a liquid.

Conduction: The transmission of something through a passage or medium without motion of the medium. Heat is conducted between adjacent parts of a medium if their temperatures are different. Moving electrons in a metal constitute electric conduction.

Conservation of energy: An important physical principle that requires that there be no change in the total energy in any energy transformation.

Control rod: A rod often made of boron or cadmium which can be inserted into the core of a nuclear reactor to control the fission reactions by selective absorption of neutrons.

Convection: Transfer of heat from one place to another by actual motion of the heated material.

Converter: A device such as an automobile engine or nuclear reactor that converts the intrinsic potential energy of an appropriate fuel into another form of energy.

Cooling tower: A tower used to collect the heated water effluent from the condenser of a steam turbine and to transfer heat to the atmosphere.

Coulomb (C): The metric unit of electric charge. In terms of the charge of the electron, the smallest unit in nature, the sum of the charges of 6¼ billion-billion electrons equals one coulomb.

Criticality: A condition in a nuclear reactor in which the neutron generation and loss rates are equal.

Critical mass: The minimum mass of fissionable material that will sustain a nuclear chain reaction.

Crankcase blowby: The leakage of gaseous or liquid hydrocarbons between a piston and its cylinder during operation of an internal combustion engine.

Crude oil: The liquid part of petroleum.

Curie (Ci):   A unit of radioactivity amounting to 37 billion (37×10^9) disintegrations per second.

Cyclone collector: A particulate-collecting device in which a whirlwind of air throws particles out of an air stream to the walls of a container on whose surface they collect or adhere.

Density: The amount of something divided by the volume it occupies. Mass density is the mass of a substance divided by the volume it occupies.

Department of Energy (DOE):   The 12th Cabinet agency. Established on October 1, 1977, it absorbed all functions of the Federal Energy Administration, Federal Power Commission, and Energy Research and Development Administration and selected energy activities of the Department of Interior, Department of Defense, Interstate Commerce Commission, Department of Commerce, and Department of Housing and Urban Development.

Deuterium: An isotope of hydrogen having one proton and one neutron in its nucleus.

Deuteron: The nucleus of the deuterium atom.

Diesel engine: An internal combustion engine named for its inventor Rudolf Diesel, German mechanical engineer and inventor, 1858–1913. Requiring no spark plugs, the diesel engine achieves ignition of a fuel–air mixture by high compression.

Direct current (d.c.):   An electric current for which charges flow in one direction.

Disintegration: A transformation involving the nucleus of an atom resulting in a less massive configuration by emission of radiation that may be either particle or electromagnetic.

Doubling time: The time it takes for something that is changing exponentially to double in amount.

Dry-bulb temperature: The temperature measured by a dry thermometer.

Effectiveness: For a heat pump, the effectiveness is the total energy deposited in an area of interest, a room, for example, divided by the energy expended to cause the energy transfer. For a refrigerator or air conditioner, the effectiveness is the energy removed from a cool region, a room or the inside of a refrigerator, for example, divided by the energy expended to cause the energy transfer.

Efficiency: The useful output of any system divided by the total input.

Electric current: Electric charges in motion. The ampere is the measuring unit.

Electric field: Qualitatively, a region of space where an electric charge feels a force.

Electric force: A force between two objects each having the physical property of charge.

Electricity: In the popular sense, it is electric current used as a source of power.

Electric potential difference: The work done on a charge to move it between two points divided by the strength of the charge. The volt is the measuring unit.

Electrical resistance: The resistance provided by the structure of a conductor to the flow of electric charges.

Electrolysis: The decomposition of a liquid into its atomic constituents as a result of an electric current in the liquid. The electrolysis of water produces hydrogen and oxygen.

Electron volt (eV): A unit of energy. It is the energy acquired by an electron accelerated through a potential difference of one volt. $1 \text{ eV} = 1.602 \times 10^{-19}$ joules.

Electrostatic precipitator: An apparatus for the removal of suspended particles from a gas by charging the particles and precipitating them through application of a strong electric field.

Emergency core cooling system (ECCS): A system designed to safely dissipate heat from the core of a nuclear reactor in the event of a sudden loss of the normal cooling facilities.

Energy: The capacity for doing work. The joule is the metric measuring unit.

Energy conservation: Distinguished from the principle of conservation of energy, it incorporates the idea of saving, or not wasting, energy.

Entropy: A measure of the disorder (or chaos) in a system.

Environmental Protection Agency (EPA): A federal agency officially established on December 2, 1970 under a Presidential reorganizational plan. Placed in EPA were the Interior Department's Federal Water Quality Administration; the HEW Department's National Air Pollution Control Administration, Bureau of Solid Waste, and Bureau of Water Hygiene; the pesticide registration, research, and regulation functions of the Agriculture, Interior, and HEW Departments, and certain radiation functions of the Nuclear Regulatory Commission; the Federal Radiation Council; and HEW's Bureau of Radiological Health. The 1970 Clean Air amendments also created an Office of Noise Abatement within EPA.

Epilimnion: The upper stratum (layer) of a lake or sea.

Equilibrium: A condition involving no change in the translational and rotational motion of a system.

Eutrophic: Designating a body of water in which the increase of mineral and organic nutrients has reduced the dissolved oxygen, producing an environment that favors plant over animal life.

Evaporation: The change of state of a liquid as it passes to vapor. The liquid cools in the process.

Exponential growth: The growth in some quantity characterized by doubling of its value at regular intervals of time. For example, if some quantity doubles its value every ten years, it is growing at an exponential rate.

Extrapolate: To infer information by extending known information. Information obtained from a graph extended beyond known values is an example of extrapolation.

Fast neutrons: A loosely defined classification of neutrons according to speed (or energy). A neutron with energy greater than 10 keV is called a fast neutron.

Feedback: The return of a portion of the output of any process or system to its input.

Fertile nucleus: A nucleus which serves as the target for generating fissionable fuel in a nuclear breeder reactor.

First law efficiency: Recognizing explicitly the first law of thermodynamics, first law efficiency is identical to the popular meaning of efficiency, i.e. the useful output of an energy converter divided by the input energy.

First law of thermodynamics: An extension of the principle of conservation of energy to include heat and internal energy.

Fission: The splitting of an atomic nucleus into fragments, two of which are usually much more massive than the remaining fragments. Energy is released in the process.

Fission fragment: An atom, usually radioactive, that is formed from the fission (or splitting) of a heavy nucleus.

Flue gas desulfurization: Using chemical reactions to remove sulfur oxides from the gases produced in the burning of coal.

Fluorocarbon: A type of molecule having fluorine and carbon atoms as constituents.

Fly ash: Fine particles that are generated from noncombustible products in the burning of coal.

Fossil fuel: Fuel such as coal or petroleum derived from remnants of plants and organisms of a past geological age.

Freight efficiency: A measure of the amount of freight that can be moved some distance by a given mode of transportation for an expenditure of a certain amount of fuel (energy). It is defined as the number of tons of freight moved multiplied by the number of miles obtained per gallon of gasoline used.

Fusion: In nuclear terms, a nuclear reaction in which nuclei are combined (fused) to form other nuclei. Energy is released in the process.

Gamma ray: A quantum (or photon) of electromagnetic radiation. It is distinguished from other electromagnetic radiation by its energy. The energy limits are loosely defined, but typically photons with energies greater than 100 keV are called gamma rays.

Garboil: A heavy oillike, combustible fluid that is produced from trash and garbage.

Gas cooled fast breeder reactor (GCFBR): A fast breeder reactor which uses a gas to transfer the heat from the reactor core.

Gaseous diffusion: The diffusion or migration of a gas through a semiporous membrane. Because the rates of diffusion of different types of molecules depend on the masses of the molecules, the process is used to separate different molecular species such as $^{235}UF_6$ and $^{238}UF_6$.

Genetic effect: A damaging effect to a person's genes which could produce an abnormality in the offspring. Genetic damage can be caused by nuclear radiation.

Geometric mean: For N values of something of interest, the geometric mean is the Nth root of the product of all N values.

Gravitational collector: A device for collecting noncombustible products released in the burning of coal. It utilizes the gravitational attraction between the earth and the particles to effect the separation.

Gravitational field: Qualitatively, a region of space where a mass feels a force due to an interaction with another mass.

Gravitational force: A force between two objects each having the physical property of mass.

Gravitational potential energy: Potential energy of a mass due to an elevated position above the earth's surface. Numerically, gravitational potential energy is the product of weight and height ($E_p = mgh$). The metric units are joules.

Greenhouse effect: An effect that produces a warming of a system by trapping electromagnetic radiation within the system. A common greenhouse warms primarily by trapping infrared radiation within the structure.

Gross National Product (GNP): The total market value of all the goods and services produced by a nation during a specified period.

Half-life: The time required for one-half of a given number of radioactive atoms to disintegrate.

Heat: A form of energy transfer associated with the motion of atoms and molecules.

Heating degree days: Used as a factor in determining the heating requirements for a building, it is the multiplicative product of the average inside-outside temperature difference and length of a heating period. Temperature difference is measured relative to a base temperature of 65 °F.

Heating factor: A numerical factor used to relate heating requirements in a given location to those in Central United States.

Heating zone: A geographical zone where heating requirements are roughly the same for a given type of structure.

Heat plume: The hot water discharge of a steam turbine into a body of water is confined to a "feather-shaped" region called a heat plume.

Heat pump: A device used to supply heat to a building by pumping energy from the cooler exterior. In principle, it operates like a household refrigerator.

Heat sink: A system to which heat can flow.

Heavy water reactor (HWR): In a heavy water reactor, the hydrogen component of the water molecules in the moderator is of the second heaviest (in mass) isotopic form ($^2_1H_1$, i.e. deuterium).

Hertz (Hz): The hertz is a unit of frequency equal to one cycle per second. Named for Heinrich Hertz, German physicist, 1857–1894.

High-level radioactive waste: Intensely radioactive materials, primarily fission fragments and actinides, produced by nuclear reactions in a nuclear reactor.

Horsepower (hp): A unit of power equivalent to 746 watts.

Humidity: A qualitative term pertaining to the water vapor in air.

Hydrocarbon: A chemical compound containing only hydrogen and carbon.

Hydroelectric power: Electric power that is produced from the conversion of the kinetic energy of water.

Hydrogen economy: An economy that is based on energy produced when hydrogen is burned in an oxygen atmosphere.

Hypolimnion: The lower stratum (layer) of a lake or sea.

Income energy: Energy provided to supplement or replace the energy locked up within the earth. The sun is Earth's source of income energy.

Inertial confinement: Taking advantage of the inertial property of mass to confine nuclear fusion reactions in a pellet of frozen deuterium and tritium, for example.

Insulation: Material used to restrict heat flow between two regions, for example, between the interior and exterior of a house.

Internal combustion engine: An engine, such as an automotive gasoline piston or rotary engine, in which fuel is burned within the engine proper rather than in an external furnace as in a steam engine.

Internal energy: Thermodynamically, it is the kinetic and potential energy of the atomic and molecular constituents of a system.

Isothermal process: A thermodynamic process for which the temperature does not change. Melting and boiling are isothermal processes.

Isotopes: Atoms with nuclei having the same number of protons, but a different number of neutrons.

Joule (J): The joule is a metric unit of work or energy. The work done by a force of one newton moving an object one meter in the direction of the force. Named for James Prescott Joule, British physicist, 1818–1889.

Kelvin (K): Unit of temperature based on a freezing point of water of 273 and a boiling point of water of 373. Named for Lord Kelvin, William Thomson, British physicist, 1824–1907.

Kilowatt-hour (kW-hr): A unit of energy equal to a power of one kilowatt (1000 W) acting for one hour.

Kinetic energy: Energy due to motion. Numerically, it is one-half the product of the mass of a body times the square of its velocity. ($E_k = \frac{1}{2} mv^2$).

Lapse rate: The variation of temperature with respect to height above the earth.

Laser: A word coined from the first letters of the main words of *"light amplification by stimulated emission of radiation."* It is a device that produces a beam of electromagnetic radiation having very uniform wavelength. In some lasers, the beam is very intense.

Light water reactor (LWR): In a light water reactor, the hydrogen component of the water molecules in the moderator is of the lightest (in mass) isotopic form ($^1_1H_0$).

Liquid metal fast breeder reactor (LMFBR): A nuclear breeder reactor using fast neutrons to produce the fissionable fuel and a liquid metal, usually sodium, as the heat transfer medium.

London smog: A heavily polluted atmosphere consisting mainly of particulates and sulfur dioxide emitted from the combustion of coal and petroleum products.

Loss of cooling accident (LOCA): An accident in a nuclear reactor that evolves from an accidental loss of coolant.

Low-level radioactive waste: Radioactive material produced in the routine operation of a nuclear reactor. It must be handled carefully but is substantially less problematical than the high-level radioactive wastes.

Magnetic field: Qualitatively, a region of space where the pole of a magnet or a moving charge feels a force.

Magnetic field line: An imaginary line in a magnetic field along which a tiny compass needle aligns.

Magnetohydrodynamics (MHD): The science concerned with the motion of electrically conducting fluids through electric and magnetic fields.

Mass: The measure of an object's resistance to acceleration.

Microgram per cubic meter ($\mu g/m^3$): One microgram (10^{-6} g) of a given substance in a one cubic meter volume of air.

Micrometer (μm): One millionth (10^{-6}) of a meter. Sometimes called a micron.

Model: A mathematical description of a physical system.

Moderator: A mechanism used to reduce the speed of neutrons in a nuclear reactor.

Molecule: A bound system of two or more atoms.

Natural gas: The gaseous component of petroleum. It is primarily methane (CH_4) and is commonly used as a household and industrial fuel.

Neutrino: A massless, electrically neutral particle emitted in beta decay.

Neutron: An electrically neutral subatomic particle. Protons and neutrons are the constituents of the atomic nucleus.

Newton (N): A newton is a metric unit of force. The force required to accelerate one kilogram, at one meter per second, each second. Named for Sir Isaac Newton, English scientist, 1642–1727.

Nitrogen oxides (NO_x): In energy considerations, molecules produced by the oxidation of atmospheric nitrogen in the combustion of fossil fuels.

Nuclear parks: A concentration of several nuclear power plants into a localized region.

Nuclear reactor: A device for converting nuclear energy released from nuclear fission reactions to heat.

Nucleon: A proton or a neutron.

Nucleus (of an atom): The positively charged central region of an atom. It is composed of neutrons and protons and contains nearly all the mass of the atom.

Ocean thermal energy converter (OTEC): A thermodynamic system that derives thermal energy from the ocean and produces mechanical energy that powers an electric generator.

Octane rating: A numerical rating of the ability of a gasoline to reduce an audible knock (noise) in an internal combustion engine.

Offshore siting: The location (siting) of a nuclear power plant in the ocean offshore from the continent.

Ohm: The ohm is the unit of electrical resistance. If a potential difference of 1 volt across some electrical element causes a current of 1 ampere, the electrical resistance is 1 ohm ($R = V/I$).

Ohm's law: If the electrical resistance of a device, a toaster element, for example, does not depend on the current in the device, then it is said to obey Ohm's law.

Oil shale: A brownish rock containing a solid hydrocarbon material called kerogen that can be converted to a crude oil product.

Ozone (O_3): A highly reactive molecule containing three atoms of oxygen.

Ozone layer: An atmospheric layer of ozone located at altitudes between 20,000 and 30,000 meters.

Pair production: The production of an electron and a positron from the interaction of a gamma ray photon with an atom. To produce an electron-positron pair, the photon must have a minimum energy of 1.022 MeV.

Particulate: Existing in the form of minute separate particles.

Parts-per-million (ppm): A measure of the concentration of a substance. On a molecular basis, a 1 ppm concentration of a given molecule means that one of every one million molecules is a molecule of the specified type.

Passenger efficiency: A measure of the number of people that can be moved some distance by a given mode of transportation for an expenditure of a certain amount of fuel (energy). It is defined as the number of passengers moved multiplied by the miles obtained per gallon of gasoline used.

Peaking unit: An auxiliary electric power system that is used to supplement the prime power system during peak periods of electricity demand.

Petroleum: A natural flammable mixture of solid, liquid, and gaseous hydrocarbon products found principally beneath the earth's surface. Gasoline, fuel oil, lubricating oil, and paraffin wax, as examples, are refined from petroleum.

Photochemical oxidants: Primarily ozone, but also including other chemical compounds created by smog conditions.

Photochemical (Los Angeles) smog: A heavily polluted atmosphere consisting mainly of eye-irritating chemicals produced from the interaction of automobile exhaust gases with sunlight.

Photoelectric effect: In nuclear physics, the absorption of a photon by an atom and a subsequent ejection of an atomic electron.

Photon: A quantum of electromagnetic energy.

Photosynthesis: The process by which plants convert electromagnetic (solar) energy, carbon dioxide, and water to carbohydrates and oxygen.

Photovoltaic cell: Commonly called a solar cell, it is a device that converts electromagnetic (solar) energy directly to electric energy.

Polarization: Denoting alignment or the assumption of two, usually conflicting, positions. Separation of oppositely charged particles is called polarization.

Pollutant: A constituent of some substance (air, water, and land, for example) that alters the value of the substance.

Potential difference: The work required to move an amount of charge between two positions divided by the size of the charge ($V = W/Q$).

Potential energy: Energy that is potentially convertible to another form of energy, usually kinetic.

Power: Rate of doing work. The watt is the metric measuring unit.

Pressure: The force on a surface divided by the area over which the force acts. ($P = F/A$). Typical units are newtons/m^2 and pounds/in^2.

Pressurized water reactor (PWR): A nuclear reactor in which the water in contact with the reactor core is allowed to become pressurized to prevent boiling and thereby allows a higher operating temperature.

Price–Anderson Act: Enacted in 1957, the act provides insurance for the owners (at the owner's expense) of a nuclear power plant against liability claims made in the event of a nuclear accident.

Primary air quality standards: Levels of air quality that are judged necessary, with an adequate margin of safety, to protect the public health.

Proton: A subatomic particle containing one unit of positive charge. Protons and neutrons are the constituents of the atomic nucleus.

Proven energy reserve: A source of energy known to be available at current prices and extractable with current technology.

Pumped-storage system: An elevated reservoir of water in which the water has been pumped into the reservoir. Ultimately, the water is channeled through water turbines that drive electric generators.

Radiation: The emission and propagation of waves (such as light and sound) or particles (such as beta and alpha).

Radiation absorbed dose (rad): A unit of energy absorbed from ionizing radiation. One rad is equal to 0.00001 joule per gram of irradiated material.

Radioactivity: The emission of particles or electromagnetic radiation from unstable atomic nuclei.

Rate: The change in some quantity divided by the time required to produce the change.

Relative biological effectiveness (RBE): A measure of the capacity of a specific ionizing radiation to produce a specific biological effect.

Relative humidity: The amount of water vapor in the air at a specific temperature divided by the maximum amount of water vapor that the air can contain at that temperature.

Roentgen equivalent man (rem): A biological unit of absorbed ionizing radiation that accounts for biological damage. It is the radiation dose measured in rads multiplied by the relative biological effectiveness (RBE).

R-value: A measure of the resistance to the flow of heat through a medium. Generally expressed in $ft^2 \cdot hr \cdot °F$ per Btu.

Secondary air quality standards: Levels of air quality that are judged necessary to protect public welfare. For example, by protecting property, materials, and economic values.

Second law efficiency: Recognizing explicitly the second law of thermodynamics, second law efficiency compares the actual efficiency of an energy converter with the maximum efficiency guaranteed by the second law of thermodynamics.

Second law of thermodynamics: The law can be formulated in several ways. Two practical formulations are 1) heat does not flow spontaneously from a cooler to a hotter object and 2) in a cyclical device, heat cannot be transformed wholly to work.

Settling velocity: The velocity (or speed) with which a particle in the atmosphere settles to the ground.

Smog: A word derived from smoke and fog describing a polluted atmosphere that has resulted from the combustion of fossil fuels.

Solar cell: A popular name for a photovoltaic cell.

Somatic effect: A biologically damaging effect to the human body. Somatic damage can be caused by nuclear radiation.

Specific heat: The amount of heat required to raise the temperature of a unit mass (or weight) of a substance by one degree. Typical units are calories per gram per degree Celsius and Btu per pound per degree Fahrenheit.

Speed: Distance travelled divided by the time required to achieve that distance. It is distinguished from velocity by disregarding any reference to the direction travelled.

Stratified charge engine: Basically a conventional, reciprocating, internal combustion engine with two vertically connected combustion chambers above each cylinder. Ignition of a fuel mixture rich in gasoline vapor is instigated in the upper chamber. The flame progresses to the lower chamber and ignites a somewhat leaner fuel mixture.

Subsidence: The sinking or subsiding of an air mass.

Sulfur dioxide (SO_2): A stable molecule consisting of one atom of sulfur and two atoms of oxygen. It is emitted as an unwanted by-product in the combustion of coal.

Superconductor: A material that presents no resistance to the flow of electrons if the temperature of the material is maintained at a sufficiently low temperature, typically about the temperature of liquid helium (4.2 K).

Surface tension: A molecular force at the surface of a liquid that restricts the escape of molecules from the liquid.

Synergism: A situation in which the combined action of two or more agents acting together is greater than the sum of the actions of the agents acting separately.

Tar sands: Viscous, oily, liquid hydrocarbon products intermixed in the pore spaces of sandstone.

Temperature: A measure of the sensation of hotness or coldness referred to a standard scale such as the height of a column of mercury in a tube.

Temperature inversion: An atmospheric condition in which the temperature increases with increasing altitude.

Tetraethyl lead: A lead compound, $(C_2H_5)_4Pb$, added to gasoline to increase the octane rating.

Thermal conductivity: A property of a medium—solid, liquid, or gas—measuring the ability to conduct heat. Typical units are (calories/second) per cm • °C or (Btu/second) per ft • °F.

Thermal energy: The random kinetic energy of an atom or molecule that is numerically equal to $3/2 \, kT$ where k is Boltzmann's constant and T is the kelvin temperature.

Thermal pollution: The addition of heat to the environment, usually the water environment, to the extent that the value of the environment is altered.

Thermal radiation: The electromagnetic radiation emitted by any object maintained at a temperature above zero kelvins.

Thermal reactor: A device used in conjunction with an internal combustion engine to complete the combustion of hydrocarbons and carbon monoxide.

Thermal stratification: The layering of a body of water into sections of nearly constant temperature.

Thermodynamics: The science of the relationship between heat and other forms of energy.

Thermostat: A temperature-controlling device that signals a heating or cooling supply to add or remove heat depending on whether or not the temperature is to be raised or lowered.

Tidal power plant: An electric power plant that uses the incoming and outgoing kinetic energy of tides to drive water turbines that drive electric generators.

Tokamak: Derived from the Russian words for toroidal magnetic chamber, it designates a type of nuclear fusion reactor.

Transformer: A device used to change the size of an alternating current.

Tritium: An isotope of hydrogen having one proton and two neutrons in its nucleus.

Triton: The nucleus of the tritium atom.

Troposphere: The lower region of the earth's atmosphere extending up to an average altitude of about 40,000 ft (eight miles).

Turbidity: In meteorology, any condition in the atmosphere that reduces its transparency to radiation, especially visible radiation.

Turbine: A device in which the kinetic energy of a fluid is converted to rotational kinetic energy of a shaft by impulses exerted on vanes attached to the shaft.

Unproven energy reserve: A source of energy thought to be available because of geologic information and oil-exploration experience.

Velocity: The displacement of an object divided by the time required to achieve the displacement. It is distinguished from speed by accounting for the direction of the displacement.

Viscous force: A frictional force exerted on an object moving in a fluid.

Visual range: The distance, under daylight conditions, at which the apparent contrast between an object chosen as the target of observation and its background becomes equal to the threshold of contrast of the observer.

Volt (V): The volt is the unit of potential difference. If 1 joule of work is required to move 1 coulomb of charge between two positions, the potential difference between the positions is 1 volt ($V = W/Q$).

Voltage: A popular expression for potential difference.

Wankel engine: An internal combustion engine in which the kinetic energy of the exploding gas is converted directly to rotational kinetic energy.

Watt (W): The watt is the metric unit of power. A rate of doing work of one joule per second is a watt ($P = W/t$).

Wave: A disturbance or oscillation propagated with a definite velocity from point to point in a medium.

Weight: The force of gravity on a mass. Numerically, weight is equal to the product of mass and acceleration due to gravity ($W = mg$). Typical units are newtons (metric) and pounds (British).

Wet-bulb temperature: The temperature measured by a thermometer covered with a cloth moistened with water.

Work: The product of the distance an object is moved times the force operating in the direction that the object moves.

X-ray: A quantum (or photon) of electromagnetic radiation. It is distinguished from other electromagnetic radiation by its energy. The energy limits are loosely defined but typically photons with energies between about 1 and 100 keV are called X-rays.

INDEX

Thermal energy, 133, 143, 159–160, 292, 295, 328, 340, 350
  in oceans, 280
  storage, 264–265, 271–273
Thermal enrichment, 158
Thermal pollution, 117, 156–157, 333
  concerns for, 128
  effects on aquatic life, 157
  effects on water quality, 156–157
  from geothermal sources, 295
  origin of, 156
  from solar electric power plants, 272
Thermal radiation, 188–189, 192, 271
Thermometer, 130
Thermostat, 343
Theta pinch, 331
Thorium, fertile nucleus, 318
Threshold hypothesis, radiation damage, 240
Tidal energy, 304–308
  projects, 304–305, 307
Tidal range, 306
Tides, 8, 305
  neap, 306
  spring, 306
Toffler, Alvin, author of *Future Shock,*
    quotes, 31, 44
Tokamak, 330, 333
Transformer, 111–112
  use in transmitting electric power, 112–113
Transportation, and air pollution, 52–53, 74
  energy requirements, 352–354
  opportunities for energy saving, 351–356
Trash and garbage, energy source, 8, 292, 297
  projects, 298–299
Tritium, 207, 327
  beta decay of, 211–212
  containment problems, 333
  nuclear fusion fuel, 334
  production in nuclear fusion reactors, 333
Troposphere, 178
Turbidity, 183–184

Ultraviolet radiation, 185, 193–194
  biological effects, 193
  percent in sunlight, 264
Uniform motion, 32
Union Electric Company, trash energy system, 298
Unproven oil reserve, 14
  lifetime, U.S., 17
Uranium, 226, 250, 295, 318, 320, 324
  actinide production from, 233, 243
  fission characteristics, 212, 225–226
  natural abundance, 226, 244, 250
  separation methods, 226–227
  use in nuclear breeder reactor, 318–319

Uranium hexafluoride, 226–227
Uranium oxide, fuel element, 228

Vaporization, 158
Velocity, 31
  settling for particulates, 61
  and speed, distinction, 30
Viscous force, 61
Visual range, 183
Volt (V), 103
Voltage, 104, 112

Wankel rotary engine, 87–89
Water, coolant in nuclear reactor, 229
  dissociation into hydrogen and oxygen, 279–280
  heavy, 230
  moderator in nuclear reactor, 229
  problems in nuclear breeder reactor, 320
  solar heating of, 267
  thermal effects on quality, 156
  turbine, principle, 292
  warm, uses from power plants, 158
Watt (W), 41, 106
Watt-hour, 106
Watt-year, energy unit, 48 (Problem 16)
Wave, 185
Wave speed, 185, 188
Wavelength, 185
  distribution in thermal radiation, 189
  electromagnetic radiation, 209
Weight, 33, 35
Wet-bulb temperature, 163–164
Wet-type cooling tower, 167
Wind energy, 8, 277–278
Wind turbine, in the ocean, 279
  power from, 277
Work, 35, 37
  done by an expanding gas, 128, 132
  by moving electric charges, 102
  negative, 37
  net, 35, 141
  and potential difference, 103
  rate of doing, 41
  relation to force, 35
  units, 37
Work-energy principle, 37
Wood, heat content, 281

X-ray, 188, 209
  cell damage by, 236
  diagnostic, doses, 237
  radiation doses for somatic effects, 238
  relative biological effectiveness, 236